T0328611

Thermal Power Plant

Thermal Power Plant
Pre-Operational Activities

Dipak K. Sarkar

ELSEVIER

AMSTERDAM • BOSTON • HEIDELBERG • LONDON • NEW YORK • OXFORD
PARIS • SAN DIEGO • SAN FRANCISCO • SINGAPORE • SYDNEY • TOKYO

Elsevier
Radarweg 29, PO Box 211, 1000 AE Amsterdam, Netherlands
The Boulevard, Langford Lane, Kidlington, Oxford OX5 1GB, United Kingdom
50 Hampshire Street, 5th Floor, Cambridge, MA 02139, United States

Library of Congress Cataloging-in-Publication Data
A catalog record for this book is available from the Library of Congress

British Library Cataloguing-in-Publication Data
A catalogue record for this book is available from the British Library

ISBN: 978-0-08-101112-6

For information on all Elsevier publications
visit our website at https://www.elsevier.com/

Working together
to grow libraries in
developing countries

www.elsevier.com • www.bookaid.org

Publisher: Joe Hayton
Acquisition Editor: Cari Owen
Editorial Project Manager: Alex White
Production Project Manager: Poulouse Joseph
Cover Designer: Victoria Pearson

Typeset by SPi Global, India

Dedicated to my grandparents
Nalini Prabha-Rajani Kanta
and
Surabala-Surendra Nath

Contents

PART 1 PREOPERATIONAL CLEANING OF VARIOUS SUB-SYSTEMS   43

PART 2 ACTIVITIES THAT MAKE CRITICAL EQUIPMENT READY TO PUT THEM IN SERVICE   193

Preface

I worked in the premier engineering consultancy firm of India, M/S Development Consultants Private Limited (DCPL), for more than four decades. During this long tenure I had the opportunity to work in innumerable thermal power plants of various sizes and configurations. The most important part of this association was my exposure to chronological development in power plant technology, both within and outside India. Against this backdrop, when I was executing various projects I observed that there is a dearth of published books wherein various aspects of preoperational activities are consolidated. This was the starting point in getting motivated to write this book, *Thermal Power Plant—Preoperational Activities*, in which I tried my utmost to incorporate the state-of-the-art technology applicable to these activities.

Design of a thermal power plant is a desktop study, while operation of the plant falls exclusively within the purview of field engineering. In between design and operation there are certain areas, generally known in the industry as preoperational activities. The procedure for execution of these activities is developed during desktop study, but is executed in field only. So these activities act as a go-between for design and operation to ensure unruffled power generation as far as the end user is concerned. Thus, this book may be construed as complemental to my previous book, *Thermal Power Plant—Design and Operation*, published by Elsevier.

This book aims to address some of the essential preoperational activities which are extremely important to carry out in line with the practice followed in the industry globally. Smooth, trouble-free, and economic operation of thermal power plants can be ensured if preoperational activities are carried out with the utmost care in order to establish that prior to the start-up of various systems, equipment, and/or the plant as a whole, they would be ready in all respects in accordance with manufacturers' recommendations, and at the same time fulfill all requirements of applicable statutory guidelines. The main purpose of addressing the essential features of preoperational activities is to ensure economic generation from a plant. Hence, the primary focus of this book is on professional engineers.

The contents of the book are such that the book should not be treated as a conventional textbook used in technical institutes. For design engineers, this book would act as a reference to help them develop project-specific precommissioning manuals of a new plant. It would also fulfill the needs of commissioning engineers, suppliers, and utility operators during the execution period of new plants. While successful completion of preoperational activities is essentially to be accomplished and certified prior to the start-up of new plants, areas addressed in this work are applicable to running plants as well. In running plants, this book would be a tool to help operation and maintenance engineers/suppliers (contractors) execute cleaning/testing activities (eg, overhaul, critical inspection, major repair) successfully to comply with regulatory requirements.

Degradation in the performance of major prime-movers—namely steam turbines, gas turbines, and diesel engines—of running units is a common phenomenon in the industry. In order to arrest the severity of degradation, it is generally recommended to undertake a major overhaul of these prime-movers following a specified period of operating hours. In the case of steam turbines, this period is usually 6 years, while for gas turbines and diesel engines, permissible operating hours that are usually recommended by manufacturers are 48,000 and 8000, respectively. Steam generators are generally inspected after about 12–24 months of operation to meet the statutory requirements of the boiler inspectorate. Any defect observed during inspection needs to be attended to before the boiler inspectorate extends permission to restart the steam generator.

In many countries initiatives have been undertaken to improve the efficiency and environmental performance of their existing thermal power plants through refurbishment, upgradation, rehabilitation, and modernization activities. Before putting these old plants into service on completion of routine overhauls, on completion of renovation and modernization, following a major repair, or after long shutdown of critical equipment or of the plant, it is compulsory to carry out preoperational activities to meet statutory requirements; otherwise, these plants would not be in a position to supply uninterrupted power.

Before discussing details of these activities, it is essential that readers are conversant with an outline of a thermal power plant or that readers' knowledge of a thermal power plant is refreshed for the convenience of understanding various systems and equipment. Hence, Chapter 1 addresses "General Description of Thermal Power Plants." In addition, a generic description of systems and equipment pertaining to specific types of preoperational activities or thermal power plants is discussed substantially in various chapters.

Successful completion of a preoperational cleaning activity or a prestart-up activity depends on how consciously each activity is carried out, such that a plant does not face any untoward incident during normal running, lest generation gets perturbed. This is ensured by adhering to the guidelines laid down by an internationally recognized quality management system (eg, ISO 9000). For the convenience of readers to adopt a foolproof quality management system, guidelines on "Quality Assurance (QA) and Quality Control (QC) (Applicable to Preoperational Activities)" are discussed in Chapter 2. Prior to conducting any preoperational activity it is extremely essential to observe certain precautionary measures, and also to ensure that certain activities/items are completed beforehand. These two aspects are addressed under "Precautions" and "Prerequisites" in Chapters 3–17. The QA team must be responsible for fulfilling these aspects. Thereafter the QC team takes over to ensure successful completion of each preoperational activity, following the steps of developing preparatory arrangements, operating procedures to be followed, availability of required materials (chemicals/water/any special gadgets), availability of safety equipment, and so on.

Looking toward commissioning of a new plant or recommissioning of a running plant from availability of electric power to commercial operation, the effective order of preoperational activities that are generally followed in the industry are presented here:

1. Hydraulic test of steam generator
2. Airtightness/leakage test of the furnace, air, and flue gas paths of steam generator
3. Alkali flushing of preboiler system
4. Flushing of fuel oil piping system
5. Blowing of fuel gas piping system
6. Steam generator initial firing and drying out of insulation
7. Chemical cleaning of steam generator
8. Flushing of lube oil piping system
9. Steam/air blowing of main steam, cold reheat, hot reheat, and other steam pipe lines
10. Floating of steam generator safety valves
11. Clean airflow test of pulverizers
12. Condenser flood test and vacuum-tightness test
13. Generator drying out and airtightness test
14. Filling of generator with hydrogen and protection stability test of generator
15. Completion test of the power station

While some of the aforementioned activities pertain to the preoperational cleaning of piping systems, the remaining activities are conducted to ensure the integrity of critical equipment. Hence, preoperational activities, as laid down in this book, are addressed under Part 1 and Part 2, as categorized here:

Part 1: Preoperational Cleaning of Various Sub-Systems

Preoperational cleaning of various piping systems of modern thermal power plants assumes considerable importance because of the high-quality demand of flowing fluid, be it steam, water, oil, or gas, through different pipe lines. During the process of manufacturing, transportation, storing, and erection of various piping systems, in spite of taking the best precautionary measures, a certain amount of dirt, mill scale, oil, grease, and so on finds its way into these systems. These unwanted constituents need to be cleaned prior to putting into service "erected new piping" or "replaced old piping."

In running units, deposits may grow inside pipe lines due to improper water treatment or from process contamination. Corrosion inside the pipe lines of operating units may take place either from an improperly controlled pH of demineralized makeup water or from concentration of boiler water salts.

In order to get rid of these undesirable elements from the piping systems of both new units and running units, as far as practicable, preoperational cleaning of piping systems is carried out. Based on the type of flowing fluids through various piping systems, different types of cleaning activities, as adopted in the industry, are discussed in the following chapters:

   i.   Chapter 3: Alkali flushing of preboiler system
  ii.   Chapter 4: Flushing of fuel oil piping system
 iii.   Chapter 5: Blowing of fuel gas piping system
 iv.   Chapter 6: Chemical cleaning of boiler
  v.   Chapter 7: Flushing of lube oil piping system
 vi.   Chapter 8: Steam/air blowing of main steam, cold reheat, hot reheat, and other steam pipe lines

Part 2: Activities that Make Critical Equipment Ready to Put Them in Service

This section addresses activities which are carried out to make critical equipment ready prior to putting it in service and to establish the integrity of this equipment, along with its associated systems. Activities covered in this section are:

   i.   Chapter 9: Hydraulic test of steam generator
  ii.   Chapter 10: Airtightness/leakage test of the furnace, air, and flue gas paths of steam generator
 iii.   Chapter 11: Steam generator initial firing and drying out of insulation
 iv.   Chapter 12: Floating of steam generator safety valves
  v.   Chapter 13: Clean airflow test of pulverizers
 vi.   Chapter 14: Condenser flood test and vacuum-tightness test
 vii.   Chapter 15: Generator drying out and air-tightness test
viii.   Chapter 16: Filling of generator with hydrogen and protection stability test of generator
 ix.   Chapter 17: Completion test of the power station

As a prequel to Chapter 17, "Brief Description of Performance Guarantee Tests" is addressed in Appendix A.

Safety is a fundamental necessity for operating any plant. Hence, Appendix B lays down "General Safety Guidelines."

When a plant is in operation, under maintenance, or kept under mothballed condition, certain valves/areas/systems purposefully need to be kept isolated. Any attempt to violate such isolation, even inadvertently, may lead to harm, injury, or major disaster of the plant and personnel. In order to obviate such an inadvertent attempt of violation, warning tags of various types are applied on isolated valves/areas/systems. Typical "Tagging Procedures" delineating various warning content are therefore presented in Appendix C of this book.

In accordance with current global practice, SI units have been used throughout the book. Nevertheless, for the convenience of readers, conversion factors from SI units to the metric system of units to the imperial and US system of units are addressed in Appendix D.

Reader suggestions for the improvement of the contents of this book are welcome, and would be acknowledged gratefully by the author.

Dipak K. Sarkar
June 30, 2016

Acknowledgments

The overwhelming response I received from power engineers for my book *Thermal Power Plant—Design and Operation* prompted me to write this book. I am sincerely thankful to the power engineering community for motivating me in this regard. I also am indebted to my colleagues and friends in the industry for their advice on bridging the gap between the design and operation of a thermal power plant in the form of a book on preoperational activities. I am particularly grateful to Mr. Samiran Chakraboty, retired chief engineer, erstwhile M/S ACC Babcock Limited (ABL), currently M/S Alstom India Limited, who firmly expressed the necessity of a book on preoperational activities of a thermal power plant for the benefit of members of start-up, commissioning, and operation engineering departments.

I also am indebted to M/S Development Consultants Private Limited (DCPL) because while working in this organization I received exposure to a variety of technologies chosen by different manufacturers adopted in different countries, the benefits of which were reaped while writing this book.

Mr. S.K. Saha, general manager DCPL, an outstanding operation engineer, shared his long experience that helped me in preparing the manuscript. I gratefully acknowledge Mr. Saha's support.

While I was in DCPL I learnt many aspects of electrical system and electrical protection system from my the then colleagues Mr. P.S. Bhattacharya and Mr. S.K. Bhattacharya. Knowledge gathered thus was subsequently honed at the field. Chapters 15 and 16 might not have been developed by me without my interaction with both the Bhattacharyas. I am specially indebted to them.

I thank my son, Krishanu, daughter, Purbita, and son-in-law, Sudip, who were the source of encouragement in writing this book. My wife Anita's inspiration needs special mention; without her ceaseless support I could not have finished the book on time.

I am grateful to Elsevier for allowing me to borrow materials on system description from my previous book, *Thermal Power Plant—Design and Operation*. I also am indebted to the publishing and editorial team at Elsevier—Sarah Hughes, Joe Hayton, Cari Owen, Alex White, Lucy Beg, Poulouse Joseph and Victoria Pearson—for their support and guidance.

Dipak K. Sarkar
May 27, 2016

List of Acronyms/Abbreviations

μS	Microsiemens
a, abs	Absolute
A	Ash (content in coal)/ampere
ABMA	American Boiler Manufacturers Association
A/C	Air/cloth
AC	Alternating current/air conditioning
ACF	Activated carbon filter
ACW	Auxiliary cooling water
ad	Air dried
AFBC	Atmospheric fluidized bed combustion
AFR	Air-fuel ratio
AH	Air heater
AHS	Ash handling system
a.k.a.	Also known as
ANSI	American National Standards Institute
APC	Auxiliary power consumption
API	American Petroleum Institute
APS	Automatic plant start-up and shutdown system
AQC	Air quality control (ESP/bag filter and FGD)
ar	As received
AS	Auxiliary steam
ASME	American Society of Mechanical Engineers
ASTM	American Society for Testing & Materials
atm	Atmosphere
AVR	Automatic voltage regulator
AVT	All volatile treatment
AWWA	American Water Works Association
b	Bar
B	Billion
BA	Bottom ash

B&W	The Babcock & Wilcox Company
BDC	Bottom dead center
BEI	British Electricity Institute
BF	Base factor
BFBC	Bubbling fluidized bed combustion
BFP	Boiler feed pump
BHRA	British Hydraulic Research Association
BIS	Bureau of Indian Standards
BMCR	Boiler maximum continuous rating
BMS	Burner management system
BOOS	Burner out of service
BOP	Balance of plant
BP	Booster pump
BPVC	Boiler and pressure vessel code
BSI	British Standards Institution
Btu	British Thermal Unit
BWR	Boiling water reactor
C	Carbon/celsius/centegrade
Ca	Calcium
CA	Compressed air/citric acid
CAA	Clean Air Act, U.S.A.
CAAA	Clean air act amendments
CBD	Continuous blow down
cc	Cubic centimeter
CC	Combined cycle
CCCW	Closed cycle cooling water
CCGT	Combined cycle gas turbine
CCPP	Combined cycle power plant
CE	Combustion Engineering Inc./collecting electrode
CEA	Central Electricity Authority, India
CEGB	Central Electricity Generating Board
CEN	(ComitéEuropéen de Normalisation)-European Committee for Standardization
CEP	Condensate extraction pump
CFBC	Circulating fluidized bed combustion
cfm	Cubic feet per minute
CFR	Cleaning force ratio
CHF	Critical heat flux
CHP	Combined heat and power

CHS	Coal handling system
CI	Combustion inspection of gas turbine
C.I.	Compression ignition
cm	Centimeter
CO	Carbon monoxide
CO_2	Carbon dioxide
cP	Centipoise
CPCB	Central Pollution Control Board, India
CR	Compression ratio
CR/CRH	Cold reheat
CSA	Canadian Standards Association
CSB	Chemical Safety and Hazard Investigation Board, U.S.A.
CV	Calorific value/control valve
CW	Circulating (condenser cooling) water
cwt	Hundredweight
D	Drain/diameter
D, d	Day
DAF	Dry ash free
dB	Decibel
DAS	Data acquisition system
DC	Direct current
DCA	Drain cooler approach
DCS	Distributed control system
DE	Discharge electrode
deg	Degree
DIN	DeutschesInstitutfürNormung
DM	De-mineralized
dmmf	Dry mineral matter free
DMW	De-mineralized water
DNB	Departure from nucleate boiling
DO	Dissolved oxygen
DSI	Duct sorbent injection
EA	Excess air
ECS	Environmental control systems
EDI	Electrical de-ionization unit
eff/EFF	Efficiency
EHS	Environmental, health and safety
EIA	Environmental impact assessment
emf	Electromotive force

EMV	Effective migration velocity
EPA	Environmental Protection Agency, U.S.A
EPRI	Electric Power Research Institute, U.S.A
EPRS	Effective projected radiant surface
ESI	Economizer sorbent injection
ESP	Electrostatic precipitator
ESV	Emergency stop valve
EU	European Union
EX	Extraction
F	Fahrenheit
FA	Fly ash
FAC	Flow accelerated corrosion
FBC	Fluidized bed combustion
FBR	Fast breeder reactor
FC	Fixed carbon (in coal)
FD	Forced draft
FEGT	Furnace exit gas temperature
FGD	Flue gas desulfurization
FGR	Flue gas recirculation
FIG	Figure
FFH	Factored fired hours
FO	Furnace oil
FSI	Furnace sorbent injection
ft	Foot/feet
FW	Feed water
FWH	Feed water heater
fpm	Feet per minute
g	Gram/gauge/acceleration due to gravity (1 kg m/Ns^2)
G	Gallon/giga
GB	China National Standard
GCB	Generator circuit breaker
GCR	Gas cooled reactor
GCS	Gas conditioning skid
GCV	Gross calorific value
GE	General Electric Company
GGH	Gas to gas heater
GHG	Greenhouse gas
GJ	Giga joule
GLR	Generator lock-out relay

GOST	GosudartsvennyeStandarty, Russia
GPHR	Gross plant heat rate
gpm	Gallons per minute
gr	Grain
GT	Gas turbine/generator transformer
GTMCR	Gas turbine maximum continuous rating
h	Hour
H	Hydrogen
H, h	Enthalpy
HAF	Hydroxyacetic-formic acid
HAP	Hazardous air pollutants
HAZ	Heat affected zone
HCSD	High concentration slurry disposal system
HEI	Heat Exchange Institute, U.S.A.
HFO	Heavy fuel oil
Hg	Mercury
HGI	Hardgrove grindability index
HGPI	Hot gas path inspection
HHV	Higher heating value
HI	Hydraulic Institute, Inc., U.S.A.
hp	Horse power
HP	High pressure
HPLP	High pressure & low pressure
HRSG	Heat recovery steam generator
HR	Heat rate
HR/HRH	Hot reheat
HSD	High speed diesel
HSI	Hybrid sorbent injection
HT	High tension/hemispherical temperature (of ash)
HV	High volatile/high voltage
HVAC	Heating-ventilation and air conditioning
HWR	Heavy water reactor
Hz	Hertz (frequency)
IA	Instrument air
I&C	Instrumentation and control
IBR	Indian Boiler Regulations
ICE	Internal combustion engine
ICS	Integrated control system
ID	Induced draft/inside diameter

IDT	Initial-deformation temperature (of ash)
IEC	International Electrotechnical Commission
IEEE	Institute of Electric and Electronic Engineers
IFC	International Finance Corporation
IGCC	Integrated gasification combined cycle
IM	Inherent moisture (content in coal)
Imp.	Imperial
in	Inch/inches
IPB	Isolated phase bus
IP	Intermediate pressure
IPCC	Intergovernmental Panel on Climate Change
IPP	Independent power producer
IR	Infrared radiation/insulation resistance
IS	Indian Standards
ISO	International Standards Organization
ISO Condition	Pressure: 1.013 kPa, temperature: 288 K, relative humidity: 65% (pertain to atmospheric condition)
IV	Interceptor valve
J	Joule
JO	Jacking oil
k	Kilo
K	Kelvin/potassium
kA	Kilo ampere
kcal	Kilocalories
kg	Kilogram
kg-mole	Kilogram-mole
kg/s	Kilogram per second
kJ	Kilojoule
km	Kilometer
KOD	Knock out drum
kPa	Kilopascal
kV	Kilovolt
kW	Kilowatt
kWh	Kilowatt hour
l	Liter
lb	Pound
LBB	Local breaker back-up
LDO	Light diesel oil
LEA	Low excess-air

LFO	Light fuel oil
LH_2	Liquid hydrogen
LHV	Lower heating value
LMTD	Log mean temperature difference
LNB	Low-NOx burner
LNG	Liquefied natural gas
LO	Lube oil
LP	Low pressure
LPG	Liquefied petroleum gas
LSHS	Low sulfur heavy stock
LT	Low tension
LV	Low volatile
m	Meter
M	Moisture (content in coal)/million
MAF	Moisture and ash free
max	Maximum
MB	Mixed bed unit
MCC	Motor control center
MCR	Maximum continuous rating
MEP	Mean effective pressure
MFR	Master fuel relay
MFT	Master fuel trip
mg	Milligram
Mg	Magnesium
MGD	Million gallons per day
m/h	Miles per hour
MI	Major inspection of gas turbine
min	Minute/minimum
MJ	Megajoule
ml	Milliliter
mm	Millimeter
MM	Mineral matter (content in coal)
MMT	Minimum metal temperature
mol	Molecular
MOEF	Ministry of Environment & Forests, India
MOP	Main (main shaft driven) oil pump
MOT	Main oil (main lube oil) tank
MPa	Megapascal
mph	Miles per hour

m/s	Meter/second
MS	Main steam
mV	Milli volt
MW	Mega watt
N	Newton/nitrogen
Na	Sodium
NCV	Net calorific value
NEMA	National electrical manufacturers association
NFPA	National Fire Protection Association, U.S.A.
Nm^3	Normal cubic meter
NO	Nitric oxide
NO_2	Nitrous oxide
NO_x	Nitrogen oxides
NPHR	Net plant heat rate
NPSH	Net positive suction head
NRV	Non-return valve
NTP	Normal temperature and pressure (273 K and 101.3 kPa)
O	Oxygen
O&M	Operation and maintenance
OCC	Open circuit characteristic
OD	Outside diameter
OEM	Original equipment manufacturer
OFA	Over-fire air
OH	Operating hours
OLTC	On-line tap changer
OPEC	Organization of petroleum exporting countries
OSHA	Occupational Safety & Health Administration
OT	Oxygen treatment
oz	Ounce
P	Power/poise
P, p	Pressure
Pa	Pascal
PA	Primary air
PAC	Powdered activated carbon
PC	Pulverized coal
PCC	Power control center
PF	Power factor/pulverized fuel
PFBC	Pressurized fluidized bed combustion
PG	Performance guarantee

pH	Negative log of hydrogen ion concentration
PI	Polarization index
P&ID	Process and instrumentation diagram
PLF	Plant load factor
PM	Particulate matter
ppb	Parts per billion (mass)
PPE	Personal protective equipment
ppm	Parts per million (mass)
ppmv	Parts per million by volume
PRDS	Pressure reducing and de-superheating
Press.	Pressure
PRV	Pressure reducing valve
PSF	Pressure sand filter
psi	Pounds per square inch
PTC	Performance test code
PWR	Pressurized water reactor
QA	Quality assurance
QC	Quality control
R	Rankine/universal gas constant
rad	Radian
rev	Revolution
RH	Reheater
RHO	Reheater outlet
RO	Reverse osmosis
rpm	Revolution per minute
RTD	Resistance temperature detector
RW	Raw water
s	Second
S	Sulfur
S, s	Entropy
SA	Secondary air/service air
SAC	Strong acid cation unit/Standardization Administration of the People's Republic of China
SBA	Strong base anion unit
SCA	Specific collection area
SCAH or SCAPH	Steam coil air pre-heater
SCC	Short circuit characteristic/submerged chain conveyor
scf	Standard cubic feet
scfm	Standard cubic feet per minute

SCR	Selective catalytic reduction
SDA	Spray drier absorber
sec	Second
SG	Steam generator
SH	Superheater
SHO	Superheater outlet
SI	Systeme International D'Unites/International System of Units
S.I.	Spark ignition
SLD	Single line diagram
SM	Surface moisture (content in coal)
SNCR	Selective non-catalytic reduction
SO_2	Sulfur dioxide
SO_3	Sulfur trioxide
SO_x	Sulfur oxides
sp. gr.	Specific gravity
SPM	Suspended particulate matter
SS	Stainless steel
s.s.c.	Specific steam consumption
ST	Steam turbine/station transformer/softening temperature (of ash)
STP	Standard temperature and pressure (288 K and 101.3 kPa)
SWAS	Steam water analysis system
t	Ton/tonne
T	Temperature/turbine
TAC	Tariff Advisory Committee, India
TDC	Top Dead Center
TDS	Total dissolved solids
TEMA	Tubular exchanger manufacturer association
Temp.	Temperature
TG	Turbo-generator
TLR	Turbine lock-out relay
TM	Total moisture (content in coal)
TMCR	Turbine maximum continuous rating
tph	Tonnes per hour
TPSC	Toshiba Power Services Corporation, Japan
T/R	Transformer rectifier set
TSP	Total suspended particulates
TSS	Total suspended solids
TTD	Terminal temperature difference
U, u	Internal energy

UAT	Unit auxiliary transformer
UBC	Unburned carbon
ULR	Unit lock-out relay
UPS	Uninterrupted power supply system
UV	Ultra violet
V	Volt
V, v	Volume
VDI	Verlag des VereinsDeutscherIngenieure
VM	Volatile matter (in coal)
vol	Volume
vs	Versus
VWO	Valve wide open condition
W	Watt/work
WAC	Weak acid cation unit
WB	World Bank
WBA	Weak base anion unit
WFGD	Wet flue gas desulfurization
wg	Water gauge
WHRB	Waste heat recovery boiler
wt	Weight
yd	Yard/yards
YGP	Yancy, geer and price (Index)
yr	Year

General Description of Thermal Power Plants

1.1 Introduction

Electricity was first supplied to the public back in the year 1881. It was a hydroelectric-generating station built on Niagara Falls supplying power to New York State. Tokyo Electric Lighting was the first in Asia to begin supplying electricity to the public in the year 1886. First time in Europe, London was electrified in 1888. Generation of electricity in India commenced from the year 1899 with the electrification of the city of Kolkata having a generating capacity of 1000 W.

A power-generating plant is an industrial facility for the generation of electric power. At the center of nearly all power stations is a generator, which is a rotating machine that converts mechanical energy into electric energy. The energy source harnessed to turn the generator varies widely. It depends chiefly on fuels and on the types of technology.

Most power stations in the world burn fossil fuels, that is, coal, oil, and natural gas, and some use nuclear power. In order to abate greenhouse gas (GHG) emission, in addition to the above there is an increasing use of cleaner, renewable sources of energy such as hydroelectric (addressed under a separate section below), solar, wind, and wave.

The following paragraphs briefly describe various types of power plants, as shown in Fig. 1.1. "Thermal power plant" is discussed in detail separately.

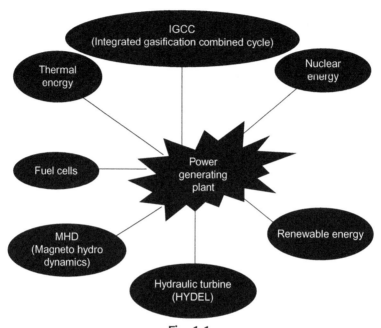

Fig. 1.1
Various types of power-generating plants.

1.1.1 Nuclear Power Plant [1]

Energy generation by a nuclear fuel takes place either by the process of nuclear fission of heavy fissile elements in a nuclear reactor, resulting in chain reactions, or by the process of nuclear fusion, in which simple atomic nuclei are fused together to form complex nuclei, as in the case of fusion of hydrogen isotopes to form helium. The process of nuclear fusion is also known as thermonuclear reaction, which is difficult to control even on date. As a result, the main source of nuclear energy is available at the present time mainly from nuclear fission.

In the heart of a "nuclear power plant" there is a nuclear reactor, wherein a controlled chain reaction of nuclear fission of heavy elements takes place. The most common fissile radioactive heavy metals are the naturally occurring isotope of uranium, U^{235}, artificial isotope of uranium, U^{233}, and artificial element plutonium, P^{239}. In a nuclear reactor, plutonium is produced from naturally occurring isotope of uranium, U^{238}, and U^{233} is produced from naturally occurring element thorium, Th^{232}.

The nuclear energy thus liberated is converted into heat that is removed from the reactor by a coolant, eg, liquid sodium. Hot liquid sodium is then passed through another heat exchanger where water is circulated as a coolant agent, which absorbs heat, resulting in generation of steam. This steam generator emits virtually no carbon dioxide, sulfur, or mercury. Nevertheless, a major concern of a nuclear power plant is that the area surrounding the nuclear reactor is potentially radioactive. Further nuclear wastes, if not disposed of taking special care, may cause a devastating effect on living beings and inanimate objects, including the environment.

Nuclear reactors are of various types, that is, pressurized-water reactor (PWR), boiling-water reactor (BWR), gas-cooled reactor (GCR), heavy-water reactor (HWR), and fast-breeder reactor (FBR).

1.1.1.1 Pressurized-water reactor (PWR)

A PWR power plant is composed of two loops in a series, the coolant loop, called the primary loop, and the water-steam or working fluid loop. The coolant picks up reactor heat and transfers it to the working fluid in the steam generator. The steam is then used in a Rankine cycle to generate electricity (Fig. 1.2).

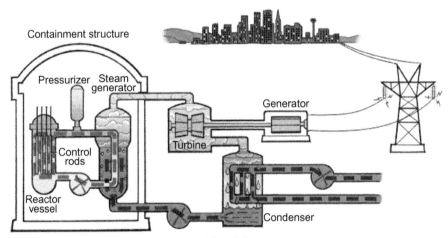

Fig. 1.2

Flow diagram of a pressurized-water reactor. *Source: From Fig. B.2, P 555. D.K. Sarkar, Thermal Power Plant—Design and Operation, 2015, Elsevier; Amsterdam, Netherlands.*

1.1.1.2 Boiling-water reactor (BWR)

In BWR, the coolant is in direct contact with the heat-producing nuclear fuel and boils in the same compartment in which the fuel is located. Liquid enters the reactor core at the bottom, flows upwards, and, when it reaches at the top of the core, it gets converted into a very wet mixture of liquid and vapor. The vapor is then separated from the liquid in a steam separator and flows through a turbine to generate power (Fig. 1.3).

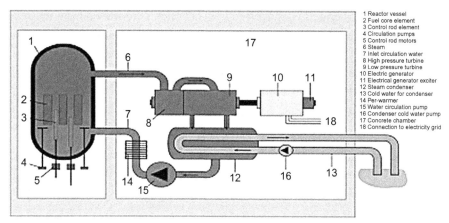

Fig. 1.3

Flow diagram of a boiling-water reactor. *Source: From Fig. B.3, P 556. D.K. Sarkar, Thermal Power Plant–Design and Operation, 2015, Elsevier; Amsterdam, Netherlands.*

1.1.1.3 Gas-cooled reactor (GCR)

A GCR is cooled by a gas. The gas absorbs heat from the reactor; this hot coolant then can be used either directly as the working fluid of a combustion turbine to generate electricity or indirectly to generate steam. There are two different types of GCR. One type utilizes both natural- and enriched-uranium fuels with CO_2 as coolant and graphite as moderator. Another one uses enriched fuels, helium as coolant, and heavy water as moderator (Fig. 1.4).

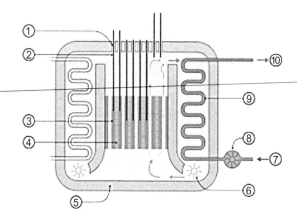

1. Charge tubes   2. Control rods   3. Graphite moderator   4. Fuel assemblies
5. Concrete pressure vessel and radiation shielding   6. Gas circulator   7. Water
8. Water circulator   9. Heat exchanger   10. Steam

Fig. 1.4

Sectional view of a typical gas-cooled reactor. *Source: From Fig. B.4, P 557. D.K. Sarkar, Thermal Power Plant–Design and Operation, 2015, Elsevier; Amsterdam, Netherlands.*

1.1.1.4 Heavy-water reactor (HWR)

In HWR, heavy water (D_2O) is used as coolant-moderator with natural-uranium fuels, instead of enriched uranium. Since heavy water is pressurized, it can be heated to higher temperatures without boiling. Production cost of D_2O, however, is very high compared with the cost of normal water (H_2O) (Fig. 1.5).

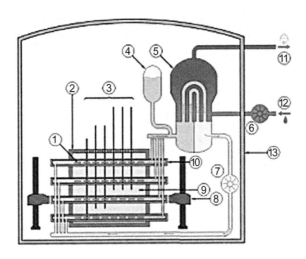

Key

1	Fuel bundle	8	Fueling machines
2	Calandria (reactor core)	9	Heavy water moderator
3	Adjuster rods	10	Pressure tube
4	Heavy water pressure reservoir	11	Steam going to steam turbine
5	Steam generator	12	Cold water returning from turbine
6	Light water pump	13	Containment building made of reinforced concrete
7	Heavy water pump		

Fig. 1.5

Sectional view of a typical heavy-water reactor. *Source: From Fig. B.5, P 558. D.K. Sarkar, Thermal Power Plant–Design and Operation, 2015, Elsevier; Amsterdam, Netherlands.*

1.1.1.5 Fast-breeder reactor (FBR)

FBR is so named because of its design to breed fuel by producing more fissionable fuel than it can consume. In FBR, neutrons are not slowed down to thermal energies by a moderator. Coolant and other reactor materials moderate the neutrons. Reactors are cooled by sodium (Fig. 1.6).

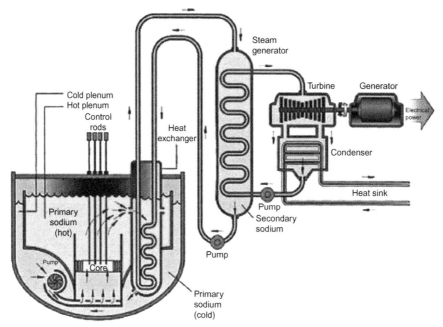

Fig. 1.6

Sectional view of a typical fast-breeder reactor. *Source: From Fig. B.7, P 559. D.K. Sarkar, Thermal Power Plant–Design and Operation, 2015, Elsevier; Amsterdam, Netherlands.*

1.1.2 Hydel Power Plant [1]

Hydroelectric or Hydel power plants comprise of hydraulic turbines, which can be of either vertical shaft or horizontal shaft. The preference for a horizontal shaft lies with the impulse-type while a vertical shaft, with the reaction types. The hydraulic turbine converts the potential energy of supplied water into mechanical energy of a rotating shaft, which in turn drives a generator to produce electricity.

In the impulse turbine (Pelton Wheel), the static head is completely transformed into a velocity head in the guide vane. This type is of relatively low (specific) speed, suitable for higher heads. Impulse turbines receive their water supply directly from the pipe line (Fig. 1.7).

In the reaction turbine, the static head is partly transformed into a velocity head in the guide vane. The radial flow type (Francis Turbine) is of relatively medium (specific) speed, suitable for medium heads. In Francis Turbine, high-pressure water enters the turbine with radial inflow and leaves the turbine axially. The high-pressure water, while passing through guide vanes, rotates the shaft to produce power (Fig. 1.8).

The propeller type (Kaplan Turbine) is of relatively high (specific) speed, suitable for low heads; thus, it is essential to pass large flow rates of water through the Kaplan turbine to produce power. Incoming water enters the passage in the radial direction and is forced to exit in the axial direction that in turn rotates the shaft for producing power (Fig. 1.9).

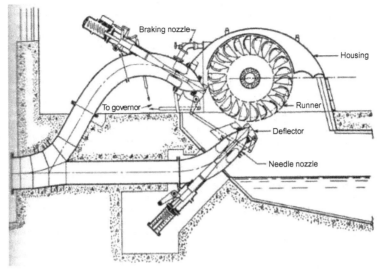

Fig. 1.7

Pelton wheel. *Source: From Fig. 15, P 9–221, Chapter 9: Hydraulic Turbines. Theodore Baumeister and Lionel S. Marks. Mechanical Engineers' Handbook (Sixth Edition). McGraw-Hill Book Company, Inc. 1958.*

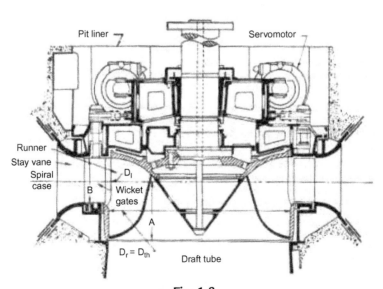

Fig. 1.8

Sectional view of a Francis turbine. *Source: From Fig. 1, P- 9-208, CH-9: Hydraulic Turbines. Theodore Baumeister and Lionel S. Marks. MECHANICAL ENGINEERS' HANDBOOK (Sixth Edition). McGraw-Hill Book Company, Inc., 1958.*

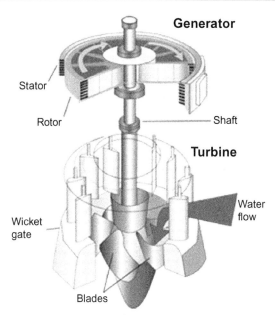

Generator

Stator

Rotor

Shaft

Turbine

Water flow

Wicket gate

Blades

Fig. 1.9

Sectional view of Kaplan turbine. *Source: From Fig. A.13. P 550. D.K. Sarkar, Thermal Power Plant–Design and Operation, 2015, Elsevier; Amsterdam, Netherlands.*

Note

The specific speed of a hydraulic turbine is defined as the speed of a geometrically similar hydraulic turbine that would develop 1 kW power under 1 m head of water column.

$$\text{Specific speed (rpm), } N_S = \frac{NP^{1/2}}{H^{5/4}}$$

where, $N=$ the normal working speed (rpm), $P=$ power output of the turbine (kW), and $H=$ the net or effective head (m).

1.1.3 Renewable Energy Power Plant [1]

Renewable energy resources draw natural energy flows of the earth, that is, solar, wind, geothermal, ocean thermal, ocean wave, ocean tidal, biomass, and storage energy. They recur, sometimes periodically, almost inexhaustible and are free for the taking. They are clean, barring biomass, almost free from causing environmental pollution, and sometimes even noise-free.

1.1.3.1 Solar energy

The total quantity of solar energy incident upon the earth is immense, but the energy is very diffuse, cyclic, and doesn't work at night without a battery-type storage device. It also suffers from atmospheric interference from clouds, particulate matter, gases, and so on. Solar radiation may either be converted into thermal or photovoltaic energy to generate electricity. To harness full potential of solar energy, a vast area of land is required to collect enough energy to electrify a community. As a result, solar technologies are very expensive.

1.1.3.2 Wind energy

Wind power is pollution free and its source of energy is also free. Winds are stronger and more consistent in offshore and high altitude sites than those at land. Wind power is suitable in areas with generally steady winds. The movement of air is utilized to run wind mills or wind turbines to produce electricity.

1.1.3.3 Geothermal energy

Geothermal energy is primarily energy from the core of the earth. The natural heat in the earth has manifested itself for thousands of years. It is recoverable as steam or hot water. Natural steam that spouted from the earth is utilized in a geothermal power generating station. Geothermal power is cost effective, reliable, sustainable, and environmentally friendly.

1.1.3.4 Ocean thermal energy

Seas and oceans absorb solar radiation resulting in ocean currents and moderate temperature gradients from the water surface downward. Since the surface of water receives direct sunlight, it is warmer, but below the surface the ocean is very cold. This temperature gradient is utilized in a heat engine to generate power. The temperature difference between the surface and down below has to be at least 21 K.

1.1.3.5 Ocean wave energy

When wind blows over the surface of oceans, large waves with high kinetic energy are generated. This energy is used in turbines to generate power.

1.1.3.6 Ocean tidal energy

Ocean tides, caused by lunar and gravitational forces, result in the rise and fall of waters with ranges that vary daily and seasonally. The potential energy of such tidal waves can be trapped to generate power.

1.1.3.7 Biomass

Biomass is another source of renewable energy, which is abundantly available. It is an organic matter produced by plants. It includes wood waste and bagasse that can be used for direct combustion in a furnace to boil water; steam thus produced is used in a conventional turbo generator to produce electricity. Trees like poplar, sycamore, and eucalyptus may be chipped and pulverized for burning in a power plant. Animal and human waste may be used to produce methane, which can be used directly in combustion turbines to generate electricity.

Biomass is produced from solar energy by photosynthesis. As such, it absorbs the same amount of carbon in growing as it releases when consumed as a fuel. So it does not add to carbon dioxide in the atmosphere.

1.1.3.8 Storage energy

The objective of this source is to store energy when the electric demand is lower than the generating capacity and to release the stored energy when the demand is higher. Thereby, it ensures a reliable, efficient, and economic supply of electricity. Various sources of stored energy are described below.

In *pumped hydro*, large quantities of water are transferred and stored from a lower to an upper reservoir during the lean period of power demand. When the demand increases, the stored water is released to the lower reservoir. Thus the potential energy of the water is utilized to drive a pump turbine.

Compressed air storage comprises compressing and storing of air in reservoirs. The stored energy is then released during periods of peak demand by expansion of air through an air turbine.

The *flywheel rotor* is physically connected to a motor-generator set. In the charging mode, during off-peak periods, the motor adds energy to the flywheel. During periods of peak demand, the flywheel rotor coasts down to drive the generator.

In the discharge mode of a *battery*, direct current is generated to serve various purposes. In the charge mode, the battery can be restored to its original condition by reversing the direction of the current obtained from a DC generator or by an alternator equipped with a rectifier.

In a *superconducting magnet*, a coil is made of a material in a superconducting state. Once it is charged, the current will not decay and the magnetic energy can be stored indefinitely. The stored energy can be released back to the network by discharging the coil.

Chemical reactions are of reversible type. The heat of reaction of reversible chemical reactions is used to store thermal energy during endothermic reactions and to release it during exothermic reactions.

1.1.4 Integrated Gasification Combined Cycle (IGCC)

In a IGCC, coal is turned into gas—synthesis gas (syngas). It then removes impurities from the coal gas before it is combusted. This results in lower emissions of sulfur dioxide, particulates, and mercury. It also results in improved efficiency compared to conventional pulverized coal. The plant is called "integrated" because its syngas is produced in a gasification unit in the plant. The gasification process produces heat, and this is reclaimed by steam in "waste heat boilers." Steam turbines use this steam in a combined cycle.

1.1.5 Magnetohydrodynamic (MHD) Generator

The MHD generator or dynamo transforms thermal energy or kinetic energy directly into electricity. MHD generators are different from traditional electric generators in that they can operate at high temperatures without moving parts. The exhaust of a plasma MHD generator is a flame, still able to heat the boilers of a steam power plant.

In the MHD generator, the solid conductors are replaced by a gaseous conductor, an ionized gas. If such a gas is passed at a high velocity through a powerful magnetic field, a current is generated, which can be extracted by placing electrodes in suitable position in the stream (Fig. 1.10).

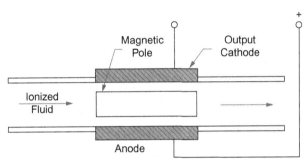

Fig. 1.10

MHD generator. *Source: From https://newenergytreasure.files.wordpress.com/2014/01/ciht-mhd-1.jpg.*

1.1.6 Fuel Cell Power Plant

A fuel cell generates electrical power by continuously converting the chemical energy of a fuel into electrical energy by way of an electrochemical reaction (Fig. 1.11). Fuel cells typically utilize hydrogen as the fuel and oxygen (usually from air) as the oxidant in the electrochemical reaction. The reaction results in electricity, along with by-product water and heat. The fuel cell itself has no moving parts, making it a quiet and reliable source of power. With their high power generating efficiency (40–60%), ease in installing near consumers, and applicability for cogeneration (heat and electricity), fuel cells are expected to achieve substantial energy conservation. In addition, fuel cells can utilize natural gas, methanol, or coal gas as fuel, serving as a driving force to encourage the use of oil-alternative energy sources. These plants, however, suffer from generating capacity limitation.

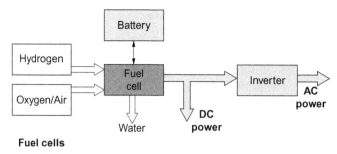

Fuel cells

Fig. 1.11

Fuel cell power plant. *Source: The Electropaedia–Hydrogen Fuelled Electricity Generation. http://www. mpoweruk.com/hydrogen_fuel.htm.*

1.1.7 Thermal Power Plant [1]

In contrast to all types of power plants described in the foregoing paragraphs, a thermal power plant utilizes the "heat of combustion" of fossil fuels. Fossil fuels are hydrocarbons, primarily coal and liquid petroleum or natural gas, formed from the fossilized remains of dead plants and animals by exposure to heat and pressure in the earth's crust over hundreds of millions of years. These fuels consist of a large number of complex compounds of five principal elements: carbon (C), hydrogen (H), oxygen (O), sulfur (S) and nitrogen (N). All fuels contain also mineral matter (A) and moisture (M) to a certain extent. However, there are just three combustible elements of significance in a fuel; these are carbon, hydrogen, and sulfur—of which carbon is the principal combustible element. Fossil fuels generate substantial quantities of heat per unit of mass or volume by reacting with an oxidant in a combustion process.

Two major constituents of a thermal power plant are a prime mover, which develops mechanical work, and a generator or an alternator, in which mechanical work is converted into electrical energy. Whatever may be the type of prime movers—steam turbine, gas turbine, or diesel engine—a generator is common to all of them for developing electrical energy.

Generators of various plants may have different capacities, but their voltage and frequency essentially have to be identical and constant and shall be in conformance with the voltage and frequency of the high-voltage (132/220/400/700 kV) grid/bus, to which output power from a generator is transmitted through generator transformers, isolators, and generator circuit breakers for further distribution to different consumers. A typical electrical distribution system is shown in Fig. 1.12.

Fig. 1.13 depicts various types of thermal power plants, which are generally described under the following paragraphs.

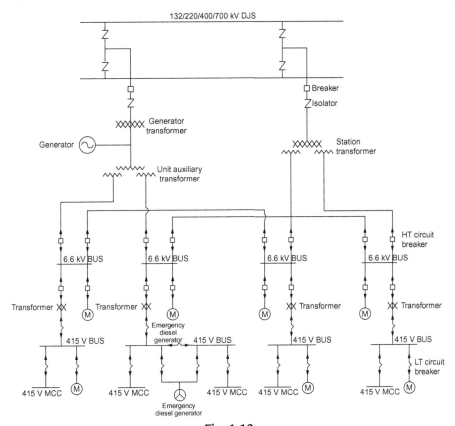

Fig. 1.12

Typical electrical distribution system. *Source: From Fig. 9.32, P 352. D.K. Sarkar, Thermal Power Plant—Design and Operation, 2015, Elsevier; Amsterdam, Netherlands.*

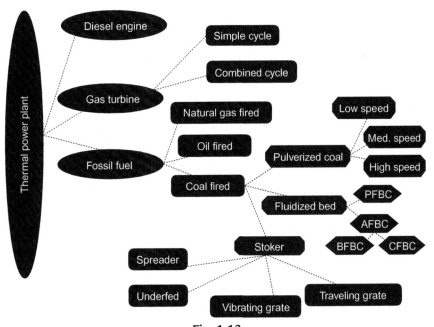

Fig. 1.13

Types of thermal power plants.

1.2 Steam Power Plant [1]

A steam power plant constitutes a steam generator (also termed as boiler), a steam turbine, a generator (also known as alternator), condenser, heaters, pumps, fans, and other auxiliaries. In the steam generator, chemical energy available in fossil fuel (coal, fuel oil, and natural gas) is converted to heat energy by combustion. The heat thus liberated is absorbed by low-temperature water to become high-energy steam in the steam generator. The thermal energy of the steam is converted to mechanical energy and then to electrical energy in the steam turbine and generator. The typical flow path of working medium, which is steam-water, in a steam power plant is shown in Fig. 1.14.

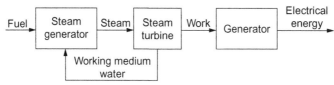

Fig. 1.14

Typical flow path of working medium. *Source: From Fig. 2.1, P 40. D.K. Sarkar, Thermal Power Plant–Design and Operation, 2015, Elsevier; Amsterdam, Netherlands.*

Air is fed into the furnace for combustion of fuel-forming products of combustion or flue gas. Flue gas passing over heating surfaces at various zones gets cooled and is discharged to atmosphere through a stack. The heat released by burning fuel is absorbed in different heat transfer surfaces—superheater, reheater, and economizer, to heat the working fluid and in the air heater to heat the ambient air prior to entering into the furnace.

The thermodynamic cycle that acts as the backbone of steam power plants is the *Rankine cycle*, which is described below.

The Rankine cycle is an ideal thermodynamic cycle involving the following processes:

  i. steam generation in a boiler at constant pressure
 ii. isentropic expansion in a steam turbine
iii. condensation in a condenser at constant pressure
iv. pressurizing condensate to boiler pressure by isentropic compression

The following states/processes are represented in *P–V* and *T–S* diagrams of the ideal Rankine cycle (Fig. 1.15):

- State 1: Condition of saturated vapor at temperature T_1 and pressure p_1.
- Process 1–2: The vapor then expands through the turbine reversibly and adiabatically (isentropically) to temperature at T_2 and pressure at p_2.
- State 2: The exhaust vapor is usually in the two-phase region at temperature T_2 and pressure p_2.
- Process 2–3: The low-pressure wet vapor, at temperature T_2 ($=T_3$) and at pressure p_2 ($=p_3$), is then liquefied in the condenser into saturated water and reaches the state 3.
- State 3: Condition of saturated liquid at temperature T_3 and pressure p_3.
- Process 3–4: The saturated water, at the condenser pressure p_3, is then compressed isentropically by the boiler feed pump to subcooled liquid at pressure p_4.
- State 4: Condition of subcooled liquid at temperature T_4 and pressure p_4.
- Process 4–1: The subcooled liquid at state 4 is heated at constant pressure p_4, in the "economizer" section of the steam generator, to a saturated liquid at state 5. The saturated liquid is further heated in the "evaporator" section of the steam generator to saturated vapor at constant temperature and constant pressure to its initial state (T_1 and p_1).

Efficiency of the ideal Rankine cycle is quite low, which may be improved by "superheating" and "reheating" the steam and by raising the temperature of feedwater at economizer inlet by "regenerative" heating of the feedwater, that is, by transferring heat from steam part of the cycle to the feedwater.

In Fig. 1.15, state 1′ corresponds to superheat condition of steam and state 2′ is the exhaust vapor phase after expansion of superheated steam in steam turbine.

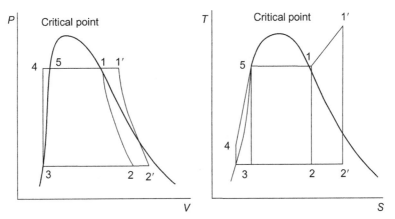

Fig. 1.15

Ideal Rankine cycle. *Source: From Fig. 1.9, P 16. D.K. Sarkar, Thermal Power Plant—Design and Operation, 2015, Elsevier; Amsterdam, Netherlands.*

1.2.1 Steam Generator

Based on its use, location, and so on, a steam generator may be of various types as discussed below.

1.2.1.1 Fire-tube steam generator

In a fire-tube steam generator, the furnace and the grate, on which coal is burnt, are located underneath the front end of the shell (Fig. 1.16). The gases pass horizontally along its underside to the rear, typically reverse direction, and pass through horizontal tubes to stack at the front. The fire tubes could be placed in a furnace horizontally or vertically or in an inclined plane.

Steam-generating capacity and outlet steam pressure of these steam generators are limited; hence, they are typically used for low-pressure steam. This type of steam generator was used on virtually all steam locomotives. Fire-tube steam generators are also typical of early marine applications and small vessels. They find extensive use in the stationary engineering field, typically for low-pressure steam use such as heating a building. These steam generators are susceptible to explosions.

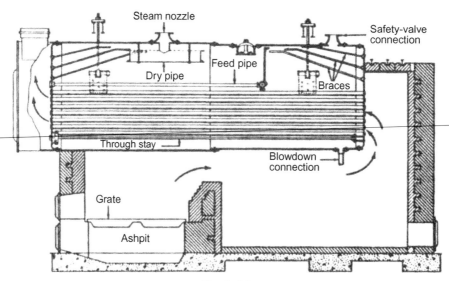

Fig. 1.16
Sectional view of a typical fire-tube steam generator. *Source: From Fig. 3–2, P 82. M. M. El-Wakil. Powerplant Technology. McGraw-Hill, Inc., 1984.*

1.2.1.2 Water-tube steam generator

Flue gases in a water-tube steam generator pass outside the tubes, thereby tubes get heated externally by the fire; water is circulated through the tubes for evaporation. Baffles are installed across the tubes to allow crossflow of flue gases to ensure maximum exposure of the tubes (Fig. 1.17).

In large utility steam generators, water-filled tubes make up the walls of the furnace called water-walls to generate steam and saturated steam coming out of boiler drum reenter the furnace through a superheater to becoming superheated.

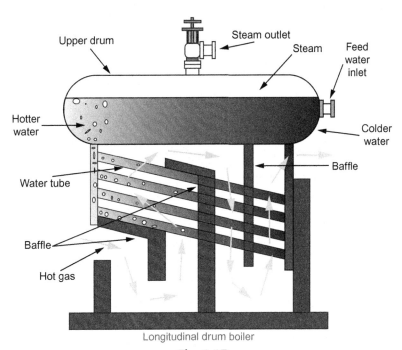

Fig. 1.17

Sectional view of a typical water-tube steam generator. *Source: Water Tube Boiler, Operation and Types of Water Tube Boiler. http://www.electrical4u.com/water-tube-boiler-operation-and-types-of-water-tube-boiler/.*

1.2.1.3 Natural circulation steam generator

A steam generator in which motion of the working fluid in the evaporator is caused by a thermosiphon effect on heating the tubes is called a "natural circulation steam generator." In this steam generator, water flows from drum through downcomers to the bottom of the furnace and does not receive any heat; water then moves up through risers or evaporator tubes, absorbs heat of combustion of fuel, becomes a mixture of water and steam, and returns to the drum (Fig. 1.18). Water inside riser tubes gets heated; thus, the density of ascending mixture of hot water and steam in riser tubes is much lower than the density of descending water in downcomers. This difference in density between relatively cool water and a mixture of hot water and steam causes the water to circulate in this steam generator. All natural circulation boilers are drum-type steam generators.

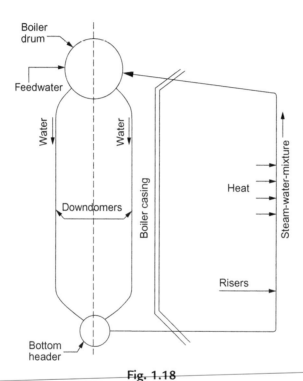

Fig. 1.18

Natural circulation steam generator. *Source: From Fig. 2.5, P 48. D.K. Sarkar, Thermal Power Plant–Design and Operation, 2015, Elsevier; Amsterdam, Netherlands.*

1.2.1.4 Forced-circulation steam generator

When the operating pressure of the working medium is 18 MPa or higher, the density difference between water in downcomers and water-steam mixture in risers is so small that it is difficult to maintain natural circulation. In such cases, the fluid flow in steam generator tubes is ensured by incorporating pumps in the downcomer circuit (Fig. 1.19). Hence, the name is *assisted or forced circulation*. Downcomer tubes from the steam generator drum are connected to a common header that serves as the suction header of the pumps. Discharge piping of these pumps is then connected to another header where riser tubes are issued.

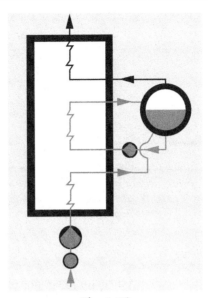

Fig. 1.19

Forced circulation steam generator. *Source: From Fig. 3–6, P 3–12, Boilers. Editor – Carl Bozzuto. CLEAN COMBUSTION TECHNOLOGIES (5th Edition). Alstom.*

1.2.1.5 Subcritical steam generator and supercritical steam generator

At the thermodynamic critical point of a steam-water cycle, the pressure of steam-water is 22.12 MPa, the temperature is 647.14 K, and the density is 324 kg m^{-3}. Steam generators that operate below the critical point condition are known as *subcritical steam generators*, while one operating above the critical point condition is termed a *supercritical steam generator*. While a subcritical steam generator could be of either drum type or once-through type, a supercritical steam generator is essentially of once-through type.

1.2.1.6 Drum-type steam generator

All drum steam generators are subcritical steam generators. A drum steam generator can be either the natural circulation type or the forced/assisted circulation type. The drum acts as a reservoir for the working medium. The lower portion of the drum with feedwater is called the water space, and the upper portion of the drum occupied by steam is called the steam space.

1.2.1.7 Once-through steam generator

These types of steam generators do not possess any boiler drum. Feedwater in this type of steam generator enters the bottom of each tube only once, continuously gets converted to steam, and discharges as steam from the top of the tube. As a result, there is no distinct boundary between the economizer, evaporator (located within the furnace), and superheater (Fig. 1.20). These steam generators can be operated either at subcritical or at supercritical pressures.

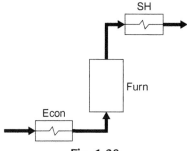

Fig. 1.20

Once-through steam generator. *Source: Figure 26 ©, P 5–18. Stultz SC, Killo JB, editors. STEAM ITS GENERATION AND USE. The Babcock and Wilcox Company. 41st Edition. 2005 Courtesy of The Babcock & Wilcox Company.*

1.2.1.8 Stoker-fired steam generator

In early stages of coal firing, the stoker-fired steam generator provided the most economical method for burning coal in almost all industrial steam generators rated less than 28 kg s^{-1} of steam. This boiler was capable of burning wide range of coals, from bituminous to lignite.

In a stoker-fired steam generator, coal is pushed, dropped, or thrown onto a grate to form a fuelbed. Coal may be fed to stokers either from above, *overfeed*, or from below, *underfeed*. Under the active fuelbed, there is a layer of fuel ash that, along with air flow through the grate, keeps metal parts at allowable operating temperature.

Traveling grate (Fig. 1.21), vibrating grate, and spreader are different type of stokers.

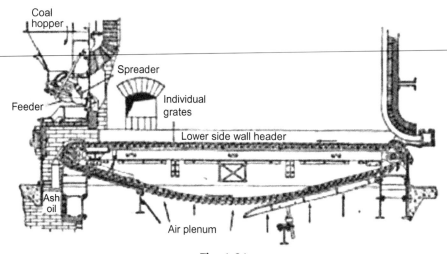

Fig. 1.21

Traveling-grate overfeed stoker. *Source: From Fig. 11, P 9–18, Section 9, Power Generation. Theodore Baumeister and Lionel S. Marks, Mechanical Engineers' Handbook (Sixth Edition), McGraw-Hill Book Company, Inc., 1958.*

1.2.1.9 Fluidized-bed steam generator

Fluidized-bed combustion ensures burning of solid fuel in suspension, in a hot inert solid bed material of sand, limestone, refractory, or ash, with high heat transfer to the furnace and low combustion temperatures (1073–1223 K). Fluidized-bed combustion is comprised of a mixture of particles suspended in an upwardly flowing gas stream.

By accelerating the upward flow of the gas stream, a situation can be reached when the buoyant force and drag force of particles overcome their gravity force and start floating in the gas; thereby, the combination of solid particles and gas exhibits fluidlike properties and ensures burning of solid fuel in suspension. Fluidized-bed combustor is amenable to firing a variety of solid fuels—coal, petro-coke, biomass, wood, coal washery rejects, and so on—with varying heating value, ash content, and moisture content in the same unit. In this steam generator, pollutants in products of combustion are reduced concurrently with combustion.

There are two different types of atmospheric fluidized-bed combustors: *bubbling fluidized bed* (BFB; see Fig. 1.22) and *circulating fluidized bed* (CFB; see Fig. 1.23).

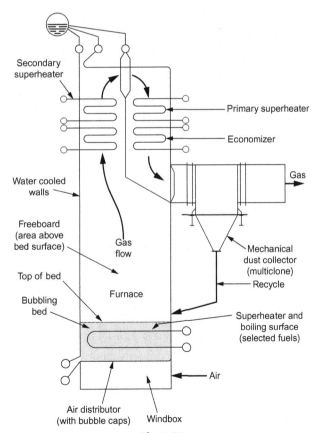

Fig.1.22

Gas flow path of a typical BFBC boiler. *Source: From Fig. 3, P 17–2, Fluidized-Bed Combustion Stultz SC, Kitto JB, editors. Steam Its Generation and Use (41st Edition). Courtesy of The Babcock & Wilcox Company.*

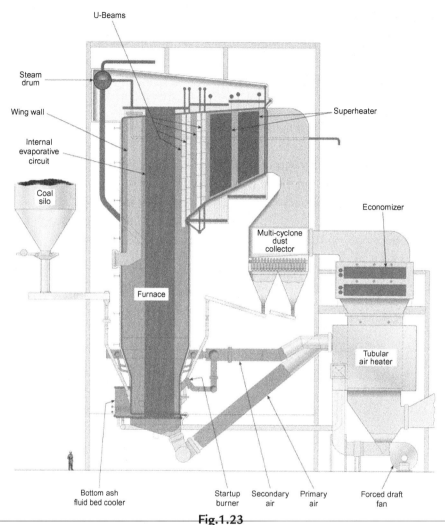

Fig.1.23

Sectional view of a typical CFBC boiler. *Source: From http://www.babcock.com/products/Pages/Circulating-Fluidized-Bed-Boiler.aspx.*

1.2.1.10 Pulverized coal-fired steam generator

In pulverized coal firing, fine particles of coal are fired in suspension. It resulted in more complete combustion of coal and higher system efficiencies. The process of pulverization of coal is carried out in two stages. In the first stage, raw coal is crushed to a size of not more than 15–25 mm. The crushed coal is then transferred to grinding mills, ie, pulverizers, where it is ground to a fine particle size.

The air that transports coal to the burner is called the primary air, while secondary air is introduced around or near the burner (Fig. 1.24). The burners impart a rotary motion to the coal-air mixture in a central zone and the secondary air around the nozzle—all within the burner. The rotary motion provides premixing for the coal and air along with some turbulence.

Based on their operating speed, pulverizers/mills are classified as low-, medium-, and high-speed mills.

Note

Detail treatment on pulverizers and their classification are discussed in Chapter 13.

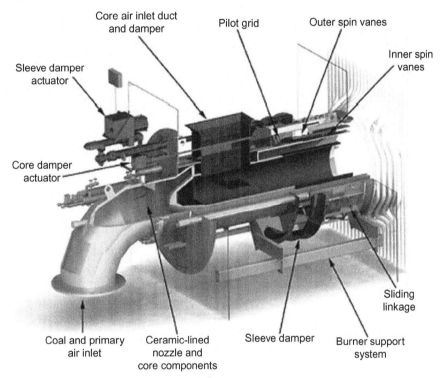

Fig.1.24

Sectional view of a coal burner. *Source: From http://www.babcock.com/products/Pages/AireJet-Low-NOx-Coal-Burner.aspx.*

1.2.2 Steam Turbine

In a steam power plant, high-energy steam, generated in a steam generator, expands through a steam turbine and undergoes changes in pressure, temperature, and heat content. During this process of expansion, the high-energy steam also performs mechanical work. The plant also consists of a condenser, where steam rejects heat in cooling water and returns to the original state (Fig. 1.25).

In a steam turbine, the inlet and exhaust pressures and the inlet temperature of the steam determine the theoretical energy available at the turbine inlet.

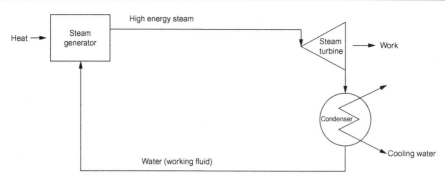

Fig.1.25

Schematic diagram of thermal power plant. *Source: From Fig. 6.1, P 190. D.K. Sarkar, Thermal Power Plant–Design and Operation, 2015, Elsevier; Amsterdam, Netherlands.*

Depending on its type, direction of steam flow, number of pressure stages, number of cylinders, and so on, steam turbines are classified as impulse, reaction, condensing, noncondensing, automatic extraction, controlled extraction, regenerative extraction, mixed-pressure, reheat, and so on, as discussed below. In practice a specific turbine may be described as a combination of two or more of these classifications.

1.1.2.1 Impulse turbine (Fig. 1.26)

An impulse turbine is comprised of a stage of stationary nozzles followed by a stage of moving blades. In this turbine, the potential energy of steam is converted into kinetic energy in nozzles. A complete process of expansion of steam with consequent pressure drop occurs only across

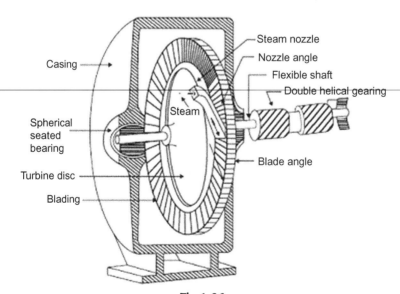

Fig.1.26

De Laval's impulse turbine. *Source: From Fig. 213, P412, CHXIV The Theory Of The Steam Turbine, D.A. Wrangham, Theory and Practice of Heat Engine, Cambridge University Press, 1960.*

these nozzles, with a net increase in steam velocity across the stage. Moving rotor blades absorb the kinetic energy of high-velocity steam jets and convert it to mechanical work, resulting in rotation of the turbine shaft.

1.1.2.2 Reaction turbine (Fig. 1.27)

A reaction turbine is constructed of rows of stationary blades and rows of moving blades. The stationary blades act as nozzles. The expansion of steam in the reaction turbine occurs both in stationary or guide blades and in rotating or moving blades such that the pressure drop is about equally divided between stationary and rotating blades. Upon expanding across the moving blades, steam comes out as a high-velocity steam jet that does the mechanical work of rotating the turbine shaft.

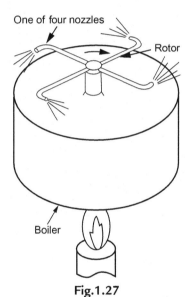

Fig.1.27
Pure reaction turbine. *Source: What are Reaction turbines? https://www.quora.com/What-are-Reaction-turbines.*

1.1.2.3 Axial turbine

In this turbine, steam flows in a direction parallel to the axis of the turbine.

1.1.2.4 Radial turbine

Steam flow in this turbine takes place in a direction perpendicular to the axis of the turbine.

1.1.2.5 Single-stage turbine

These turbines are mostly used for driving centrifugal compressors, blowers, and so on.

1.1.2.6 Multi-stage turbines

They are used primarily for generation of power and are made in a wide range of power capacities varying from small to large.

1.1.2.7 Single-shaft tandem compound turbine

Rotors of all the cylinders of this turbine are mounted on one and the same shaft and coupled to a single generator.

1.1.2.8 Multishaft cross-compound turbine

In this turbine, the rotor shaft of each cylinder is placed parallel to each other.

1.1.2.9 Turbine with throttle governing

In a turbine with throttle governing, fresh steam enters through one or more simultaneously operated control or throttle valves and the steam flow is controlled by the degree of opening of these valves. The upstream steam pressure at throttle valves will be constant, irrespective of the turbine loading, but the pressure after throttle valves will reduce as the valve opening reduces. Throttle governing is not suitable for part load operation or for fluctuating load demand.

1.1.2.10 Turbine with nozzle governing

In a turbine with nozzle governing, fresh steam enters through two or more consecutively opening control valves, which in turn is varied according to the load on the turbine. Nozzle-governed machines are preferable where the loading regime involves prolonged operation under part load conditions, because the turbine efficiency is higher at part loads due to reduced throttling losses.

1.1.2.11 Turbine with bypass governing

In this turbine, steam is normally supplied through a primary valve to meet the economic load. When the load demand exceeds the economic load, steam is also directly fed to one, two, or even three intermediate stages of the turbine.

1.1.2.12 Condensing turbine

Exhaust steam from that turbine, at a pressure less than atmospheric, is directed to the condenser. Steam is also extracted from intermediate stages of the turbine either for feedwater heating or for industrial and heating purposes.

1.1.2.13 Back-pressure turbine

In this turbine, extraction steam from intermediate stages as well as steam from turbine exhaust is supplied at various pressure and temperature conditions for meeting industrial process requirement and heating purposes.

1.1.2.14 Stationary steam turbine with constant speed of rotation

This turbine is primarily used for driving generators (alternators) to generate power.

1.1.2.15 Stationary steam turbines with variable speed of rotation

This is meant for driving turbo-blowers, air circulators, pumps, and so on.

1.1.2.16 Nonstationary steam turbines with variable speed of rotation

This turbine is usually employed in steamers, ships, and so on.

1.3 Gas Turbine Power Plant [1]

A *gas turbine*, also called a combustion turbine, is a rotary engine that extracts energy from a flow of combustion gas. It has an upstream compressor coupled to a downstream turbine and a combustion chamber in-between (Fig. 1.28). In the compressor, atmospheric air is sucked in at point 1 in the figure; consequently air is pressurized. This pressurized air then enters the combustion chamber, point 2, wherein fuel is sprayed for combustion to take place. Energy is released when air is mixed with the fuel (either liquid or gaseous) and ignited in the combustor. As a consequence of the combustion at constant pressure, the temperature of air vis-à-vis the mixture is increased. High-temperature products of combustion of fuel and air are directed over the turbine blades, point 3, get expanded, thereby spinning the turbine and at the same time powering the compressor. Finally, the gases are passed through additional turbine blades, generating additional thrust by accelerating the hot exhaust gases by expansion back to atmospheric pressure, point 4, and performing useful work. Energy is extracted in the form of shaft power that is used to power electrical generators, aircraft, trains, ships, tanks, and so on.

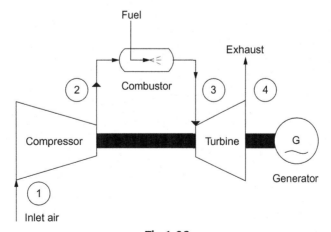

Fig.1.28
Simple cycle gas turbine power plant. *Source: Towards Energy Conversion in Qatar, http://file.scirp.org/ Html/5-2650044_40880.htm.*

Gas turbines are ideal for electricity generation in periods of peak electricity demand, since they can be started and stopped quickly, enabling them to meet energy demand. They are smooth running, and their completion time to full operation is the fastest compared to other power-generating plants.

When running in simple cycle mode, the heat rate of a gas turbine is very high compared to the heat rate of other prime movers. The heat rate can be substantially improved by operating the gas turbine in combined cycle mode (Fig. 1.29). In a combined cycle plant, high heat content in gas turbine exhaust gas is recovered in a heat recovery steam generator (HRSG) to generate high-pressure, high-temperature steam, which is then passed through a steam turbine to produce additional power.

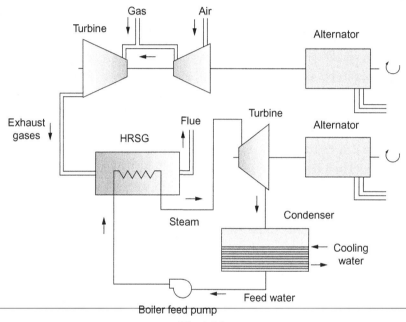

Fig.1.29

Combined cycle gas turbine power plant. *Source: An Overview of Combined Cycle Power Plant. http://electrical-engineering-portal.com/an-overview-of-combined-cycle-power-plant. Reproduced with permission from EEP.*

Principle of operation of a gas turbine is based on the thermodynamic cycle known as the Brayton cycle or the Joule cycle. Fig. 1.30 represents *P–V* and *T–S* diagrams of an ideal Brayton cycle. The cycle is comprised of two isentropic and two isobaric processes as described below.

- The compressor draws in ambient air and compresses it isentropically from position 1 to position 2 at pressure P_2.
- The compressed air is discharged into a combustion chamber, where fuel is burned, heating that air at constant pressure from position 2 to position 3.

- The heated, compressed product of combustion (gas) then creates energy as it expands through a turbine isentropically from position 3 to position 4.
- The gas is then cooled from position 4 to position 1 at constant pressure P_1.

In a closed cycle, cooling of gas takes place within a heat exchanger; but in an open cycle, gas is discharged to the atmosphere.

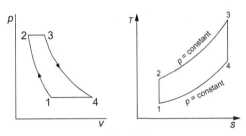

Fig.1.30

Ideal Brayton cycle or Joule cycle. *Source: From Fig. 7.7 and 7.8, P 245. D.K. Sarkar, Thermal Power Plant–Design and Operation, 2015, Elsevier; Amsterdam, Netherlands.*

1.4 Diesel-Generating Station [1]

A diesel engine is an internal combustion engine that uses the heat of compression of air to initiate ignition of fuel; a process by which fuel is injected after the air is compressed in the combustion chamber causing the fuel to self- ignite in the presence of high temperature compressed air.

Compared to the thermal efficiency of steam or gas turbines of equivalent size, the diesel engine has the highest thermal efficiency (50% or more) due to its very high compression ratio. In spite of diesel's high-efficiency, a diesel power plant is generally used as an emergency supply station because of the high cost of diesel oil.

Most diesel engines have large pistons and therefore draw more air and fuel, which results in a bigger and more powerful combustion. This is effective in large vehicles such as trucks, diesel locomotives, and generators.

Working in a diesel engine comprises four distinct processes: induction and cooling, compression, combustion, and expansion and exhaust (Fig. 1.31). The thermodynamic cycle in which these processes are represented is the ideal diesel cycle and *P–V* and *T–S* diagrams which are shown in Fig. 1.32.

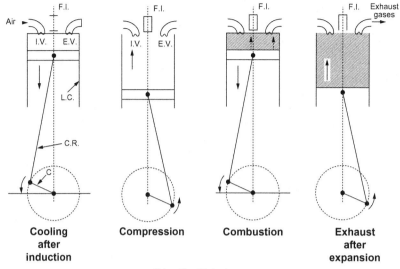

F.I. = Fuel injecter

I.V. = Inlet valve

E.V. = Exhaust valve

Fig. 1.31

Working processes of diesel engine. *Source: From Fig. 8.4, 8.5, 8.6 and 8.7, P 292 & 293. D.K. Sarkar, Thermal Power Plant—Design and Operation, 2015, Elsevier; Amsterdam, Netherlands.*

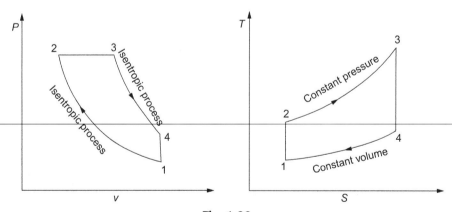

Fig. 1.32

Ideal diesel cycle. *Source: From Fig. 8.2 & 8.3, P 290 & 291. D.K. Sarkar, Thermal Power Plant—Design and Operation, 2015, Elsevier; Amsterdam, Netherlands.*

The cycle is described as:

- Process 1–2 is an isentropic compression process. The air is compressed isentropically through a volume ratio v_1/v_2. During this process, work is done by the piston compressing the working fluid.

- Process 2–3 is a reversible constant-pressure heat addition process. Heat is supplied by combustion of the fuel while the air expands at constant pressure to volume v_3 during this process.
- Process 3–4 is an isentropic expansion process to the original volume v_1. Work is done by the working fluid expanding on to the piston, which produces usable torque.
- Process 4–1 is a reversible constant volume heat rejection (cooling) process. During this process, heat is rejected by exhausting air until the cycle is completed.

Reference

[1] D.K. Sarkar, Thermal Power Plant—Design and Operation, Elsevier, Amsterdam, Netherlands, 2015.

CHAPTER 2

Quality Assurance and Quality Control (Applicable to Preoperational Activities)

Chapter Outline

2.1 Introduction

Webster's dictionary defines quality as "a degree of superiority" or "a grade of excellence." This is a generic definition of quality found in the dictionary in order to characterize quality with respect to excellence, fineness, or grade of excellence, eg, "silks of fine quality," "food of poor quality," "quality man (a person of high social position)," "quality (percentage by weight) of vapor in saturated steam," and so on. Qualitative analysis in *Chemistry* means analysis of a substance in order to ascertain the nature of its constituents [1].

The definition of quality above, however, does not apply to areas while delivering a product or adopting a process or carrying out an activity. In these cases, ISO 9000 quality systems are more relevant since per these systems many of the terms of quality have specific meanings and applications rather than generic definitions. ISO 9000:2015 defines "quality" as "the degree to which a set of inherent characteristics fulfills a set of requirements" [2]. If those characteristics meet all requirements (that is, characteristics are in "conformity" with requirements), high or excellent quality is achieved; but, if those characteristics do not meet all requirements (that is, there lies "nonconformity" or "partial conformity" between characteristics and requirements), a low or poor level of quality is achieved. However, while applied to preoperational activities all characteristics must fulfill all the requirements without any concession, lest plant operating personnel face nagging interruption during running a plant with consequent loss of revenue.

Thermal Power Plant. http://dx.doi.org/10.1016/B978-0-08-101112-6.00002-2

During the execution of a preoperational activity, a team of personnel of the supplier (contractor) gets involved to deliver their part of the service. So, to ensure that the quality of work or service rendered by the supplier will meet specified requirements of the customer, it is imperative that coordinated action among these personnel is established beforehand, so that on completion of that service, the supplier can show through objective evidence that he has fulfilled what he has agreed to the customer.

Note

In ISO 9001:2015 the term "supplier" (contractor) has been replaced with a new terminology "external provider." Since, ISO 9001:2008 will remain effective up to the next 3 years, the term "supplier" is retained in this chapter.

From the foregoing, it may be opined that the supplier must aim to enhance customer satisfaction; at the same time he must assure that applicable statutory and regulatory requirements have been met. Such confidence can be extended by a supplier only if he adheres to an established "quality assurance (QA)" and "quality control (QC)" program. An adequate QA/QC program helps in improving transparency, consistency, and completeness of the service provided by the supplier that in turn would enhance customer's confidence. While every person of the supplier's team associated with a service or an activity must be conscious in maintaining QC, customer's confidence can further be strengthened if supplier's representatives, engaged in QA program, act as an independent authority within supplier's organization; they will be not involved directly in the execution of a service or an activity, thereby conveying the message that they are working impartially.

The customer would feel assured about the quality of work undertaken and services rendered by the supplier if he adopts and casts his procedure in line with the guidelines laid in ISO 9000 Quality Management Systems. Potential benefits of adopting this standard are [3]:

(a) the ability to consistently provide products and services that meet customer and applicable statutory and regulatory requirements;
(b) facilitating opportunities to enhance customer satisfaction;
(c) addressing risks and opportunities associated with its context and objectives;
(d) the ability to demonstrate conformity to specified quality management system (QMS) requirements.

2.1.1 Quality Assurance

QA is a system or set of procedures intended to ensure that the delivery of a service or the quality of a product under development (before work is complete, as opposed to postexecution) is assessed and compared with specified requirements; it is implemented in a quality system and is focused on providing confidence that quality requirements will be met.

While QC is focused on fulfilling process outputs, that is, end results, QA comprises administrative and procedural activities, following which requirements for a product, service, or activity will be fulfilled. QA is a method of preventing mistakes or avoiding problems that may crop up during execution of a service, and that in a way is able to predict that the end result will fulfill the desired requirements of the customer. Thus QA is applied to the preproduction stage of a physical product or to the preexecution stage of a service.

As a part of the QA program, following "general precautionary measures" applicable to all preoperational activities must be observed for safe, smooth, and trouble-free execution of each activity. In addition "specific precautionary measures" that must be observed prior to executing individual preoperational activity are addressed in the concerned chapter.

General precautionary measures

  i. Fire-fighting system must be ready in all respects around the area of activity, and concerned personnel must be trained to operate the system.
  ii. It shall be ensured that the area of activity is well ventilated and adequately illuminated.
  iii. Area surrounding the test arena should be cordoned off with fluorescent tape/rope.
  iv. The area of activity must be free from any obstacle, including temporary piping, valves, ladders, scaffoldings, to ensure unrestrained movement of personnel associated with this activity.
  v. This area with associated appurtenance should not be the source of any hazard when an activity is getting executed.
  vi. Suitable warning signs to restrict entrance of unauthorized personnel inside cordoned area shall be displayed. Authorized personnel, associated with any activity, shall be provided with "access passes."
  vii. Authorized personnel must have proper knowledge of all attributes of an activity.
  viii. Display "no smoking," "danger," "keep off," and other safety tags (in regional and English languages) at various places around the area of activity.
  ix. Tanks, pumps, fans, compressors, blowers, valves, dampers, gates, associated equipment and systems, which are in service or are energized, shall be provided with proper safety tags (Appendix C) so as to obviate any inadvertent operation or to obviate any injury to equipment and personnel.
  x. Proper communicating system shall be established among all concerned personnel through walky-talkies, a public address system, or any other suitable means.

2.1.2 Quality Control

QC is defined as a system or a set of procedures intended to verifying and maintaining a desired level of quality in a product or service by careful planning, use of proper equipment, continued inspection, and corrective action, as and when required, in order to fulfill agreed quality requirements. A QC program involves thoroughly examining and testing services pertaining to specific preoperational activity. The basic intention of adopting a QC process by the supplier is to meet specific requirements of the customer, thereby ensuring that the products or services provided by the supplier are dependable, satisfactory, safe, and economic.

In order to implement an effective QC program, the supplier must first develop a procedure applicable to a specific preoperational activity, which will produce satisfactory results. Once the procedure is approved by the customer, the engineer, and, if required, the original equipment manufacturer (OEM), the supplier may execute the activity when the "front" is made available to him. During the execution period of a specific activity, the QC process must remain ongoing to ensure that the activity is carried out in line with the approved procedure, so that occurrence of any trouble could be detected and resolved by adopting suitable remedial measures without much difficulty, lest failure to get the desired result would compel the supplier to repeat the activity at no extra charge until the customer is satisfied with the result.

2.2 ISO 9000:2015—Quality Management Systems—Fundamentals and Vocabulary

The intrinsic objective of adopting ISO 9000 Quality Management Systems by a supplier is to ensure the following:

  i.   the QC system is in place and effective;
 ii.   the QA system is implemented at the supplier's end;
iii.   the supplier is in a position to deliver quality products and services;
 iv.   the supplier's product will fulfill customer satisfaction;
  v.   the supplier's products and processes improve continually to meet customer needs and expectations, which may change with time. It is more so because of ceaseless pressures from suppliers' competitors and advancements on technical fronts.

The customer may not feel high satisfaction even after all contractual requirements are fulfilled by the supplier. Hence, from the supplier's point of view he must always make an all-out effort to get a document from the customer certifying that "supplier has met all specified requirements," which the supplier may use in the future as his credential to other prospective customers. At the same time, the supplier must analyze customer requirements, improve his services for the satisfaction of both current and prospective customers, and keep his services under control. Thus adopting ISO 9000 will help in developing the supplier's business growth.

2.2.1 Definitions

It is clarified in Section 2.1 above that meaning of a word as given in a dictionary is quite different from that adopted in ISO Quality Management Systems. Hence, in line with the ISO, the following definition is relevant to the words applicable to preoperational activity [2].

Characteristic It is "a distinctive feature or property of an object, which can be inherent or assigned and can be qualitative or quantitative."

Competence Competence of a supplier means he is able to apply knowledge and skill to achieve specified results and is qualified to do the job.

Complaint When a customer is not satisfied with a product or a service, his expression of dissatisfaction is lodged as a complaint. The supplier must respond to such complaint without any rider.

Concession A concession is a special approval that is granted by a customer for a specific use and limited time to release a nonconforming product or service by the supplier.

Conformity It means "fulfillment of a requirement."

Contract A contract is a binding agreement between a customer and a supplier.

Corrective action This is the process of eliminating the causes of existing nonconformities and ensuring at the same time prevention of further recurrence of the same nonconformity.

Customer A customer is a recipient of a product or a service provided by a supplier. A customer could be an individual or an organization. He must identify and define his needs before procuring a specific job or service.

Customer satisfaction Customer satisfaction is a feeling or perception. It is a matter of degree to which a customer's specified requirements are fulfilled by the supplier.

Defect When a product or service fails to meet specified or intended requirements, this is termed as a defect. It is one of the various types of nonconformity.

Documented information The documented information is controlled and maintained and includes all the information that a customer or a supplier needs to operate and record.

Feedback The term "feedback" refers to comments or opinions expressed by a customer about a product or a service.

Grade Grade is the status given to different quality requirements of products/services having identical functional use.

Improvement Improvement is a set of activities that a supplier carries out in order to get better results.

Infrastructure This refers to the entire system of facilities, equipment, technologies, and support services that a supplier needs to carry out his obligations.

Knowledge Knowledge is a collection of information that is construed as true with a high level of certainty.

Management system A management system is a set of interrelated or interacting elements to formulate policies and objectives of an organization; it also establishes that policies are followed and objectives are achieved.

Monitoring Monitoring is the mode of determining the status of an activity, a process, or a system at different stages or at different times through supervision, check, and observation.

Nonconformity Nonconformity means failure of fulfilling a requirement.

Object It can be either imaginary (services, ideas) or tangible (products, tools).

Objective It is the result a customer or a supplier intends to achieve.

Objective evidence Objective evidence is depicted by end results, measurements, observations, and so on.

Opportunity Opportunities can arise as a result of a situation favorable to achieving an intended result (eg, attract customers, develop new products/services, reduce wastes, improve productivity). A positive deviation arising from a risk can provide an opportunity, but not all positive effects of risk result in opportunities.

Output An output is the result of a process and can be either tangible or intangible. There are four generic categories of output; these are services, software, hardware, and processed materials.

Performance Performance refers to measurable results that are achieved through activities, processes, products, services, and so on.

Process A process is either a set of activities that are interrelated or that interact with one another. A process uses resources to transform inputs into outputs.

Product A product is the result of a process. A product may be either a tangible output, eg, hardware, processed materials, or an intangible output, eg, services, or software. A product of a preoperational activity is both tangible and intangible.

Quality Quality is defined as the degree to which a set of inherent characteristics fulfills a set of requirements. A quality is said to be high or excellent when those characteristics meet all requirements, it is termed as low or poor when those characteristics do not meet all requirements.

In a contractual environment, meeting specified needs or guaranteed parameters is termed as fulfillment of the quality.

Quality objective A quality objective is derived from an organization's quality policy so that the organization can achieve its intended quality result.

Quality policy A quality policy should be consistent with the overall policy of the organization and express top management's commitment to the QMS. The object of the quality policy must be achievable. It should provide also the framework to set quality objectives. The language of quality policy must be easy to understand.

Regrade Improvement of the grade of a nonconforming product/service is made in order to make it conform to the specified requirement.

Repair Repair refers to removal of a defect.

Requirement A requirement of a customer is "a need or expectation either stated (specified requirement by customer) or implied (as, for example, expectation from the end result on completion of a preoperational activity)." Needs may change with time for the same customer or needs may vary from customer to customer.

Rework The process by which nonconformity of a product or service is removed is called rework.

Risk Risk is the effect of uncertainty, and any such uncertainty can have positive or negative effects. Risk-based thinking is essential for achieving an effective QMS.

Service A service is an intangible output, eg, advice, warranty, repair, delivery, and transportation, provided by the supplier to the customer. It is the result of a process.

Statutory requirement A statutory requirement is defined by a legislative body and is obligatory on the part of supplier to fulfill these requirements.

Supplier A supplier provides a product or a service to a customer. The supplier could be an individual or an organization. Before extending a service, supplier must address how he will fulfill the expectations or needs of a customer and also will meet applicable statutory and regulatory requirements by establishing and maintaining documented procedures.

System A system is defined as a set of interrelated or interacting elements.

Validation Validation uses objective evidence to confirm that the requirements which define an intended use or application have been fulfilled.

Verification Verification uses objective evidence to confirm that specified requirements of a product or a service have been met.

2.3 ISO 9001:2015—Quality Management Systems—Requirements

Note

1. The following passages provide only informative sections of ISO 9001:2015 and do not contain the complete content of this standard. To view the complete content, readers are advised to refer to the published standard itself.
2. Customers would feel assured about the quality of work undertaken and services rendered by suppliers if suppliers establish and integrate this standard into their QMS [4].
   This standard brings quality management and continual improvement into the heart of an organization.

ISO 9001:2015 specifies requirements for a QMS when an organization:

(a) needs to demonstrate its ability to consistently provide products and services that meet customer and applicable statutory and regulatory requirements; and

(b) aims to enhance customer satisfaction through the effective application of the system, including processes for improvement of the system and the assurance of conformity to customer and applicable statutory and regulatory requirements.

All the requirements of ISO 9001:2015 are generic and are intended to be applicable to any organization, regardless of its type or size, or the products and services it provides. The approach of this standard is "Management of the processes and the system as a whole can be achieved using the PDCA (Plan-Do-Check-Act) cycle with an overall focus on risk-based thinking aimed at taking advantage of opportunities and preventing undesirable results." The structure of this approach is shown in Fig. 2.1. From this figure it is evident that clauses 4–10 are consolidated in the PDCA cycle. Hence, for general guidance of the reader or the supplier, these clauses are briefly described below [5].

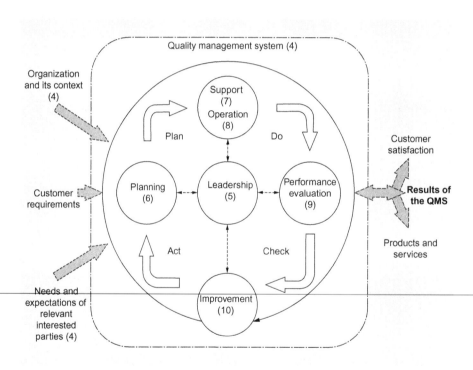

Fig. 2.1
Structure of ISO 9001:2015 in the PDCA cycle. *From Fig. 2, Clause 0.3.2, ISO 9001:2015 – Quality Management Systems – Requirements. Copyright remains with ISO.*

Clause 4: Context of the organization. This clause gives an organization the opportunity to identify and understand the factors and parties in their environment that support the QMS.

Clause 5: Leadership. The purpose of this clause is to demonstrate leadership and commitment by leading from the "top management," which may be an individual or a group of people.

Clause 6: Planning. The organization will need to plan actions to address both risks and opportunities, how to integrate and implement the actions into its management system processes, and evaluate the effectiveness of these actions. Another key element of this clause is the need to establish measurable quality objectives. The last part of the clause considers planning of changes which must be done in a planned and systemic manner.

Clause 7: Support. This clause ensures that there are the right resources, people, and infrastructure to meet the organizational goals.

Clause 8: Operation. This clause deals with the execution of the plans and processes that enable the organization to meet customer requirements and design products and services.

Clause 9: Performance evaluation. Performance evaluation covers requirements for monitoring, measurement, analysis, and evaluation; this clause also deals with what needs to be measured, methods employed, when data should be analyzed and reported on, and at what intervals.

Clause 10: Improvement. This clause stipulates that organizations should determine and identify opportunities for improvement, such as improved processes, products, and services, to enhance customer satisfaction.

2.4 Procedure

A written QA/QC procedure is a fundamental element of a QA/QC program. Implementing QA/QC procedure requires resources, expertise, and time. Hence, this program outlines QA/QC activities to be performed, the personnel responsible for these activities, and the schedule for completing these activities [6].

In order to implement the QA/QC program pertaining to each preoperational activity effectively, the supplier must first identify a "leader" of this program along with his contact details. Other members of the QA/QC team, who will act under the guidance of the leader, shall be also identified.

Responsibilities of the QA/QC leader are given below:

i.   act as the "single point contact" between the supplier and the customer;
ii. clarify and communicate QA/QC responsibilities to each member of his team;
iii. develop and maintain a QA/QC checklist applicable to each preoperational activity. During the preparation of this checklist, time management shall also be addressed;

Note

Each chapter on preoperational activity describes its applicable QA/QC checklist.

iv.   make a list of apparatus/instruments required for testing and analysis of chemicals, samples, specimens, and so on;

v.   make a list of safety gadgets as applicable to a specific preoperational activity;

vi.   ensure that requisite quantity of clarified water, DM water, chemicals, oil, gas, air, and so on, are available;

vii.   arrange compressors, blowers, pumps, hydraulic jacks, gagging devices, applicable instruments, and so on, as required;

viii.   verify the completion of each item of this checklist before executing the preoperational activity;

ix.   ensure that a specific preoperational activity is executed in accordance with the approved written procedure, thus fulfilling the specified requirement of the customer;

x.   on completion of a specific preoperational activity, record and maintain a protocol jointly signed by the customer, the engineer, the OEM (whenever required), and the supplier certifying that the activity is executed successfully.

References

[1] Webster's Encyclopedic Unabridged Dictionary of the English Language, Gramercy Books, New York/Avenel, New Jersey, 1996.

[2] ISO 9000:2015—Quality Management Systems—Fundamentals and vocabulary.

[3] ISO 9001:2015—Quality Management Systems—Requirements.

[4] S.K. Sarkar, S. Saraswati, Construction Technology, Oxford University Press, New Delhi, 2008.

[5] Moving from ISO 9001:2008 to ISO 9001:2015 Transition guide. http://www.bsigroup.com/LocalFiles/en-GB/iso-9001/Revisions/ISO-9001-FDISTransition-Guide-FINAL-July-2015.pdf.

[6] Template 3: Description of QA/QC Procedures. https://www3.epa.gov/.

PART 1

Preoperational Cleaning of Various Sub-Systems

Alkali Flushing of Preboiler System

3.1 Introduction

The preboiler system of a steam power plant constitutes condensate, feedwater, and feedwater heater drains systems. Alkali flushing of these systems of both new and running units is carried out to make these systems free of undesirable elements, lest these elements will contaminate the flowing fluid, viz. condensate and feedwater, and delay either commissioning of new steam generators or re-commissioning of running steam generators.

In new units undesirable elements like dirt, oil, grease, rust, welding slag, and so on, may develop and remain in various piping systems during manufacturing, transportation, storing, and erection stages [1].

In running units, deposits may grow inside feedwater pipe lines due to improperly controlled water treatment or process contamination (Fig. 3.1) [2]. These deposits impair heat transfer and may lead to corrosion originating from concentration of steam generator water salts.

Fig. 3.1

Internal deposits resulting from poor water treatment. *From Fig. 3, P 42–3, Chapter 42: Water and Steam Chemistry, Deposits and Corrosion STEAM Its Generation and Use (41st Edition). Courtesy of The Babcock & Wilcox Company.*

Corrosion inside pipe lines of a preboiler system of operating units may also take place from de-mineralized makeup water (which is slightly acidic; pH ranges from 6.8 to 7.0) if pH of the flowing fluid falls below the recommended limit (Tables 3.1 and 3.2) due to inadequate dozing of suitable inhibitor, eg, ammonia (NH_3), which is dozed if non-copper based alloy material, like SS, Ti, and so on, is used in pipe lines/tubes of water-steam circuit; for copper based alloy material, however, morpholine (C_4H_9NO), cyclohexylamine ($C_6H_{13}N$), and so on, is dozed as inhibitor. Corrosion products from feedwater pipe lines may get carried over to steam generators and deposited in steam generator tubes. In the event that corrosion rate is severe, feedwater pipe lines may rupture (Fig. 3.2) due to thinning of pipe lines [2]. Excessive dozing of inhibitor is also very harmful and may cause caustic gouging (Figs. 3.3 and 3.4) and damage pipe lines.

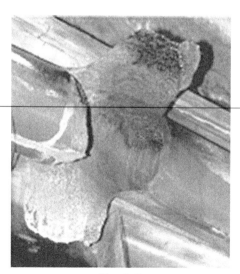

Fig. 3.2

Rupture of feedwater pipe. *From Fig. 14, P 42–16, Chapter 42: Water and Steam Chemistry, Deposits and Corrosion STEAM Its Generation and Use (41st Edition). Courtesy of The Babcock & Wilcox Company.*

Fig. 3.3
Caustic gouging.

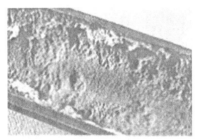

Fig. 3.4
Caustic gouging.

The discussion above establishes the importance of alkali flushing of a preboiler system. Hence, irrespective of whether the steam generator is new or old, having a clean preboiler system would ensure availability of the recommended chemical condition of condensate, feedwater, and steam generator water for smooth, trouble-free operation of a unit.

In modern high pressure, high-temperature large-drum-type steam generators and once-through steam generators purity of condensate and feedwater that shall be maintained is presented in Tables 3.1 and 3.2. The purity of steam generator water of high-pressure, high-temperature drum-type steam generators shall conform to the requirements laid down in Table 3.3. In once-through steam generators, since there is no provision of blow down, purity of feedwater must conform to the purity of superheated steam (Table 6.1).

Note

Normally extraction steam system is cleaned mechanically. This system, however, may be also included in the alkali flushing scheme if recommended by the turbine manufacturer.

Table 3.1 Analysis of condensate

Contaminants	Unit	Normal Condition (Maximum)
Total dissolved solids	$\mu g\ kg^{-1}$	100
Silica (as SiO_2)	$\mu g\ kg^{-1}$	20
Chloride (as Cl)	$\mu g\ kg^{-1}$	10
Sodium (as Na)	$\mu g\ kg^{-1}$	10
Iron (as Fe)	$\mu g\ kg^{-1}$	20
Copper (as Cu)	$\mu g\ kg^{-1}$	3
pH	–	9.0–9.6
Specific conductivity, at 298 K	$\mu S\ cm^{-1}$	4–11
Conductivity (after cation exchanger), at 298 K	$\mu S\ cm^{-1}$	0.2

Table 3.2 Analysis of feed water at economizer inlet

Contaminants	Unit	Drum Boiler	Once-Through Boiler	
		Normal Condition (Maximum)		
		All-Volatile Treatment	All-Volatile Treatment	Oxygenated Treatment
Total hardness (as $CaCO_3$)	$\mu mol\ L^{-1}$	3	3	1
Dissolved oxygen	$\mu g\ kg^{-1}$	7	7	50–150
Iron (as Fe)	$\mu g\ kg^{-1}$	20	20	5
Copper	$\mu g\ kg^{-1}$	5	5	1
Silica (as SiO_2)	$\mu g\ kg^{-1}$	20	20	20
Sodium + Potassium (as Na + K)	$\mu g\ kg^{-1}$	5	5	5
Hydrazine	$\mu g\ kg^{-1}$	10–50	10–50	10–50
pH	–	9.2–9.6	9.2–9.6	8.0–8.5
Specific conductivity, at 298 K	$\mu S\ cm^{-1}$	4–11	4–11	2.5–7.0
Conductivity (after cation exchanger), at 298 K	$\mu S\ cm^{-1}$	0.2	0.2	0.2
Oil Content	$mg\ kg^{-1}$	0.3	0.3	0.3

$\mu g\ kg^{-1} = ppb$; $mg\ kg^{-1} = ppm$.

Table 3.3 Analysis of steam generator water of drum-type steam generators

Contaminants	Unit	Normal Condition (Maximum)
Total dissolved solids	$mg\ kg^{-1}$	15
Suspended solids	$mg\ kg^{-1}$	1
Silica (as SiO_2)	$mg\ kg^{-1}$	8

$mg\ kg^{-1} = ppm$.

3.2 Description of Preboiler System [3]

As discussed above, the preboiler system of a steam power plant constitutes condensate, feedwater, and feedwater heater drains systems. Each of these systems is described in the following paragraphs.

Condensate system (Fig. 3.5): Steam after expansion in the low pressure (LP) turbine is condensed in the condenser. The main condensate is collected in the hotwell, which is then pumped by condensate extraction pumps (CEPs) to the deaerator through steam jet-air ejectors (not required if the vacuum pump is used), gland steam condenser (GSC), and LP feedwater heaters, where condensate is successively heated by extraction steam (Fig. 3.6). During this process, condensed steam collected from different heaters/coolers in the cycle joins the main condensate flow, is heated in the low-pressure heaters, and is finally stored in the feedwater storage tank of the deaerating heater or deaerator. A common minimum flow recirculation line (in large plants each CEP may be provided with a dedicated recirculation line) is provided downstream of the GSC to the hotwell to protect the running pump/s in the event that throughput to deaerating heater falls below a minimum flow limit. Each of the LP feedwater heaters is provided with individual isolation valves as well as a bypass valve.

Ammonia/morpholine is dozed in the condensate line to maintain pH and conductivity of condensate water (Table 3.1). Dissolved oxygen (DO) in condensate downstream of the hotwell should not exceed 15 μg kg^{-1}.

Feedwater system (Fig. 3.5): Feedwater flows from the deaerator feed storage tank, which is the beginning of the feedwater system, to the suction of boiler feed pumps (BFPs) through booster pumps (BPs) (if provided). High-pressure discharge from BFP flows through the high pressure (HP) feedwater heaters before entering the economizer through the feed regulating/control station, which regulates feedwater flow to the steam generator to maintain boiler drum level. Each BFP is provided with a minimum flow recirculation line for protection of the pump, which ensures the minimum flow through the pump when throughput to the steam generator is stopped or falls below the minimum flow limit. Each of the HP feedwater heaters is provided with individual isolation valves as well as a bypass valve. Hydrazine is dozed at the suction of BFPs to maintain pH and DO of feedwater (Table 3.2).

The HP feedwater heaters receive extraction steam from the cold reheat line and IP (intermediate pressure) turbine (Fig. 3.6). The steam and feedwater flow through HP heaters in a counter-flow arrangement. Feedwater successively flows through subcooling (for final cooling of drain condensate), condensing (for condensing steam), and separate de-superheating (for de-superheating extraction steam) zones provided in the HP heaters and absorb heat from the extraction steam as it passes through heaters.

The feedwater system supplies properly heated and conditioned water to the steam generator and maintains the boiler drum level in accordance with the steam generator load. This system also sends water to the steam generator superheater attemperator, reheater attemperator, auxiliary steam de-superheaters, HP/LP bypass de-superheaters, and so on.

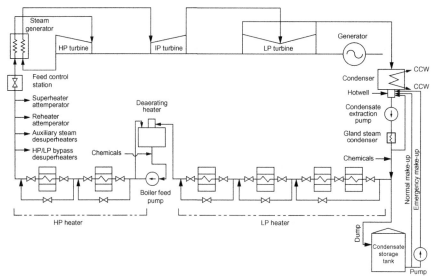

Fig. 3.5

Condensate, condensate make-up and dump system and feedwater system.

Feedwater heater drains system (Fig. 3.6): The condensed steam accumulated in each closed feedwater heater shell needs to be separated from its heat source, the extraction steam, and be cascaded to the next lower-pressure heater as the heater drain. In the event that the next lower-pressure heater is not available or the level in the heater under consideration reaches a high level, the drain from the affected HP heater should be diverted to the deaerating heater and the drain from the affected LP heater diverted to the condenser. An alternate drain to the condenser from each heater is also provided. All heater drains are the gravity type.

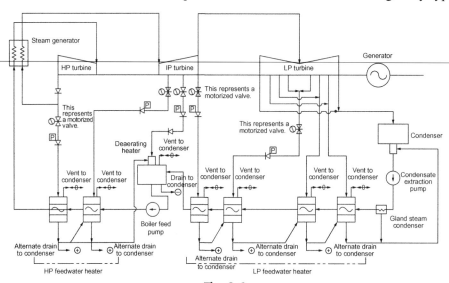

Fig. 3.6

Extraction steam, heater drains, and vents system.

3.3 Precautions

Over and above "general precautionary measures" described under Section 2.1.1 in Chapter 2, the following specific precautionary measures must be observed before executing alkali flushing of a preboiler system:

i. Alkali solution is injurious to health. When it comes in contact with the body, alkali solution may burn skin; alkali fumes may cause injury to eyes. Hence, all probable safety measures shall be exercised by operating personnel while handling alkali solution. They must keep on wearing alkali proof hand-gloves, face mask, and protective clothing/apron as well as protective glasses for eyes when working with caustics and as long as the cleaning process is continued. If contact is made with caustics, operating personnel must immediately flush with fresh water;

ii. In the area surrounding the chemical-dissolving tank, circulating pumps should be adequate so that any spillage of chemicals from the system does not damage any permanent equipment;

iii. During the process of alkali flushing, dirt and debris from various parts of the system get collected into the chemical-dissolving tank and may in turn infiltrate and choke the suction filters of the running chemical-circulating pumps. Choking of the suction filter could be ascertained by observing a fall in pump discharge pressure. On occurrence of such an event, the standby chemical-circulating pump/s shall be put into service; then the running pump is to be stopped and its suction filter is to be cleaned and placed in position. This pump will then act as a standby pump;

iv. Emergency stop push buttons must be provided in the vicinity of the flushing area for stopping chemical-circulating pumps as may be necessary;

v. Electrical installation shall be at a safe distance from pumps and care shall be taken to ensure protection against rain.

3.4 Prerequisites

Prior to carrying out the alkali flushing process, a thorough inspection of systems under purview is essential to ascertain the following (Table 3.4):

Table 3.4 Areas/items to be checked

Sl. No.	Areas/Items	Ok (√)
1	Before erecting the piping systems, protective coating used on internal surfaces is removed as far as practicable, as a safeguard against corrosion during transport and storage.	
2	Prior to erecting all temporary piping systems, they are thoroughly cleaned mechanically.	
3	All piping systems, whether permanent or temporary, are erected and supported properly along with associated valves. Protocol signed jointly by the customer, the engineer, and the supplier (contractor), certifying completion of erection of piping systems, is verified.	

Continued

Table 3.4   Areas/items to be checked—cont'd

Sl. No.	Areas/Items	Ok ($\sqrt{}$)
4	All welded and flanged joints are secured.	
5	All permanent pipe lines are hydrotested at 1.5 times the design pressure. Protocol signed jointly by the customer, the engineer, and the supplier (contractor), certifying acceptability of the hydro test, is verified.	
6	Trays are removed from inside the deaerator.	
7	Thermal insulation with lagging is provided on the following: i. Condensate extraction pump (CEP) discharge pipe line up to deaerator; ii. Deaerator; iii. Boiler feed pump (BFP), suction pipe lines; iv. BFP recirculation pipe lines; v. BFP discharge pipe line up to economizer; vi. High-pressure and low-pressure feedwater heater drain pipe lines	
8	Construction materials, in regard to temporary scaffolding, wooden planks, welding rod ends, cotton wastes, and so on, are removed from the condenser shell, tube bundles, hotwell, and deaerator, along with the feedwater storage tank.	
9	The condenser shell, tube banks, and hotwell are made free of accumulated dirt by water-washing prior to boxing up the condenser.	
10	The tube side of the feedwater heaters, air ejector aftercoolers (if provided), drain coolers, and gland steam condenser, are bypassed.	
11	Instruments installed on the permanent piping systems are isolated.	
12	Main condensate flow measuring orifice plate is removed from the condensate piping close to the deaerator, and a suitable spool piece is installed in that place.	
13	Flow nozzles in each BFP recirculation line and in the main feed water line are removed and replaced with suitable spool pieces.	
14	The BFP, booster pump (BP) (if provided), and CEP are isolated from the flushing circuit and bypassed by erecting temporary pipe lines across them.	
15	The main condensate flow control valve and feedwater heater drip control valves are kept closed. Bypass to these control valves are kept open.	
16	Feed control station control valves are kept closed, keeping their bypass valve fully open.	
17	Erection of temporary chemical-cleaning pumps is complete.	
18	A temporary tank for dissolving chemicals, along with associated fittings and attachments, is erected.	
19	Temporary piping fitted with temporary valves is erected.	
20	Temporary instruments are in place.	
21	Temporary platforms are provided wherever required.	
22	Estimated quantity of clarified water (Section 3.7.2) is available.	
23	Estimated quantity of DM water (Section 3.7.3) is available.	
24	Estimated quantity of chemicals (Section 3.7.4) is available.	
25	Apparatus required for chemical analysis (Section 3.4.1) is arranged.	
26	Safety gadgets (Section 3.4.2) are arranged.	
27	Verify that safety tags, eg, "no smoking," "danger," "keep off," and so on (in regional and English languages) are displayed wherever required.	
28	Verify also that tanks, pumps, valves, associated equipment, and systems, which are in service or are energized, are provided with proper safety tags.	

3.4.1 Typical List of Apparatus

Apparatus that are typically required for chemical analysis during alkali flushing of a preboiler system are listed below.

Size and quantity of each of these items would vary from project to project and shall have to be assessed by the chemist associated with cleaning activities:

1. titrating flasks
2. standard flasks
3. polyethylene bottles
4. measuring cylinders
5. pipettes
6. funnels
7. colorimetric cylinders
8. platinum crucible
9. photo colorimeter for determination of phosphate and silica
10. pH meter for determining pH of the solution
11. special pH papers of range 0–12
12. conductivity meter for determining conductivity of the solution
13. electric hot plate

3.4.2 Typical List of Safety Gadgets

The following equipment, generally used for safety of operating personnel during alkali flushing, are arranged:

1. gum boots of assorted sizes
2. rubber gloves
3. asbestos gloves
4. rubber or polythene aprons
5. hard hats of assorted sizes
6. side-covered safety goggles with plain glass
7. gas masks
8. first-aid box
9. safety sign boards
10. safety tags

3.5 Preparatory Arrangements

a. A temporary tank will be required for dissolving chemicals for carrying out the alkali flushing process. This tank shall be an atmospheric tank and serve as the suction tank for temporary chemical-circulating pumps. Chemical solutions supplied by the circulating pumps will drain back to this tank, thus establishing a continuous circulation. The tank shall be provided with suitable overflow, drain and sampling connections, vent outlets, and so on. Since it would be prudent to use the same tank for carrying out "alkali flushing of the preboiler system" and "chemical cleaning of the steam generator" (Chapter 6), this tank may be located at a convenient place between the steam generator and the TG hall at an elevated level from the ground;

b.   This tank shall be provided with a bubbler tube for admitting heating steam into the solution;

Notes

1. For an extension unit, this steam may be sourced from an existing auxiliary steam system.
2. For a grass-root unit, steam may be supplied from a temporary, saturated steam-producing steam generator.
3. Alternately, temperature of clarified water may be raised by continuous recirculation of water, keeping all drain valves closed. This procedure, however, is time consuming and may extend up to 8 h.

c.   Four (4) temporary chemical-circulating pumps, each of 50% capacity (preferable) with suction filters shall be installed at the ground level just outside the space below the chemical-dissolving tank. Alternately, $2 \times 100\%$ temporary chemical-circulating pumps with suction filters may be also used.

Note

For further clarification on selecting the number of pumps, internal surface of pump casing, and material of impeller, the reader may refer to "chemical cleaning of steam generator" (Chapter 6).

d.   In order to complete the circuit for circulation of flushing fluid, some temporary piping connecting the chemical-dissolving tank and the permanent piping systems should be erected and kept ready for putting into service.

e.   It is apparent that while carrying out alkali flushing, some temporary valves, pressure gauges, thermometers, and so on, are required to be installed;

Notes

1. These temporary connections are unit specific and vary from project to project. Hence, layout of temporary piping shall have to be jointly decided at site among the owner, the engineer, and the contractor responsible for successful completion of above cleaning activity.
2. Size and quantity of temporary piping materials, including elbows, tees, valves, supports, and so on, shall be assessed at the design stage. These requirements, however, may vary based on

the exact location of the tanks and pumps at site as well as the location of terminal points of the sources of electric power supply, clarified water, and DM water and the location of effluent discharge and draining points. Final quantity of piping materials including associated equipment shall be procured accordingly.

3. To facilitate operation of temporary valves and monitoring of pressure and temperature of circulating fluids, temporary platforms may have to be erected as required.

---

f. Quantity of chemicals, reagents, various apparatus, and safety gadgets to be required for the process should be ascertained beforehand and an adequate quantity must be kept available.

g. An adequate quantity of clarified and DM water must be assessed prior to undertaking the process and shall be made available.

h. The following instruments shall be installed:

  – Pressure gauges at the discharge of chemical-circulating pumps;
  – Thermometer at the chemical-dissolving tank.

Note

The chemical generally required for alkali flushing is trisodium phosphate (Na_3PO_4), which is sometimes supplemented with caustic soda (NaOH). The detergent used during this process is usually "Teepol," "Lissapol," or "Dodenol."

3.6 Operating Procedure

Depending on the configuration of the alkali flushing system layout, the system may be split into three or more circuits for the convenience and effectiveness of the flushing process. Some of these circuits that are typically used in the industry are described below:

i. temporary chemical-cleaning tank, temporary piping, deaerator with its feedwater storage tank, CEP recirculation lines, condensate pipe line starting from downstream of CEPs up to deaerator inlet and bypassing condensate polishing unit (CPU), GSC, and LP feedwater heaters (Fig. 3.7);

ii. temporary chemical-cleaning tank, temporary piping, deaerator with its feedwater storage tank, BFP suction lines and recirculation lines, BFP discharge lines up to economizer inlet bypassing HP feedwater heaters, and feed control station (Fig. 3.8);

iii. temporary chemical-cleaning tank, temporary piping, HP feedwater heater cascade drip lines and emergency drip lines up to deaerator and condenser, and LP feedwater heater cascade drip lines and emergency drip lines up to condenser (Fig. 3.9).

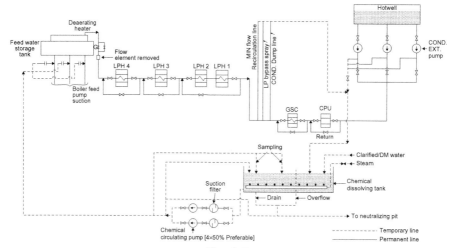

Fig. 3.7
Alkali flushing of condensate system.

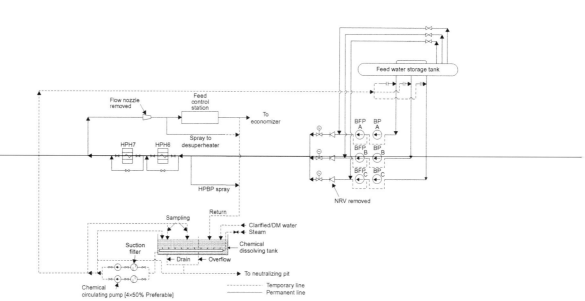

Fig. 3.8
Alkali flushing of feedwater system.

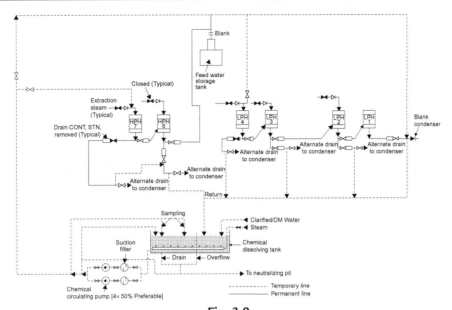

Fig. 3.9
Alkali flushing of feedwater heater drip system.

To reduce total time of completing the process successfully and to economize total requirement of chemicals used, the alkali flushing process of each circuit is sequentially divided into "cold water flushing," "hot water flushing," "hot alkali flushing," "rinsing," and "DM water flushing" processes.

Fill the chemical-dissolving tank with clarified water just above the middle of the tank. Check operation of each chemical-circulating pump by circulating water through recirculation line back to the tank.

Close all drain valves connected with the selected circuit. Fill the circuit gradually with clarified water by running any of the pumps. Open all vent valves of the selected circuit. Once it is ensured that the circuit is completely filled up, close vent valves. Carry out hydraulic test of the selected circuit including temporary pipe lines to ascertain that there is no leakage from the system. Ensure that the circuit is completely filled up.

Cold water flushing: On successful completion of the hydraulic test, start the cold water flushing process that involves circulation of water from the discharge of temporary chemical-circulating pumps through the selected circuit and temporary piping back to the tank.

Open all drain valves connected with the selected circuit. Keep on replenishing with clarified water into the tank. Continue the cold water circulation and draining process until the drain samples are visibly clear.

Hot water flushing: In continuation of the above process, close drain valves of the same circuit and fill up the system and the tank with clarified water. Close the clarified water supply valve. Start the temporary chemical-circulating pumps. Admit steam to the tank to raise the temperature of the circulating water to 363–368 K.

Continue circulation; check from time to time the quality of the effluent from the drain lines. During this process, water may have to be drained from the dissolving tank to maintain the water level in the tank. Water from the tank shall be also drained in the event that turbidity of effluent becomes very high. Water drained thus needs to be replenished again. When the effluent appears to be clear, action may be taken to stop circulation.

Repeat the above two processes for each of the circuits.

Hot alkali flushing: On completion of hot water flushing of all the circuits, check the turbidity of the water in the tank. In the event turbidity is high, drain the tank completely and refill the system with clarified water. Establish hot water circulation of the first circuit. Feed detergent in the tank. During feeding of detergent, foaming will occur and hence detergent shall be fed in stages in order to prevent formation of excessive foaming in the mixing tank. Start dozing trisodium phosphate, Na_3PO_4, in the dissolving tank in equally spaced steps. Continue recirculation without interruption. Analyze effluent samples at frequent intervals. Check chemical strength of medium and add chemicals if found necessary. Dozing rate of chemicals shall be completed in the shortest possible time in a way that achieves uniform concentration of Na_3PO_4 of about 4500 mg kg^{-1} in high-pressure, high-temperature steam generators and pH above 11.

Note

In medium- and low-pressure and low-temperature steam generators, while pH of the solution will be maintained above 11, concentration of Na_3PO_4 could be relaxed down to as low as 1500 mg kg^{-1}.

Circulate hot alkali at 363–368 K. During circulation, resort to partially draining and replenishing clarified water along with additionally dozing chemicals until the effluent becomes visibly clear. Samples shall be collected at inlet and outlet points at an interval of preferably 60 min. Samples shall be tested for mechanical impurities, pH value, and alkalinity. Flushing process may be declared complete once pH and alkalinity of samples from the supply line and return line of the circuit are reasonably the same.

The same procedure may be followed in other circuits as well. Once all the circuits are successfully flushed, stop the pumps and discontinue with the supply of heating steam. Thereafter, drain the dissolving tank and circuits completely by opening all drain valves and vent valves.

Rinsing: Refill the system following the same guideline as discussed above; flush each circuit one after another with a fresh supply of clarified water, and then drain the circuit. Repeat this process of filling, flushing, and draining of one circuit until the pH of the circulating water almost attains the pH of the fresh clarified water. Continue with the same procedure for the other circuits. On completion of rinsing of all the circuits, drain the system completely.

Open the manhole doors of the deaerator feedwater storage tank, and visually inspect the inside.

Remove all accumulated dirt, debris, and so on, from the feedwater storage tank manually; replace the manhole doors.

DM water flushing: Fill the dissolving tank with DM water, then raise the temperature of DM water to about 333 K by supplying steam to the tank. Circulate hot DM water through each circuit one by one for hot water rinsing of each circuit to remove alkalinity. Take samples from drain lines and check for phosphate, alkalinity, and pH of the effluent. Record both pH and conductivity of fresh DM water. Drain water from the circuit. Continue with filling, rinsing, and draining of the circuit until the pH and conductivity of outlet samples match with those of fresh DM water. This procedure may be repeated for all the circuits.

On completion of hot DM water flushing, a final rinse of the system may be carried out with cold DM water containing 6–10 mg kg^{-1} Na$_3$PO$_4$ through each circuit.

Note

On completion of the alkali flushing of the aforementioned three circuits, the rinsing tube side of feedwater heaters, air ejector aftercoolers, drain coolers, and GSC, which had been bypassed, as per Section 3.4 (10) above, may be flushed with cold DM water if so desired.

Drain all circuits. All temporary piping connections may be removed then and normal piping connections are restored. Control valves bypassed earlier (Section 3.4 (15&16)) may be placed in normal position.

3.7 Estimation of Clarified Water, DM Water, and Chemicals

Prior to carrying out the alkali flushing of the preboiler system, it would be prudent to estimate the exact requirement of clarified water, DM water, and chemicals. It is recommended to procure a higher quantity of these items than estimated in order to obviate any complication arising out of a short fall of any of these items during the execution of the process. To facilitate estimation of each of above items, it is advised to assess the water-holding capacity beforehand.

3.7.1 Water-Holding Capacity

The water-holding capacity of areas falling under the alkali flushing process (Table 3.5) varies from project to project, and hence shall be assessed beforehand in consultation with the manufacturer/supplier of the equipment and piping systems.

Table 3.5 Water-holding capacity

Item	Volume (m^3)
Deaerator with its feedwater storage tank Condensate pipe line, starting from downstream of condensate extraction pumps up to deaerator inlet bypassing CPU, GSC, and LP feedwater heaters Boiler feed pump suction lines and recirculation lines Boiler feed pump discharge lines up to economizer inlet bypassing HP feedwater heaters and feed control station HP feedwater heater cascade drip lines and emergency drip lines up to deaerator and condenser LP feedwater heater cascade drip lines and emergency drip lines up to condenser Temporary piping Temporary chemical-cleaning tank	

3.7.2 Clarified Water Requirement

The sequence of activities along with applicable areas, wherein clarified water will be required, is presented in Table 3.6.

Table 3.6 List of activities requiring clarified water

Activity	Areas to be Considered	Quantity (m^3)
Cold water flushing	Deaerator, condensate line, LP and HP feedwater pipe lines, feedwater heater drip lines, and all temporary pipe lines	
Hot water flushing	Deaerator, condensate line, LP and HP feedwater pipe lines, feedwater heater drip lines, and all temporary pipe lines	
Hot alkali flushing	Deaerator, condensate line, LP and HP feedwater pipe lines, feedwater heater drip lines, and all temporary pipe lines	
Rinsing	Deaerator, condensate line, LP and HP feedwater pipe lines, feedwater heater drip lines, and all temporary pipe lines	

3.7.3 DM Water Requirement

Table 3.7 identifies the area wherein DM water will be required.

Table 3.7 Areas requiring DM water

Activity	Areas to be Considered	Quantity (m^3)
DM water flushing	Deaerator, condensate line, LP & HP feedwater pipe lines, feedwater heater drip lines, and all temporary pipe lines	

3.7.4 Chemical Requirement

A typical list of chemicals to be used during alkali flushing of a preboiler system is summarized in Table 3.8. The correct amount of these chemicals shall be assessed beforehand and procured accordingly.

Table 3.8 List of chemicals

Activity	Name of Chemical	Quantity (kg)
Hot alkali flushing	Trisodium phosphate, Na_3PO_4, of concentration 4500 mg kg^{-1} in high-pressure and high-temperature steam generators; (Trisodium phosphate, Na_3PO_4, of concentration 1500 mg kg^{-1} in medium- and low-pressure and -temperature steam generators); Caustic soda, Na OH; Detergent: "Teepol," "Lissapol," or "Dodenol"	
DM water flushing	Trisodium phosphate of concentration 6–10 mg kg^{-1}	

3.8 Conclusion

On completion of successful alkali flushing of all the circuits, a preboiler system may be declared to be clean and ready to be put into service.

A protocol will be signed by all concerned (Table 3.9).

Table 3.9 Protocol (alkali flushing of preboiler system)

Circuit Under Alkali Flushing	Results of Analysis of Sample				Inference
	Samples Taken from the Solution of the Supply Line of the Circuit		Samples Taken from the Solution of the Return Line of the Circuit		
Condensate system	pH	Alkalinity	pH	Alkalinity	pH and alkalinity at supply and return lines of the circuit are the same
Feedwater system	pH	Alkalinity	pH	Alkalinity	pH and alkalinity at supply and return lines of the circuit are the same
Feedwater heater drain system	pH	Alkalinity	pH	Alkalinity	pH and alkalinity at supply and return lines of the circuit are the same
Signed by the customer		Signed by the engineer		Signed by the supplier (contractor)	

References

[1] C. Bozzuto (Ed.), Clean Combustion Technologies, fifth ed., Alstom, Windsor, CT, 2009.
[2] S.C. Stultz, J.B. Kitto (Eds.), Steam Its Generation and Use, 41st ed., The Babcock and Wilcox Company, Barberton, OH, 2005.
[3] D.K. Sarkar, Thermal Power Plant—Design and Operation, Elsevier, Amsterdam, Netherlands, 2015.

Flushing of Fuel Oil Piping System

4.1 Introduction

Reliable operation of oil-fired steam power plants, start-up, and support of (1) coal-fired steam, (2) gas turbine, and (3) diesel power plants greatly depends on the use of a clean fuel oil system free from detrimental foreign material, eg, mill scale, rust, welding beads, residue from grinding, chipping, blasting, or other maintenance activities. Liquid and solid contaminants in fuel oil may damage fuel oil pumps, burners/injectors, and various components of gas turbines and diesel engines. The presence of foreign material in fuel oil may cause a varying rate of flow of fuel oil and air to burners, causing blow-off or flashback of flame. Fuel oil

Thermal Power Plant. http://dx.doi.org/10.1016/B978-0-08-101112-6.00004-6

heavily laden with foreign material may even choke the tip of burners, thereby resulting in blow-off or pulsation of burner flame.

Contaminants may enter into the fuel oil system during manufacture, storage, field fabrication, and installation of a new unit. In running units, contaminants are generated during operation and/or get introduced during overhaul. Hence, during erection of fuel oil storage tanks, fuel oil piping, and so on, utmost precaution should be taken to keep off dirt, dust, and other undesirable debris from the fuel oil system. On completion of assembly, the whole fuel oil system, including fuel oil storage tanks, fuel oil pipe lines, strainers, heaters, and so on, shall be cleaned manually as thoroughly as possible.

Furthermore, prior to putting a fuel oil system in service, it must be flushed with an oil to remove all undesirable contaminants. A successful flush ensures that system piping and components meet acceptance criteria (Table 4.1: NAS class 10 specification with water content of <100 mg kg^{-1} (ISO 21/19/16—Section III) [1]) within a minimum length of time with a minimum of effort.

The flushing process discussed in this chapter is applicable to fuel oil systems of all types of thermal power plants—steam, gas turbine, and diesel power plants. The process deals with flushing of the entire fuel oil system exhaustively, by circulation of oil of the grade as recommended by the equipment manufacturer, by running temporary flushing oil pumps.

Table 4.1 NAS cleanliness level particle count

NAS 1638 (1964)	Based on 100 mL Sample				
	5–15 µm	15–25 µm	25–50 µm	50–100 µm	>100 µm
12	1024k	182k	32,400	5760	1024
11	512k	91,200	16,200	2880	512
10	256k	45,600	8100	1440	256
9	128k	22,800	4050	720	128
8	64,000	11,400	2025	360	64
7	32,000	5700	1012	180	32
6	16,000	2850	506	90	16
5	8000	1425	253	45	8
4	4000	712	126	22	4
3	2000	356	63	11	2
2	1000	178	32	6	1
1	500	89	16	3	1

Prior to addressing the fuel oil flushing process, certain areas common to all thermal power plants—fuel oil unloading, storing, pressurizing, and forwarding systems—are discussed below.

The fuel oil system covers receipt, storage, treatment, pressurizing, and forwarding of oil to burners/injectors of various thermal power plants at required pressure and temperature in an efficient manner, meeting all the safety requirements.

Fuel oil may be delivered to plants by road tankers or railway wagons or, if the plant is located near a port, through a pipe line from the port.

Oil delivered by road tankers or railway wagons is unloaded with the help of a set of three or four fuel oil unloading/transfer pumps, each rated for adequate capacity and head complying with the requirement of a specific plant. Unloading/transfer pumps may be of centrifugal type or screw type per the recommendation of API 676. These pumps are located at the unloading terminal points of a plant. Each of the pumps is equipped with a simplex basket strainer at the suction to transfer the oil to storage tanks. Connection is made with the tankers or wagons through screwed hoses. Pressurized fuel oil from unloading pumps is supplied through pipe lines to fuel oil storage tanks.

When the fuel oil is received through pipe lines from a port, the supply oil line from barge is to be connected to the plant fuel oil receiving line connected directly to fuel oil storage tanks. Fuel oil supply pumps, located at the port, are selected in accordance with the fuel oil unloading capacity and head as required in a specific plant.

Fuel oil storage tanks are usually vertical, cylindrical, cone-roof steel tanks designed and fabricated as per API 650. Tanks are contained within concrete dyke dimensioned to contain the total storage volume of oil from the tanks in accordance with the requirement of NFPA 30. The tanks are provided with both local level indication and remote level indication of the tank.

Each of the storage tanks is provided with foam injection facility on the shell of the tank to inject foam below the roof of the tank for fire protection. In addition, water spray nozzles and ring headers are provided at the roof and around the perimeter of the shell for cooling the tank as protection against propagation of fire on the adjacent tank.

Based on a specific plant requirement, there may be one or two fuel oil storage tanks for storing raw oil as received. In gas turbine and diesel power plants, fuel oil from raw oil storage tanks is pumped through a fuel treatment system, comprising a battery of centrifuges/purifiers, to one or two treated fuel oil storage tanks. Piping connections between raw fuel oil storage tank/s and treated fuel oil storage tank/s may be provided to facilitate utilizing both type of storage tanks to store either raw oil or treated oil.

Fuel oil storage tanks provide an in-plant storage capacity of usually 30 days of continuous operation at base load of oil-fired steam, gas turbine, and diesel power plants. However, storage capacity for less number of days of operation may be provided if decided by the plant authority.

In pulverized coal-fired steam power plants, fuel oil storage tanks will be of much lower capacity. Normally storage capacity of these tanks will fulfill 30 days of operation of a steam generator, with oil firing at 30% BMCR (boiler maximum continuous rating) for about 40–50 h of operation per month and also stabilization of flame with oil at 10% BMCR for about 8 h of operation per day.

While storing heavy fuel oil (HFO), depending on viscosity of the oil, the minimum temperature in bulk oil storage tanks is maintained at around 328 K by electric or steam heating coils provided at the bottom of the tank. Each storage tank is also provided with an electric/steam suction heater at the point of withdrawal of oil to raise the temperature of oil to about 353 K to ensure pumpable fluidity of the oil. From raw fuel oil storage tank/s, hot fuel oil is transferred to treated fuel oil storage tanks through fuel oil transfer pumps and a fuel treatment system. Hot fuel oil from treated fuel oil storage tank/s is supplied to the suction of $2 \times 100\%$ screw-type pressurizing/forwarding fuel oil pumps to deliver oil to the respective burners/injectors of a steam generator/gas turbine/diesel engine. For ensuring the required viscosity of HFO at the burner/injector end and proper atomization of fuel oil, the minimum temperature of oil should be maintained at around 353–393 K as recommended by the equipment manufacturer. For this purpose two shell and tube-type heat exchangers, each of 100% capacity, are mounted downstream of fuel oil pressurizing/forwarding pumps located in a fuel oil pump house. Each of the oil heaters, heated by steam, shall normally be kept in line to avoid amassing stagnant oil. A temperature control station at the steam inlet line will maintain the fuel oil outlet temperature.

While using light-grade fuel oils (LFO), eg, light diesel oil (LDO), high speed diesel (HSD), and so on, heating of oil is not a necessity to make them pumpable. LFO is generally used as the start-up fuel of a fuel-firing system.

The fuel oil system is generally above ground, and adequate thermal relief valves may need to be provided to protect the piping system from overpressure due to heating from solar and atmospheric radiation. In the event fuel oil piping is laid below ground, suitable cathodic protection of piping may become a necessity.

All instrumentation and control facilities, including tank level controllers, pressure/temperature gages, control valves, and so on, in the fuel oil pump house are provided for safe and reliable operation of the system.

A totalizing flow meter is provided at the transfer line to storage tanks for indication of fuel flow and to record total oil delivered to the plant.

The HFO system of gas turbine power plants and diesel power plants usually comprises all or most of the following. The HFO system of steam power plants does not require items x–xiv* below:

   i. HFO unloading/transfer pumps;

  ii. HFO treatment system;

 iii. 2 × 100% fixed roof type raw/treated HFO storage tanks with floor coil heaters and suction heaters;

 iv. 3 × 50% HFO pressurizing/forwarding pumps;

  v. reinforced, corrosion-resistant, basket-type, duplex strainer of 111 μm (140 mesh) to 970 μm (20 mesh) at the suction of each HFO pressurizing/forwarding pump (Section 5, Clause 3.23 [2]);

 vi. 2 × 100% heaters downstream of pressurizing/forwarding pumps;

vii. a recirculation line with pressure control valve provided from the discharge line of fuel oil pressurizing/forwarding pumps to the treated oil storage tank/s;

Note

The recirculation line maintains a constant pressure at the pump discharge under all conditions of the varying load demand of the plant. The discharge of the control station is connected to the return oil line from the burner end of the plant and returned to the treated oil storage tank/s.

 viii. common duplex strainer downstream of heaters;

  ix. temperature and pressure control valves;

   x. high-pressure fuel injection pumps*;

  xi. fine screens/filters at pump suction*;

 xii. injectors*;

xiii. high-pressure pipes connecting the injection pumps to the injectors*;

 xiv. low-pressure pipes*;

  xv. regulating and isolating valves;

 xvi. nonreturn valves;

xvii. relief valves;

xviii. piping and associated hangers and supports;

 xix. instrumentation and control;

  xx. associated piping, isolating valves, fittings, supports, and so on;

 xxi. steam heating and tracing system with valves and traps.

The LFO system of gas turbine power plants and diesel power plants usually consists of all or most of the following. Items vii–xi** below are not required in the LFO system of oil-fired steam power plants:

   i. LFO unloading/transfer pumps;

  ii. LFO treatment system;

 iii. 2 × 100% fixed roof type raw/treated LFO storage tanks;

iv. $3 \times 50\%$ LFO pressurizing/forwarding pumps;

v. reinforced, corrosion-resistant, basket-type, duplex strainer of 111 μm (140 mesh) to 970 μm (20 mesh) at the suction of each LFO pressurizing/forwarding pump (Section 5, Clause 3.23 [2]);

vi. A recirculation line with a pressure control valve provided from the discharge line of the fuel oil pressurizing/forwarding pumps to the treated oil storage tank/s (The recirculation line maintains a constant pressure at the pump discharge under all conditions of the varying load demand of the plant. The discharge of the control station is connected to the treated oil storage tank/s);

vii. high-pressure fuel injection pumps**;

viii. fine screens/filters at pump suction**;

ix. injectors**;

x. high-pressure pipes connecting the injection pumps to the injectors**;

xi. low-pressure pipes**;

xii. regulating and isolating valves;

xiii. nonreturn valves;

xiv. relief valves;

xv. piping and associated hangers and supports;

xvi. instrumentation and control;

xvii. associated piping, isolating valves, fittings, supports, and so on.

Fuel oil used in thermal power plants is of different types—HSD, LDO and heavy fuel oil (HFO), the typical specification of which is shown in Table 4.2.

4.2 Description of Fuel Oil System

Fuel oil pressurizing and supplying system to burners/injectors of different types of thermal power plants is not of identical configuration. So, for the convenience of readers, a fuel oil system of various types of power plants is discussed separately under the following paragraphs.

4.2.1 Fuel Oil System of Steam Power Plant (Source: Section 9.2.24, p. 346 [3])

Fuel oil is the secondary fuel of large, pulverized coal-fired steam generators. The main purpose of a fuel oil system is to facilitate the start-up of a steam generator since pulverized coal on its own is unable to ignite and requires ignition energy from an external source. Fuel oil is required also for flame stabilization at low-load (usually 30–40% BMCR) operation with pulverized coal. The system is also capable of ensuring a low-load (up to about 30% BMCR) operation of a steam generator without coal firing.

The fuel oil system generally has provision of handling two different types of oil—LDO and HFO. The functional requirement of the above two oils is:

(1) LDO is used to light up the steam generator during cold start-up;
(2) HFO serves for hot start-up of the steam generator, flame stabilization, and low-load operation.

Table 4.2 Typical specification of fuel oils

Characteristics	Heavy Fuel Oil (HFO)	Light Diesel Oil (LDO)	High Speed Diesel Oil (HSD)
Acidity, inorganic	Nil	Nil	Nil
Acidity, total mg of KOH/g, max.	–	–	0.50
Ash, % by weight, max.	0.1	0.02	0.01
Gross calorific value ($MJ\,kg^{-1}$)	41.86	41.86	–
Specific gravity at 288 K	Not limited but to be reported	0.85	–
Flash point, K, min. (Pensky-Martens Close Cup)	339	339	311 Abel Close Cup
Pour point (K)	–	285 (Winter) 291 (Summer)	279 max.
Kinematic viscosity in centistokes	370 at 323 K	2.5–15.7 at 311 K	2.0–7.5 at 311 K
Sediment, % by weight, max.	0.25	0.10	0.05
Total sediments, mg per 100 mL, max.	–	–	1.00
Total sulfur, % by weight, max.	4.5	1.8	1.0
Water content, % by volume, max.	1.0	0.25	0.5
Carbon residue (Ramsbottom), % by weight, max.	–	1.5	0.20

Note

Kinematic viscosity : 1 stoke $= 407\,s$ Redwood viscosity No. 1 at 311 K
$= 462.7\,s$ Saybolt universal viscosity at 323 K
$= 13.16$ Engler degrees
$= 10^{-4}\,m^{2}\,s^{-1}$

4.2.1.1 Light diesel oil (LDO) system (Fig. 4.1)

The LDO system consists of high-pressure screw or gear pumps and a pressure control recirculation valve, provided on a common discharge header of pumps, for maintaining constant pressure. Pumps are also provided with strainers at their suction. These pumps take suction from LDO storage tanks. Since viscosity of this oil (Table 4.2) is compatible with that required for proper firing, as recommended by manufacturers of burners, this oil does not require prior warm-up. Hence, in this system, a return oil line from the burner end is not provided.

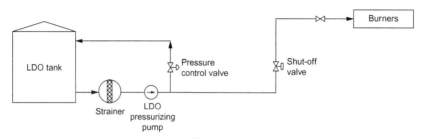

Fig. 4.1
Light diesel oil system.

4.2.1.2 Heavy fuel oil (HFO) system (Fig. 4.2)

The HFO pumping and heating system is comprised of high-pressure screw pumps and dedicated fuel oil heaters. The pumps take suction from HFO storage tanks. Strainers are provided at the suction of pumps as well as downstream of HFO heaters. A pressure control recirculation valve is provided on the common discharge header of pumps, which maintains constant pressure downstream of heaters.

Being quite viscous, HFO (Table 4.2) is required to be preheated to attain requisite viscosity at burners, for proper atomization and combustion as recommended by manufacturers. A fuel oil heater uses steam from an auxiliary steam header and is controlled by a temperature control valve at the inlet to the heater in order to maintain constant fuel oil temperature at the heater outlet.

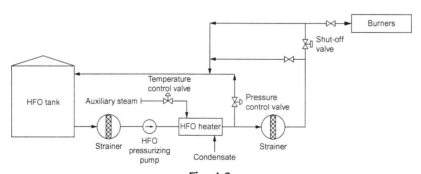

Fig. 4.2
Heavy fuel oil system.

The HFO at the required pressure and temperature is supplied to the burners. A return line, from the fuel oil header supplying the burners, is provided to route excess oil back to the oil tank and also to maintain hot oil circulation up to the boiler front for use in burners in the case of emergency. The HFO supply and return lines are insulated and either steam traced or electric traced to keep the lines warm at all times. Similarly, pumps and strainers are steam/electric jacketed.

4.2.1.3 Fuel oil burner

In the fuel oil burner, oil is atomized either utilizing pressure of fuel oil or with the help of compressed air/steam. The former type is called the mechanical-type or pressure-type atomized burners. They use 1.5–2.0 MPa oil pressure at maximum flow. The turn-down ratio (operating range) of these burners is generally low, eg, 2:1–4:1.

The limitation of the turn-down ratio of the pressure-atomized burners is largely overcome by air- or steam-atomized burners. Burner output in air- or steam-atomized burners may be varied by varying the oil pressure and air/steam pressure, correspondingly. Air/steam is supplied at a pressure of 0.5–1.2 MPa, while the maximum oil pressure is about 0.7 MPa at the burner end. The turn-down ratio of an atomized oil burner is as high as 10:1.

4.2.2 Fuel Oil System of a Gas Turbine (Source: Section 7.5.9, p. 269 [3])

Fuel oil received from an oil refinery is unloaded in raw oil storage tanks. Fuel oil from raw oil storage tanks is transferred to a fuel oil treatment system comprising a battery of centrifuges and located near these storage tanks to remove the water-soluble sodium (Na^+) and potassium (K^+) salts in the fuel oil to a level of 0.5–2 ppm as recommended by the gas turbine manufacturer for smooth trouble-free operation of gas turbines. Demineralized water is used for the water washing facility. Treated oil is then stored in treated oil storage tanks. Water discharged from the treatment plant is directly connected to the storm water system. The oil sludge from the centrifuges is removed by truck to a remote disposal area.

Oil from the treated oil storage tanks is taken to the suction of fuel oil pressurizing/forwarding pumps. A basket-type duplex strainer is provided at the suction of each of the two pumps.

These pumps supply fuel oil to the fuel oil skid of each gas turbine. This skid contains combined pump block, which forwards fuel oil to gas turbine burners at an adequate pressure required for atomization as long as the main oil shut-off valve is open. Control valves of burners are then partially opened for ignition of oil and thereafter open further as per the load demand. During shutdown of the set, all control valves and shut-off valves are closed.

Fuel oil burners are installed in a combustor or combustion chamber, which is located between the compressor and the gas turbine. Atmospheric air, upon getting compressed in the compressor, enters the combustion chamber, wherein air is mixed with fuel and gets ignited to release energy. High temperature products of combustion then enter the gas turbine to produce power.

A typical fuel oil system of gas turbines is depicted in Fig. 4.3:

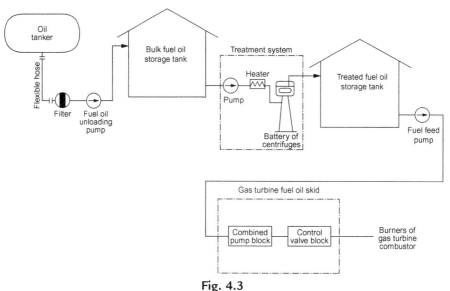

Fig. 4.3
Fuel oil system of gas turbine.

4.2.3 The Fuel Oil System of a Diesel Engine (Source: Section 8.5.1, p. 309 [3])

Fuel oil supplied from an oil refinery to a plant is unloaded in raw oil storage tanks. The fuel oil used in a diesel engine should be completely free from water and mechanical impurities. Hence, fuel oil from raw oil storage tanks is transferred to treated oil storage tanks through a fuel oil treatment system. Solid and liquid contaminants are cleaned before use to prevent damage to fuel pumps and engine components. The fuel oil pressurizing/forwarding pumps supply fuel from the treated oil storage tank/s and deliver fuel through a fine filter to the high-pressure fuel injection pumps. These pumps transfer oil to the injectors located on each cylinder.

A diesel engine is capable of burning a wide range of low-quality liquid. It should be ensured that heating of fuel oil to the required viscosity is carried out without developing thermal cracking. For the start-up of a diesel engine, either HSD or LDO is used. For normal running, a heavier grade fuel oil, eg, HFO or low sulfur heavy stock (LSHS), is generally utilized.

In a diesel engine, atmospheric air is sucked in cylinders; air is then compressed to raise its temperature to a level in which the fuel-air mixture ignites spontaneously when fuel is sprayed. The rapidly burning mixture then expands, contributing useful work vis-à-vis generating power.

Fig. 4.4 shows a typical fuel oil system of large diesel engines.

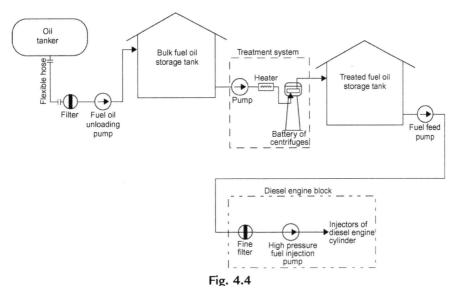

Fig. 4.4
Fuel oil system of diesel engine.

4.3 Precautions

In addition to the general precautionary measures described under Section 2.1.1: Quality Assurance, following specific precautionary measures must be observed before executing the flushing of a fuel oil piping system:

    i.   During manual cleaning of tanks, and so on, it is necessary to ensure that they are well ventilated before any person is allowed to enter them;

   ii.   Manual cleaning should be followed by a thorough visual inspection to ensure that a good cleaning job has been performed and no extraneous matters, eg, waste cloth, working tools, nuts, bolts, and/or other material, are left inside;

 iii.   Avoid contact with fuel oil, vapors, or contaminated residues;

 iv.   All valves in the fuel oil system are functionally checked for putting into service;

   v.   Extreme fire and spark protection precautions must be taken;

vi. To prevent irritation arising from any splashing of oil into eyes, eyewash facilities should be applied immediately;

vii. Emergency stop-push buttons must be provided in the vicinity of the flushing area to stop circulating oil pumps as may be necessary.

4.4 Prerequisites

Before carrying out the fuel oil flushing process, it is essential to carry out a thorough inspection of following areas/items (Table 4.3):

Table 4.3 Areas/items to be checked

Sl. No.	Areas/Items	Ok (√)
1	Before erecting the piping systems, protective coating used on internal surfaces as a safeguard against corrosion during transport and storage is removed as far as practicable.	
2	Prior to erecting all temporary piping systems, they are thoroughly cleaned mechanically.	
3	All piping systems, whether permanent or temporary, are erected and supported properly, along with associated valves. Protocol signed jointly by the customer, the engineer, and the supplier (contractor), certifying completion of erection of piping systems, is verified.	
4	All welded and flanged joints are secured.	
5	Fuel oil pipe lines from pressurizing/forwarding pumps to burner front and return lines from burner end to storage tanks are hydrotested successfully at 1.5 times of their designed pressures. Protocol signed jointly by the customer, the engineer, and the supplier (contractor), certifying acceptability of the hydrotest, is verified.	
6	Hydrotest of fuel oil storage tanks is successfully performed and the necessary protocol signed jointly by the customer, the engineer, and the supplier (contractor) is available.	
7	HFO pipe lines, strainers, oil heaters, pumps, and so on, are insulated adequately.	
8	Construction materials—temporary scaffolding, wooden planks, welding rod ends, cotton wastes, and so on, are removed.	
9	Instruments installed on the permanent piping systems are isolated.	
10	For initial purging or flushing of fuel oil pipe lines, verify availability of any of the following:	
	i. If steam purging is adopted, source of steam has required pressure, degree of superheat, and requisite quantity. Temporary piping systems are provided with proper insulation and expansion devices; ii. If air purging is utilized, the air compressor is capable of delivering an adequate quantity of air at the required pressure; iii. In the event of water flushing, the source of service water has the requisite flow and pressure	
11	Temporary platforms are provided wherever required.	
12	The estimated quantity of flushing oil (Table 4.4) is available.	
13	The oil/water separator system is in service.	
14	Materials required for fuel oil flushing (Section 4.4.1) are available.	
15	Portable fire extinguishers are provided in fuel oil storage tank/s areas and fuel oil pressurizing/forwarding pump houses.	

Continued

Table 4.3   Areas/items to be checked—cont'd

Sl. No.	Areas/Items	Ok (√)
16	Emergency eyewash facilities are available.	
17	Verify that fuel oil storage tanks and the entire fuel oil system are electrically grounded securely.	
18	Facilities are provided to dispose of any sludge, debris, strainer screens, and waste fuel resulting from fuel oil system cleaning and flushing activities.	
19	Safety gadgets (Section 4.4.2) are arranged.	
20	Verify that no smoking, danger, keep off, and other safety tags (in regional and English languages) at various places around the area of activity are provided.	
21	Verify that tanks, pumps, valves, associated equipment, and systems, which are in service or are energized, are provided with proper safety tags (Appendix C) so as to obviate any inadvertent operation or to obviate any injury to equipment and personnel.	

4.4.1 Typical List of Bill of Materials

A list of materials that are typically required for flushing of a fuel oil piping system is given below. Size and quantity of each of these items would vary from project to project and shall have to be assessed by the supplier (contractor), and duly approved by the owner, the engineer and the equipment manufacturer.

1. temporary tank for fuel oil flushing
2. $3 \times 50\%$ temporary fuel oil flushing pumps
3. temporary pipe lines
4. spool pieces
5. temporary valves
6. temporary arrangement for steam purging, or temporary arrangement for water flushing, or temporary oil-free air compressor or station instrument air for dry air purging
7. temporary fine mesh screens
8. pressure measuring instruments
9. temperature measuring instruments

4.4.2 Typical List of Safety Gadgets

Equipment that is generally used for safety of operating personnel are as follows:

1. gum boots of assorted sizes
2. rubber gloves
3. asbestos gloves
4. rubber or polythene aprons
5. hard hats of assorted sizes
6. side-covered safety goggles with plain glass
7. gas masks

8. eyewash facilities
9. first aid box
10. safety sign boards
11. safety tags

4.5 Preparatory Arrangements

In order that the flushing of fuel oil system can be completed satisfactorily within the shortest possible time, it is necessary to complete the following preparatory arrangements before the actual flushing process is undertaken. The whole system may be split into two or more circuits for the convenience of operation (Fig. 4.5), as given below:

 i. One of the circuits may commence from the unloading terminal up to the inlet to fuel oil storage tank/s;
 ii. The second circuit may start from the outlet of fuel oil storage tank/s up to the burner/injector front;
 iii. The third circuit may constitute a burner/injector recirculation line back to the fuel oil storage tank/s.

From all the circuits, flushing oil is taken back to the temporary flushing oil tank through the flushing oil return line. Thus, it will be convenient to locate the temporary flushing oil tank and pumps in the vicinity of the fuel oil storage tank/s.

4.5.1 Protection of Equipment (Clause 3.2.1 [4])

Before carrying out the flushing process, the following equipment or components of the fuel oil system are to be removed and replaced with suitable spool pieces:

 i. pressure control valve
 ii. ~~temperature control valve~~
 iii. flow meter

Before undertaking steam/air blowing or water flushing of fuel oil piping, the following precautions may be ensured:

 i. All instrument sensors shall be kept isolated;
 ii. All pumps shall be bypassed with temporary piping around them;
 iii. Strainer/filter elements need to be removed.

4.5.2 Mechanical Cleaning

It is desirable that the whole circulation system, including temporary flushing oil tank, fuel oil storage tanks, oil pipes, oil heaters, oil strainers/filters, oil purifier, and so on, has been thoroughly cleaned as far as accessibility permits to get rid of dirt, dust, and foreign materials.

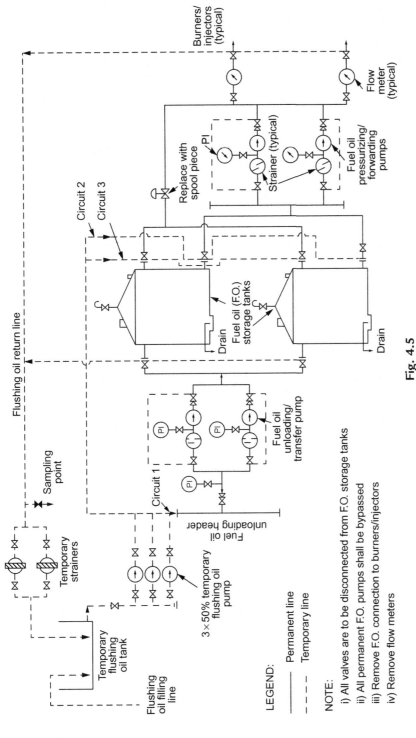

Fig. 4.5
Typical fuel oil flushing scheme.

For this reason, it may be necessary to chip and wire-brush rusty parts. Welding slag on pipe lines shall be removed by grinding or filing as far as practicable. A protective coat of paints is to be removed by puffing, brushing, and so on. Proper cleaning of the entire system will be followed by wiping the system clean by cloth. Cleaning of the above parts will ensure providing clean oil to the remaining parts of the system.

4.5.3 Erection of Temporary Tank, Piping, and Valves

In order to complete the circuit for circulation of oil, depending on the pipe layout configuration at site, temporary flushing oil tank, temporary piping with associated temporary valves connecting the permanent piping systems must be erected and kept ready for putting into service. Size and quantity of temporary piping materials including elbows, tees, valves, supports, and so on, shall be assessed at the design stage.

Sampling points are to be installed for monitoring cleanliness of oil during the progress of oil flushing.

While flushing LFO pipe lines, a suitable temporary connection is to be made with the HFO return line from the burner front to recirculate flushing oil back to the temporary flushing oil tank.

4.5.4 Steam/Air Blowing or Water Washing of Fuel Oil Piping

On completion of hydrotest of the complete fuel oil piping and system, they must be cleaned thoroughly to get rid of debris, dirt, and so on. Before cleaning the system, ensure that the temporary bypass connection to each of the permanent fuel oil pumps is provided. Remove all strainer/filter elements.

Appropriate cleaning can be achieved by carrying out steam/air blowing or water washing of fuel oil piping as described below.

I. *Steam blowing*
    Before carrying out steam blowing of fuel oil lines, it must be ensured that proper care is taken for expansion of line and support systems.
    Steam for blow-out of fuel oil piping must be at a pressure of 1.0–1.2 MPa with a degree of superheat of at least 50 K. Steam can be sourced from either station auxiliary steam system if available at site, or an auxiliary boiler can be installed temporarily.
    Temporary piping from this source is to be connected to one end of the selected circuit. Steps to be followed are:
    1. Partially open steam blow-out valve and all drains and vents of fuel oil pipe lines to warm up the piping system. After proper warming up, close the drains and vents.

Open the steam blow-out valve fully and continue blowing steam for about 3–4 h or as required until the debris coming out from other end of the piping appears visibly clean. Close the blow-out valve and allow the piping to cool for about 2–3 h. Repeat steam blowing until the exhaust debris becomes invisible to the naked eye. Then, stop blow-out of this circuit;

2. Close the blow-out valve;
3. Purge the circuit with dry compressed air for about 2 h to remove any residual condensate from the circuit;
4. Disconnect temporary piping and normalize the system.

Repeat the above steps for blowing out other circuits.

II. *Compressed air blowing*

Compressed air for blowing must be oil-free and dry at a pressure of 0.7 MPa; temperature of air should be about 20 K higher than ambient temperature. Compressed air can be sourced from either station instrument air system or air compressor to be installed temporarily. Temporary piping from this source is to be connected to one end of the selected circuit. Steps to be followed are:

1. Open instrument air header supply valve or start temporary blow-out air compressor. Open temporary blow-out valve. Open also all drain valves in the selected circuit to impart free blow through them to atmosphere for removing any liquid and/or loose material collected at low points. Likewise, open all vent valves in the selected circuit to clean them by imparting free blow to the atmosphere. Thereafter, close all drain and vent valves;
2. At the beginning of the blowing out operation, debris coming out from the exhaust end will be visible;
3. Continue blowing out the circuit until clear air comes out through the exhaust;
4. Close temporary blow-out valve and close instrument air header supply valve or stop the air compressor;
5. Disconnect temporary piping and normalize the system.

Repeat the above steps for blowing out other circuits.

III. *Water washing*

Water washing is carried out by circulating water through each circuit. Erect temporary piping from the station service water header to one end of the selected circuit connected with a temporary water washing valve. Steps to be followed are:

1. Open the temporary water washing valve along with all drain valves and vent valves of the selected circuit for removing any liquid and/or loose material collected at these valves. Thereafter close all drain and vent valves;
2. Maintain flow through the selected circuit for 2–3 h;
3. At the beginning of the flushing process, debris coming out from the exhaust end will be visible;

4. Continue with the flushing process of the circuit until the water coming out from the other end appears to be clean to the naked eye;
5. Close the temporary water flushing valve;
6. Open all drains and vents of the piping, which may retain water, to thoroughly drain out water from the system;
7. Purge the circuit with dry compressed air for about 2 h to remove any residual water from the circuit;
8. Disconnect temporary piping and normalize the system.

Repeat the above steps for water flushing of other circuits.

Note

It is evident from the above three processes that while adopting steam blowing or water flushing of fuel oil piping, it is imperative to dry out pipe lines with dry oil-free compressed air. This additional activity with arrangement for dry oil-free compressed air can be avoided if compressed air blowing of fuel oil piping is adopted.

On completion of the blowing or flushing process, strainer/filter elements should be reinstalled.

4.5.5 Installation of Fine Mesh Screen

It is recommended to install a temporary $2 \times 100\%$ simplex strainers on the flushing oil return line near the temporary flushing oil tank. During the initial stage of flushing, when the debris from the system is expected to be large, a coarse screen of 425 μm (40 mesh) may be installed inside the temporary strainers (Clause 3.2.2 [4]). These strainers must have provision of periodic withdrawal for inspection and cleaning after stopping oil circulation, if necessary. If a fine mesh screen was installed in the early stage, it is likely to become choked frequently and will need frequent inspection and cleaning. Once the content of debris in flushing oil is reduced, a fine mesh screen of 111 μm (140 mesh) (Section 5, Clause 3.23 [2]) should be placed in the temporary strainer. A temporary fine mesh screen shall also be placed over permanent strainers. These are of great value towards the end of the flushing process.

Note

A duplex 10 μm filter usually is installed immediately prior to the inlet of the fuel injection pumps to protect pump plungers and barrels from any untreated contamination or random debris remaining in the fuel. While it may appear that this final filter is not necessary due to the cleaning and treatment equipment upstream, it may be kept in mind that high-pressure diesel injection pumps are very sensitive to minute particles of debris. These materials can cause microseizures and, finally, total failure of the pump plunger and barrel (Section 5, Clause 3.31 [2]).

4.5.6 Readiness of Auxiliaries

Ensure that the following auxiliary equipment is kept ready to be put into operation prior to undertaking the oil flushing process:

i. temporary flushing oil tank to be located above ground; capacity of this tank must be more than the fuel oil system volume;
ii. $3 \times 50\%$ temporary flushing oil pumps, each rated with capacity higher than the 50% capacity of normal fuel oil pump/s; temporary pumps are to be located on the ground below the temporary flushing oil tank for getting positive suction;
iii. $2 \times 100\%$ temporary fuel oil strainers, each provided with fine mesh screens;
iv. permanent fuel oil strainers/filters fitted with temporary fine mesh screens;
v. sampling point located upstream of temporary fuel oil strainers.

4.6 Operating Procedure

The key to effective flushing of a fuel oil system is to maintain turbulent oil flow (Reynolds number > 4000) through each circuit as long as the process continues. Turbulent oil flow can be attained by adopting relatively high fluid velocity and/or lowering fluid viscosity. For achieving high fluid velocity, the capacity of each flushing oil pump is required to be higher than the capacity of normal fuel oil pumps. Lower viscosity of oil may be attained if LFO is used as the flushing medium. Use of LFO has an added advantage of doing away with the heating of flushing oil.

For the convenience of reducing total time of completing the oil flushing process, successfully fulfilling the recommendation of equipment manufacturers, the sequence of flushing activities may be followed as described below:

i. oil charge
ii. oil circulation
iii. duration of circulation
iv. inspection and cleaning
v. normalization of the fuel oil system

4.6.1 Oil Charge

Flushing oil is supplied to a temporary flushing oil tank from oil barrels/drums, located below the tank on the ground, with the help of a portable pump. The tank will be filled with flushing oil up to 75–80% of its storage capacity. Once adequate oil is poured in, one by one flushing oil circuits are to be charged by running temporary flushing oil pump/s. Top off the tank with a fresh supply of oil. The system is now ready to be put under oil flushing.

4.6.2 Oil Circulation

Begin flushing of fuel system pipe lines at low flow rates using one temporary flushing oil pump. Slowly increase flushing flow rate with additional pumps until a fuel velocity of about 3.5 m s^{-1} is achieved (Clause 3.5.1 [4]).

The circulation shall be generally continuous with stopping from time to time for inspection and cleaning of temporary strainers. Experience reveals that almost all of the foreign matters are collected in the filters or temporary strainers during the first few hours of flushing. During this time, whenever a noticeable increase in pressure drop across the strainers is observed, the strainers should be cleaned and replaced. This may occur as frequently as in 30-min intervals at the beginning. When the process is in progress, periodically open vents and drains for about 10 s each to remove stagnant debris.

Flushing shall continue until the fuel being delivered is free of construction debris to the satisfaction of the equipment manufacturer. Samples of fuel, taken from upstream of temporary fuel oil strainers, shall generally be such as to contain solid contaminants of maximum 0.53 mg dm^{-3} and free water not to exceed 10 mg kg^{-1} (Clause 3.7.2 [4]).

Bypass provided around normal fuel pumps may be now removed and fuel pumps are restored to the flushing oil circuit. Flushing oil is now allowed to pass through the fuel oil pumps and circulation of flushing oil is continued until the final cleanliness is attained, fulfilling the requirement of the equipment supplier.

4.6.3 Duration of Oil Circulation

Circulation of flushing oil through each circuit may continue for several days until the oil charge becomes sufficiently clear with no evidence of contaminants, such as lint, welding beads, other extraneous matter, and so on, on both permanent and temporary fine mesh screens and accessible parts of the system.

Size of suspended particulate matter in flushing oil must conform to NAS class 10 specification with water content of <100 mg kg^{-1} (ISO 21/19/16) (Table 4.1) [1].

Note

The flushing oil, after purification, may be reclaimed and used for flushing another installation, after it has been determined that the product is free of contaminants and still contains solvency and rust-inhibiting properties [1].

4.7 Estimation of Flushing Oil

For the determination of the flushing oil requirement, the oil-holding capacity of the system shall be ascertained beforehand (Table 4.4). Actual volume of each area is project specific and has to be supplied by the manufacturer.

Table 4.4 Areas for which oil-holding capacity is to be assessed

	Item	Volume (m³)
	Storage volume of temporary flushing oil tank	
	From temporary tank to the suction of temporary flushing oil pumps	
Circuit 1	From temporary flushing oil pump discharge to fuel oil unloading header through oil strainers, unloading/transfer pumps up to fuel oil storage tank inlet, back to temporary tank	
Circuit 2	From temporary flushing oil pump discharge to fuel oil storage tank outlet through oil strainers, pressurizing/forwarding pumps, oil heaters, fuel oil lines up to burner front, back to temporary tank	
Circuit 3	From temporary flushing oil pump discharge to burner/injector return oil lines inlet to fuel oil storage tank up to burner front, back to temporary tank	

4.8 Conclusion

The fuel oil system is now ready to be put into service for normal operation.

On completion of fuel oil flushing, a protocol (Table 4.5) is jointly signed by all concerned.

Table 4.5 Protocol (flushing of fuel oil system)

Areas		Parameters	Inference	
Temporary strainer located on the return line		Analysis of effluent	Solid contaminants of maximum 0.53 mg dm^{-3}. Number of particles in 100 mL of sample: Particle Size (μm) / Maximum Allowable Number of Particles 5 – 15 / 256,000 15 – 25 / 45,600 25 – 50 / 8100 50 – 100 / 1440 > 100 / 256	
Sampling point from burner front		Analysis of oil	Free water content $\leq$10 mg kg^{-1}	
Signed by the customer	Signed by the engineer	Signed by the steam generator manufacturer	Signed by the gas turbine/diesel engine manufacturer	Signed by the supplier (contractor)

References

[1] GE Energy, GE K 110483b: Cleanliness Requirements for Power Plant Installation, Commissioning, and Maintenance, © 2002 General Electric Company.

[2] ABS Notes on Heavy Fuel Oil, Copyright 2001 American Bureau of Shipping, ABS Plaza, 16855 Northchase Drive, Houston, TX 77060, USA.

[3] D.K. Sarkar, Thermal Power Plant—Design and Operation, Elsevier, Amsterdam, Netherlands, 2015.

[4] USACE/NAVFAC/AFCEC/NASA, UFGS-33 08 55 (July 2007), Unified Facilities Guide Specifications, Section 33 08 55, Commissioning of Fuel Facility Systems.

CHAPTER 5

Blowing of Fuel Gas Piping System

Chapter Outline

5.1 Introduction

One of the major sources of fuel for electricity generation is natural gas. Natural gas burns cleaner than other fossil fuels, such as oil and coal, and produces less carbon dioxide per unit energy released. For an equivalent amount of heat, burning natural gas produces about 30% less carbon dioxide than burning petroleum and about 45% less than burning coal [1].

Notwithstanding the above benefits of natural gas, it is seldom used as fuel in steam power plants since the cost of natural gas (typically $5.5 per MMBtu or 5.21×10^{-3} per MJ) is exorbitantly high compared to the cost of coal (typically $55 per ton or 3.24×10^{-3} per MJ). Further based on higher heating value (HHV) of fuel, efficiency of a natural gas-fired boiler (typically 85%) is lower than that of a pulverized coal-fired boiler (typically 87%).

A diesel engine cannot operate on natural gas alone. In countries where natural gas is abundant, a dual-fuel type diesel engine may utilize natural gas as a supplementary fuel to diesel oil.

Hence, for all practical purposes, natural gas as fuel is utilized in gas (combustion) turbine power plants only.

Prior to putting gas turbines into service and to ensuring their reliable operation, a fuel gas piping system must be blown to get rid of foreign materials—wooden planks, welding rod ends, cotton wastes, beverage cans, and so on, and contaminants, eg, mill scale, rust, welding beads, residue from grinding, chipping, blasting, and so on. These contaminants may cause fuel combustor clogging, injector nozzle wear, and blade particulate impingement [2].

In new units foreign materials and contaminants may sneak into the fuel gas piping system during manufacture, storage, field fabrication, and installation. Contaminants may enter into the fuel gas piping system of running units during operation and/or overhaul. Hence, during erection of a fuel gas piping system, extreme precaution should be taken to keep off dirt, dust, and other undesirable debris from this system. On completion of assembly, the whole fuel gas piping system shall be cleaned manually as thoroughly as possible. Over and above manual cleaning, this system shall be also blown with suitable blowing medium to make the piping system free from detrimental elements.

At one time it was common practice in the industry to adopt natural gas as the blowing medium. However, consequent to several severe explosion and fire events and two major catastrophes arising out of natural gas explosions at different plants in the United States in June 2009 (ConAgra Food Plant, North Carolina) and February 2010 (Kleen Energy 620 MW combined cycle power plant, Middletown, Connecticut), the US Chemical Safety and Hazard Investigation Board (CSB) strongly recommended to prohibit using natural gas as the blowing medium for the purpose of cleaning fuel gas piping since flammable gases present an explosion hazard that cannot be wholly eliminated. In response to the CSB recommendation, the National Fire Protection Association (NFPA), United States, issued a provisional standard NFPA 56 (PS) that prohibits any type of flammable gas for internal cleaning of fuel gas pipe and includes compressed air, nitrogen or helium, as well as steam and water as approved alternatives [3]. At the same time EPRI also recommended pneumatic blow cleaning processes using compressed air or compressed nitrogen in order to obviate incidental potential hazards of natural gas [2]. In addition, pigging may be adopted for cleaning of the fuel gas piping system.

During blowing out of the fuel gas piping system, the momentum ratio or cleaning force ratio (CFR) or drag of any suspended particulate matter should be greater than the CFR experienced by the particle under gas turbine maximum continuous rating (GTMCR) operation.

This would ensure that even if any debris is remaining in the system after blowing, they would not get disturbed under normal operating conditions. The CFR of a particle is defined as:

$$CFR = \left[\frac{Q_{blow}}{Q_{gtmcr}}\right]^2 \times \frac{v_{blow}}{v_{gtmcr}} \qquad (5.1)$$

where Q_{blow} = mass flow of blowing medium during blowing out (kg s^{-1});
Q_{gtmcr} = mass flow of natural gas at GTMCR (kg s^{-1});
v_{blow} = specific volume of blowing medium during blowing out (m^3 kg^{-1});
v_{gtmcr} = specific volume of natural gas under minimum permissible operating pressure at GTMCR (m^3 kg^{-1}).

The value of CFR should be between 1.2 and 2.0 or equal to the value recommended by the gas turbine manufacturer, since the onus of determining the effectiveness of blowing the fuel gas piping system mainly rests with the gas turbine manufacturer's representative.

Successful completion of the typical blowing-out process may be assessed by the permissible size and quantity of particles present in the flowing medium. Recommendations of some of the leading gas turbine manufacturers are shown in Table 5.1.

Table 5.1 Cleanliness requirements of fuel gas

Name of Gas Turbine Manufacturer	Maximum Particle Size (μm)	Particle Concentration (mg kg^{-1})
General electric	10	28
Mitsubishi	5	30
Pratt and Whitney	10	30
Rolls-Royce	20	NA
Siemens	10	20
Solar turbines	10	20

Source: Table 5-1, p. 5-2; D. Grace, Guidelines for fuel gas line cleaning using compressed air or nitrogen, EPRI, 1023628, Technical Update, 2011.

If recommended by the gas turbine manufacturer, target plates made of aluminum/copper/mild steel (softer material) and alloy steel may be installed and suitably supported at the exhaust end of the piping system. From the degree of indentation made on target plates by undesired particles in the flowing medium, the effectiveness of blowing could be assessed. Criteria for making satisfactory assessment of target plates may vary from project to project depending on the guidelines laid down by gas turbine manufacturers.

Example 5.1

A simple cycle gas turbine power plant generates maximum power at fuel gas pressure of 2.6 MPa consuming 1.2×10^6 N m^3 of fuel gas per day. Gas temperature follows ambient temperature and can go up to 314 K. If the fuel gas line is blown with compressed air at a pressure of 690 kPa and temperature of 323 K, determine the quantity of air required during blowing for achieving a CFR of 1.2. Specific volume of fuel gas is 1.109 N m^3 kg^{-1}.

If the internal diameter of the fuel gas pipe line is 300 mm, compare the velocity of flow through the fuel gas pipe line during blowing with the velocity of flow at the maximum generation of power. Determine also the sonic velocity in the piping during compressed air blowing.

Solution

Referring to Eq. (5.1) we can proceed as follows:

i. Maximum quantity of fuel gas flow,

$$Q_{gtmcr} = \frac{1.2 \times 10^6}{1.109 \times 24 \times 3600} = 12.5238 \, \text{kg s}^{-1}$$

ii. To find out specific volume of fuel gas, v_{gtmcr}, at 2.6 MPa pressure and 314 K temperature, applying the ideal gas law, we can write:

$$\frac{101.3 \times 1.109}{273} = \frac{(2600 + 101.3) \times v_{gtmcr}}{314}$$

Solving above, $v_{gtmcr} = 0.04783 \, \text{m}^3 \, \text{kg}^{-1}$.

iii. At 101.3 kPa and 323 K, specific volume of air, $v_a = 0.9174 \, \text{m}^3 \, \text{kg}^{-1}$.

Hence, at 690 kPa and 323 K, specific volume of air,

$$v_{blow} = \frac{101.3 \times 0.9174}{(690 + 101.3)} = 0.11744 \, \text{m}^3 \, \text{kg}^{-1}$$

From Eq. (5.1),

$$CFR = \left[\frac{Q_{blow}}{Q_{gtmcr}}\right]^2 \times \frac{v_{blow}}{v_{gtmcr}}$$

or,

$$1.2 = \left[\frac{Q_{blow}}{12.5238}\right]^2 \frac{0.11744}{0.04783}$$

so,

$$Q_{blow} = 12.5238 \times \sqrt{\frac{1.2 \times 0.04783}{0.11744}} = 8.7553 \, \text{kg s}^{-1}$$

iv. Velocity of air through fuel gas pipe line during blowing,

$$V_{blow} = \frac{8.7553 \times 0.11744}{\frac{\pi}{4}(0.3)^2} = 14.55 \, \text{m s}^{-1}$$

v. Velocity of fuel gas through fuel gas pipe line at maximum generation of power,

Example 5.1—cont'd

$$V_{gtmcr} = \frac{12.5238 \times 0.04783}{\frac{\pi}{4}(0.3)^2} = 8.47\,ms^{-1}$$

Since V_{blow} is much greater than V_{gtmcr}, this implies the effectiveness of the blowing process.

vi. The sonic velocity of an ideal gas is given by

$$V_c = 0.3194\sqrt{(k \times g \times P \times v)}$$

where V_c = sonic velocity (m s^{-1});
k = specific heat ratio = 1.4 (for air);
g = acceleration due to gravity (9.81 m s^{-2});
P = pressure (Pa abs);
v = specific volume of gas (m^3 kg^{-1}).
Hence, sonic velocity during blowing,

$$V_c = 0.3194\sqrt{1.4 \times 9.81 \times (690 + 101.3) \times 10^3 \times 0.11744} = 360.84\,ms^{-1}$$

Example 5.2

If the power plant of Example 5.1 receives fuel gas at pressure 2.6 MPa at the inlet to a fuel gas supply metering skid, which reduces to 1.5 MPa at the inlet to the gas turbine fuel gas skid and if other conditions of Example 5.1 remain unchanged, determine the quantity of air required along with velocity of flow during blowing.

Solution

For every 103 kPa drop in fuel gas pressure, the fuel gas temperature drops by 1 K (Source: p. 2-4 [2]).

Hence, when fuel gas pressure drops from 2.6 to 1.5 MPa, fuel gas temperature drops by about 11 K; that means fuel gas temperature at the inlet to gas turbine fuel gas skid will be 303 K (314−11).

From Eq. (5.1) it is noted that v_{gtmcr} corresponds to a minimum permissible operating pressure at GTMCR. So, specific volume of fuel gas, v_{gtmcr}, at 1.5 MPa pressure and 303 K temperature, applying ideal gas law, is

$$\frac{101.3 \times 1.109}{273} = \frac{(1500 + 101.3) \times v_{gtmcr}}{303}$$

Solving above, $v_{gtmcr} = 0.07787$ m^3 kg^{-1}.

The quantity of air required during blowing, applying Eq. (5.1), will be:

$$1.2 = \left[\frac{Q_{blow}}{12.5238}\right]^2 \frac{0.11744}{0.07787}$$

So, $Q_{blow} = 11.1713$ kg s^{-1}.

Velocity of air through fuel gas pipe line during blowing,

$$V_{blow} = \frac{11.1713 \times 0.11744}{\frac{\pi}{4}(0.3)^2} = 18.56\,ms^{-1}$$

5.2 Description of Fuel Gas System (Source: Section 7.5.8, p. 268 [1])

NFPA 85 defines natural gas as "A gaseous fuel occurring in nature and consisting mostly of organic compounds, normally methane (CH_4), ethane (C_2H_6), propane (C_3H_8), and butane (C_4H_{10}). The calorific value of natural gas varies between 26.1 and 55.9 MJ m^{-3}, the majority averaging 37.3 MJ m^{-3}" [4]. Natural gas on average contains 80–90% methane, 6–9% ethane, and 2–5% propane. Noncombustible gases present in minor quantities in natural gases are nitrogen (0.5–2.0%) and carbon dioxide (0.1–1.0%).

Prior to supplying fuel gas to the gas turbine fuel gas skid, the incoming gas is treated in a "knockout drum" (KOD) followed by a "filter separator" unit to strip the gas of all solids and liquids. Any solid particle carried with the gas stream is separated first. Liquid separation takes place subsequently. Thereafter, the fuel gas may enter into a booster compressor for boosting the gas pressure to the pressure recommended by the gas turbine manufacturer. If the pressure of fuel gas at the inlet to the fuel gas metering skid (at supply end) is considerably higher than the operating maximum gas pressure required at the inlet to the gas turbine fuel gas skid (as in Example 5.2), then the booster compressor need not be installed. Fuel gas is sent directly to gas coolers to limit fuel gas temperature to comply with the requirement of gas turbines. The cooled gas is then passed through another set of filters to remove any liquid condensed after cooling of the compressed gas before the fuel gas is conveyed to the fuel gas skid of gas turbines for further distribution to burners (Fig. 5.1).

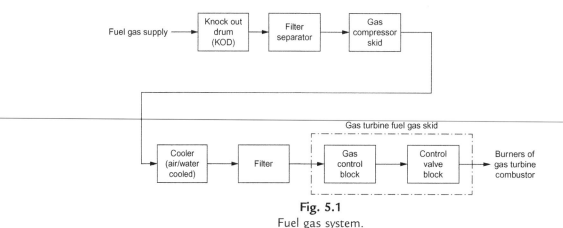

Fig. 5.1
Fuel gas system.

5.3 Precautions

Over and above the general precautionary measures described under Section 2.1.1: Quality Assurance, the following specific precautionary measures must be observed before the blowing of the fuel gas piping system:

i. Blowing of the fuel gas piping system shall be carried out in accordance with NFPA 56: Standard for Fire and Explosion Prevention During Cleaning and Purging of Flammable Gas Piping Systems;

ii. While the blowout process is underway, all activities including welding work or any other hot work, eg, gas-cutting, around the area shall be strictly prohibited;

iii. In the event that natural gas is used as the blowing medium, the following aspects need special consideration:

1. The exhaust pipe must be grounded to mitigate static electricity, the occurrence of which may cause ignition;

2. The exhaust end of the piping must be located at an elevation at a safer place;

3. A magnetic filter shall be installed between the blowout valve and the silencer to arrest metal particles, which on impact on any metallic surface located nearby the exhaust end may cause a spark to ignite the natural gas, resulting in an explosion;

4. As explained earlier, the main constituent of natural gas is methane. Hence, ensure that in a mixture with air, the content of methane does not fall within the range of 4.4–16.5% by volume, lest on ignition the mixture will explode;

5. The point of release of the gas must be carefully determined to minimize the extent of the flammable atmosphere. Hence, a complex technical evaluation of various factors is necessary, including height, location, orientation of the vent pipe, velocity and density of the natural gas being discharged, potential sources of ignition, personnel location, wind speed, and a dispersion analysis to verify that the natural gas will rapidly dissipate [5];

6. Install combustible gas monitors around the perimeter of the gas exhaust points.

iv. In the event nitrogen is used as the blowing medium, it must be kept in mind that there is a potential risk of asphyxiation. The risk may get exacerbated since the presence of nitrogen cannot be detected by sight or smell. Hence, it is essential that oxygen detection equipment must be in use to protect all personnel present in any area where nitrogen may accumulate, whether by inadvertent leak or by the blow process as designed [2];

v. While adopting the steam-blowing method, both permanent and temporary pipe lines must be suitably insulated;

vi. During the execution of the blowing-out process, the general noise level of a plant may get raised above the tolerable hearing limit of human beings (typically 65 dBA). Hence, in order to avoid unbearable noise pollution, it would be preferable to carry out the blowout operation starting from early morning, say at 06:00, up to about 19:00 on weekdays only, excluding Sundays and public holidays;

Note

Contrary to the above guideline, there is a school of thought "to perform the blowing activity on the weekend to minimize the impact to construction progress and ensure worker safety. Site personnel are limited to those who are essential to conducting the gas blow" [5].

vii.   Before inserting a target plate, the pressure in the fuel gas piping system and temporary piping is relieved and reduced to atmospheric pressure;

viii. Prior to initiating a blow, an audio-visual alarm must be sounded to clear the area, especially the discharge area;

ix.   Use ear muffs when blowing is in progress.

5.4 Prerequisites

Since blowing of the fuel gas piping system is a hazardous process, before undertaking the process a thorough inspection of the system is essential to ascertain the following (Table 5.2):

Table 5.2 Areas/items to be checked

Sl. No.	Areas/Items	Ok (√)
1	Before erecting permanent piping systems, protective coating used on internal surfaces, as a safeguard against corrosion during transport and storage, are removed as far as practicable.	
2	Prior to erecting all temporary piping systems, they are thoroughly cleaned mechanically.	
3	All piping systems, whether permanent or temporary, are erected and supported properly along with associated valves. Since the blowing medium from temporary piping systems exhausts to the atmosphere, thrust from the exhaust pipe will be transmitted to the pipe supports; hence, verify that the design of these supports are checked taking proper care.	
4	The internal of the silencer shall be thoroughly cleaned prior to its installation.	
5	The complete fuel gas piping system is hydrotested at 1.5 times the maximum operating pressure. Protocol signed jointly by the customer, the engineer, and the supplier (contractor), certifying acceptability of the hydrotest, is verified.	
6	If natural gas is used as the blowing medium, then verify the following: a. the exhaust pipe is properly grounded and the exhaust end is erected in a safe location b. combustible gas monitors are installed around the perimeter of the gas exhaust points	
7	In case nitrogen blowing is adopted, sufficient oxygen detection equipment is available to protect all personnel present.	
8	For utilizing steam as the blowing medium, ensure that all of the piping system is properly insulated.	
9	All welded and flanged joints are secured.	
10	Construction materials, including temporary scaffolding, wooden planks, welding rod ends, cotton wastes, and so on, are removed from the internals of the blowing pipes.	
11	All instruments installed on permanent piping systems are kept isolated.	

Table 5.2   Areas/items to be checked—cont'd

Sl. No.	Areas/Items	Ok (√)
12	The source of the blowing medium—auxiliary boiler or nitrogen trailer or air compressor—is arranged.	
13	Location of auxiliary boiler/nitrogen trailer/air compressor does not encroach in the area where the permanent equipment or piping is installed.	
14	Check that the following instruments are in service: i. pressure gage at the source—fuel gas pressure or steam pressure or compressed nitrogen/air pressure ii. pressure gage upstream of temporary blowout valve	
15	Relevant protections, alarms, and interlocks are checked and kept ready.	
16	Required quantity of materials (Section 5.4.1) is available.	
17	Safety gadgets (Section 5.4.2) are arranged.	
18	Temporary platforms are provided wherever required.	
19	Verify that "no smoking," "danger," "keep off," and other safety tags (in regional and English languages) are displayed at various places around the area of activity.	
20	Verify also that auxiliary boiler/nitrogen trailer/air compressor, valves, associated equipment, and systems, which are in service or are energized, are provided with proper safety tags (Appendix C).	

Note

Ensure qualified personnel only are involved in laying down the blowout guidelines and in supervising the blowout process until the normalization of all circuits.

5.4.1   Typical List of Bill of Materials

Materials that are typically required for the blowing of the fuel gas piping system are listed below. Size and quantity of each of these items would vary from project to project and shall have to be assessed by concerned personnel associated with blowing activity.

1.   pneumatically operated temporary rotary blowout valve/s;
2.   gasket sheets or rubber discs;
3.   auxiliary boiler/air compressor/nitrogen trailer/pigs;
4.   target plates;
5.   target plate holders;
6.   temporary pipe lines;
7.   spool pieces;
8.   pressure-measuring instruments.

5.4.2 *Typical List of Safety Gadgets*

The following equipment is generally used for safety of operating personnel and the area covered under blowing:

1. safety shoes of assorted sizes;
2. asbestos gloves (optional for steam blowing);
3. hard hats of assorted sizes;
4. ear muffs;
5. first aid box;
6. sufficient lighting arrangements around temporary system;
7. cordoning of the exhaust area;
8. hooters at various places to send warning sound prior to commencing each blowout operation;
9. thermal insulation of temporary piping (optional for steam blowing);
10. safety sign boards;
11. safety tags.

5.5 Preparatory Arrangements

For effective blowing of the fuel gas piping system within the shortest possible time, it is necessary to complete the following preparatory work. The whole system may be split into one or more circuits for the convenience of operation (Fig. 5.2).

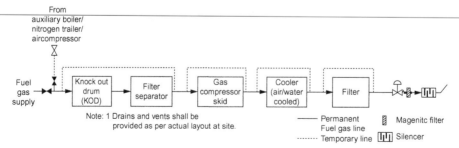

Fig. 5.2
Fuel gas piping blowing scheme.

a. In order to complete the circuit for blowing out, some temporary piping connecting the permanent piping systems to atmosphere should be erected and kept ready for putting into service;

Notes

1. These temporary connections are project specific. Hence, layout of temporary piping shall have to be jointly decided at the site among the customer, the engineer, and the supplier (contractor) responsible for successful completion of the above cleaning activity;
2. Size and quantity of temporary piping materials including elbows, tees, valves, supports, and so on, shall be assessed at the design stage. These requirements, however, may vary based on the exact layout of the temporary piping system. Final quantity of piping materials including associated equipment shall be procured accordingly;
3. Cross-section of temporary piping systems shall be at least same as the permanent piping systems.

---

b. Provide bypass around the following equipment:
    i. knockout drum and filter separator
    ii. gas compressor skid
    iii. cooler
c. Remove/replace the following items to protect them from any damage caused by exhaust debris:
    i. filter elements
    ii. control valves replaced with suitable spool pieces
    iii. thermowells
    iv. flow element/s
d. If the blowing medium is selected to be steam, then either a temporary connection from the existing auxiliary steam header is to be laid or an auxiliary boiler is to be installed;
    or
    In the event nitrogen blowing is chosen, two or three nitrogen trailers shall be arranged;
    or
    For compressed dry air blowing, two temporary oil-free rotary screw compressors, each with suction filter and silencers, should be installed;

Note

Flow capacity and required pressure of blowing medium would vary from project to project and shall be decided through pipe flow analysis and modeling jointly by the customer, the engineer, and the supplier (contractor).

---

e. The above piping systems are provided with drain valves for discharging any condensate;
f. Wherever temporary piping systems interfere with passages, ramps or gangways, suitable temporary crossings over the piping systems shall be provided;

g. The temporary blowout valve followed by magnetic filter (optional for natural gas blowing) and silencer shall be installed on the horizontal portion of the temporary piping system near the exhaust end. The final exit from temporary piping systems will be assembled with an inclination toward the sky. It shall be ensured that debris coming out from the exhaust end would not hit any structure or equipment;

h. For effective completion of blowing out, the temporary blowout valve shall be preferably pneumatically operated to ensure quick opening;

i. Target plates shall be numbered in compliance with the number of blows to be carried out. This would facilitate proper identification of target plates and assessment of chronological improvement in the blowing-out process after each blowout.

5.6 Operating Procedure

Before discussing the operating procedure of blowing the fuel gas piping system, certain aspects common to each blowing medium are highlighted below:

i. Whatever may be the blowing medium, the storage of some volume of the medium is important during blowing. The capacity of storage volume required should be in accordance with the volume of piping to be cleaned and storage capacity that is available [2];

ii. The blow process usually begins with larger bore piping, followed by smaller off-take lines [2];

iii. Blowing of each circuit is generally started with a lower pressure (as applicable to a specific blowing medium) for evaluating safety aspects of the blowing circuit. These blows are carried out very slowly and may continue for several minutes. During this phase, water and lighter debris from the piping system will be driven out. Once the presence of water in the exhaust is no longer visible, the blowing pressure may be gradually raised step by step to a pressure as recommended by the gas turbine manufacturer. The duration of blowing at each step may last 10–15 min;

iv. Industry experience reveals that most of the debris gets removed after an initial few blows. Usually after about 10–20 blows, debris coming out from the exhaust end will not be visible to the naked eye. Target plates may be installed then for analyzing the size and number of indentation formed on the surface of these plates. Blowing at maximum pressure, as recommended by the manufacturer, will continue until two or three sequential clean target plates are achieved.

v. The first few blows will be carried out bypassing major equipment as discussed under Section 5.5 "b". On successful completion of this step, the normal line will be blown for cleaning;

vi. Method of blowing could be either "packing fracture blow" [2], also known as "card board blasting" or "continuous blow" type;

Note

1. In the card board blasting method, multiple gasket sheets or rubber discs, as decided beforehand, are placed between flanges at the exhaust end of the pipe line. Once the pipe line gets pressurized, these sheets or discs will eventually rupture, releasing gush of debris from the pipe line. After each blow, ruptured sheets or discs need to be replaced with new ones;
2. In the continuous blowing method, velocity of the blowing medium is the driving force for removal of debris. On charging the fuel gas pipe lines at the required pressure, the temporary blowout valve is opened fast, kept open for a few seconds, and then closed. Due to the sudden opening of the valve, the debris inside the pipe lines will be driven out.

---

vii. Sequence of blowing as discussed in this section refers to blowing of one circuit. The same procedure may be followed while blowing other circuits;

viii. On successful completion of blowing the fuel gas piping system, if the system remains idle for more than 10 days, then pipe lines are to be preserved with nitrogen gas [2].

With above guidelines, the typical blowing out requirement of the fuel gas piping system with different blowing medium is addressed under the following paragraphs.

5.6.1 Natural Gas Blowing

Blowing of the fuel gas piping system using natural gas as the blowing medium is still followed in the industry even though "the CSB believes that using natural gas or other flammable gases to clean fuel gas piping is inherently unsafe and should be prohibited [6]."

The advantage of using natural gas as the blowing medium is that the line pressure itself, without further compression, would suffice to drive out debris from inside the fuel gas pipe line. In practice natural gas pressure at the supply end of the fuel gas metering skid is much higher than the required blowout pressure; eventually blowout pressure is established by partial opening of the fuel gas block valve at the supply end. The blowing may be started with a gas pressure of 700 kPa or as recommended by the manufacturer.

Using natural gas as the blowing medium prompts the following precautionary measures to be adhered to:

a. The exhaust end of the piping must be located at a safe, well-ventilated place;
b. Gas must be exhausted at an elevation for proper dispersion [2];
c. Exhaustive fire-fighting arrangement shall be provided to suppress any fire hazard.

While use of high pressure natural gas seems much easier due to lesser equipment involved, and slightly cheaper in installation cost, precautions needed to ensure safety are much higher. Fire hazards during the gas blow are a big concern in the industry [7].

Since methane is a potent greenhouse gas, venting natural gas to the atmosphere promotes global warming. According to the US EPA, although the effect of natural gas only lasts 9–15 years, compared to 100 years from carbon dioxide, the impact of methane is 21 times stronger than that of carbon dioxide [7].

5.6.2 Compressed Air Blowing

Of all blowing medium generally used during the blowing of fuel gas pipe lines, high-pressure oil-free dry compressed air blowing may be considered the most convenient method. Compared to natural gas blowing, compressed air blowing is definitely safer, but may require an extended time to clean the fuel gas piping system.

This method has an added advantage that air blows can be done before the gas utility has their piping available at the plant metering skid terminal point [7].

First, air pressure during blow is maintained at about 300 kPa by opening the supply end valve. Once the pipe line is pressurized, the blowout valve is opened quickly to drive out remnant water and loose debris inside piping. When the effluent from the exhaust end is observed by the naked eye to be clean and free from any water content, air pressure is gradually raised from 400 to 600 kPa, then to 700 kPa and finally to 800 kPa or as recommended by the gas turbine manufacturer.

If the card board blasting method is adopted, continue pressurizing the system until the gasket sheets or rubber discs rupture. Close the supply end valve; once the line is depressurized, install fresh gasket sheets or rubber discs. Repeat the pressurizing-depressurizing process at each starting air pressure until the pipe line is declared to be clean.

While utilizing the continuous blow method, quickly open the blowout valve at each starting air pressure and keep the valve open until the air pressure drops to about 250 kPa. This may take several seconds. Then close the blowout valve for the preparation of the next blow. The pressurization and depressurization cycles are repeated until the piping systems are observed to be clean.

The preparatory time to fill the air receiver by the air compressor will be around 8–10 min. On completion of blowing at maximum air pressure for at least four times, target plates may be installed at the exhaust end [2].

A major concern of adopting compressed air blowing is if the capacity of station instrument air compressor is not adequate enough to cater air blowing, a temporary oil-free dry air compressor along with compatible air receiver and air dryer may need to be installed for blowing activity.

5.6.3 Compressed Nitrogen Blowing

In lieu of compressed air blow, compressed nitrogen blow may be adopted as the blowing medium. Advantages of compressed nitrogen over compressed air are:

a. In this method, nitrogen trailers provide required flow capacity and blowing pressure. Hence, any additional storage area, like an air receiver, is no longer required;
b. Nitrogen itself is dry, so a separate dryer, as required for compressed air, is not a necessity.

A major disadvantage of nitrogen is there is a risk of asphyxiation. One full breath of nitrogen can render one unconscious or worse [2]. In addition, compressed nitrogen is costlier than compressed air.

The duration of compressed nitrogen blowing will be greater than that of compressed air blowing due to the larger storage capacity of the nitrogen trailer. Nitrogen gas blows are inherently safer alternatives to natural gas blows due to the use of a nonflammable gas [5].

Procedure for blowing with compressed nitrogen is similar to the procedure followed for compressed air blowing.

5.6.4 Steam Blowing

In this method steam velocity is the driving force for removal of debris from inside the piping system. Steam may be supplied at the gas metering end of the pipe line, blown through the line; then steam escapes to atmosphere through the exhaust end.

Source of steam for steam blowing could be the station auxiliary steam header (pressure and temperature are usually 1.6 MPa and 493–503 K, respectively). In case of inadequacy of steam supply from this source, a temporary auxiliary boiler may have to be installed for blowout.

If an auxiliary boiler is required to be installed, then firing in the boiler is not disturbed during blowing and blowing pressure is maintained constant until the completion of blowing.

5.6.5 Water Jet Flushing

In this method, high-pressure (about 70 MPa or higher) water jets are connected to special hoses. These hoses are inserted at one end of the fuel gas pipe line. The jet from the hose nozzle is kept forced against the pipe wall, moved around, and traversed along inside the pipe lines to dislodge loose debris and other foreign materials. After retracting the jet, debris are removed by water flushing. In this method, typical water flow ranges from 126×10^{-5} to $252 \times 10^{-5}\ \mathrm{m^3\ s^{-1}}$ [2].

5.6.6 Pigging

Pigging is a method of cleaning fuel gas pipe lines mechanically. The pipe is first pressurized with compressed air at a pressure of about 700 kPa while keeping the blowout valve closed; flexible polythene pig is then inserted through pig receiver and pig launcher (shaped as "Y") from the fuel gas supply end of the pipe line. The blowout valve is then opened. The pressure in the pipe line pushes the pig along the pipe length. When the pig is traveling, air pressure is maintained below 400 kPa. Pig picks up debris as it travels along the pipe line; on reaching the exhaust end of the pipe line pig gets collected in the "pig catcher," while debris carried with the pig blows out through the exhaust end [2]. Ensure that no one gets injured or no equipment gets damaged due to the impact of high velocity pig.

Effectiveness of cleaning by pigging is less than the effectiveness that could be achieved by using natural gas [5]. Pigging is attractive for cleaning of long distance pipe lines and may not be very convenient for cleaning of fuel gas pipe lines at plant level.

5.7 Conclusion

The acceptance criteria of a clean target plate are one having no raised edge impacts and having indentation of 3 particles no larger than 0.4 mm on the target plate [2].

Once indentation on target plates is accepted, the fuel gas piping system may be declared to be ready for further use after normalizing the permanent systems.

Table 5.3 depicts a typical protocol, which must be signed jointly by all concerned.

Table 5.3 Protocol (blowing of fuel gas piping system)

Circuit Under Blowing		Status of Aluminum/Copper/Mild Steel Target Plate	Status of Alloy Steel Target Plate
1		Fulfills the criteria "one having no raised edge impacts and having indentation of 3 particles no larger than 0.4 mm on the target plate"	Shows no further improvement
Signed by the customer	Signed by the engineer	Signed by the gas turbine manufacturer	Signed by the supplier (contractor)

References

[1] D.K. Sarkar, Thermal Power Plant—Design and Operation, Elsevier, Amsterdam, Netherlands, 2015.
[2] D. Grace, Guidelines for Fuel Gas Line Cleaning Using Compressed Air or Nitrogen, EPRI, 2011. 1023628. Technical Update.

[3] F. Durso, Jr., The making of a standard, NFPA J.®, November/December 2011. (http://www.nfpa.org/newsandpublications/nfpa-journal/2011/november-december-2011/features/the-making-of-a-standard).

[4] National Fire Protection Association, USA, NFPA 85_2004 Boiler and Combustion Systems Hazards Code, 2004.

[5] J.H. Brown, Best practices for natural gas line cleaning, 2011. http://www.powermag.com/best-practices-for-natural-gas-line-cleaning.

[6] US Chemical Safety Board (CSB), ASME urged to adopt CSB recommendation prohibiting natural gas blows at power plants.

[7] H. Phung, G. Jones, N.K. Smith, ASME: comparison between air blows and gas blows for cleaning power plant fuel gas piping systems, 2012. http://www.energy-tech.com/steam/article_57000e6d-c500-504d-8a8e-2d13ff95a410.html.

CHAPTER 6

Chemical Cleaning of a Steam Generator

Chapter Outline

6.1 Introduction

In the present day power plants, high purity steam, both main and reheat (Table 6.1), is required to ensure:

- efficient heat transfer within the steam generator,
- reduce steam generator tube failures, and
- provide turbine blades free from fouling.

These three aspects together could ensure safe operation of plants with least interruption from these fronts.

In order to receive high purity steam, a steam generator is essentially required to be chemically cleaned, to make the internal surfaces of steam-water circuit free from

Table 6.1 Analysis of superheated and reheated steam

Contaminants	Unit	Drum Boilers	Once-Through Boilers
		Normal Condition	
pH	–	9.0–9.6[a]	9.0–9.6[a]
		8.5–9.0[b]	8.5–9.0[b]
Specific conductivity, at 298 K	$\mu S\ cm^{-1}$	3–11	2.5–7.0
Conductivity (after cation exchanger), at 298 K	$\mu S\ cm^{-1}$	<0.20	<0.20
Silica (as SiO_2)	$\mu g\ kg^{-1}$	<20	<10
Sodium (as Na)	$\mu g\ kg^{-1}$	<10	<3
Iron (as Fe)	$\mu g\ kg^{-1}$	<20	<10
Copper (as Cu)	$\mu g\ kg^{-1}$	<3	–
Chloride	$\mu g\ kg^{-1}$		<3
Total dissolved solids	$\mu g\ kg^{-1}$	<50	<50

Note: $\mu g\ kg^{-1}$ = ppb.
[a] All-volatile treatment (AVT) of FW.
[b] Oxygenated treatment (OT) of FW.

i.   mill scales and oxides of iron that may develop in the manufacturing and fabrication stage of a new unit

ii. grease, oil, dirt, temporary protective coatings, and so on, which are used for preservation and may enter into the system during transport and storage

iii. weld slag, debris, and other contaminants that may remain after erection of the unit

iv. deposits, eg, metallic copper, iron oxide, aluminum, carbonates, silica and silicates, phosphate compounds, calcium and magnesium sulfates, corrosion products, sludge, and oil or process contaminants that may develop in boiler tubes of a running unit

Deposits inside steam generator tubes (Fig. 6.1) are responsible for overheating and/or corrosion of tube metal surfaces that in turn may lead to tube failure. Deposits also impair proper steam generator water circulation. Common constituents of deposits are calcium, magnesium, silica, aluminum, iron, and sodium. In running steam generators, the amount of deposits can be evaluated by using chordal thermocouples (Fig. 6.2). These thermocouples located on certain critical tubes can indicate a tube metal temperature increase caused by excessive internal deposits and also can measure the effect of any steam generator deposition on the resistance to heat transfer. Thereby, the operator is alerted, identifying the need for cleaning, lest tube failure will be imminent [1,2].

In steam generator steam is continuously generated from evaporation of water with gradual increase in concentration of dissolved salts. Eventually when there is excessive deposition of dissolved salts the ionic product of these salts exceeds their solubility limit. Thus salts, which are insoluble in steam generator water, start precipitating. Scale is formed when these precipitates are hard and adhere to the inner wall of steam generator tubes (Fig. 6.3). They are difficult to remove, even with the help of hammer and chisel. Scale may form as calcium silicate ($CaSiO_3$), magnesium silicate ($MgSiO_3$), or as sodium iron silicate. Some forms of scale are so stubbornly unyielding that they resist any type of removal—mechanical or chemical [3].

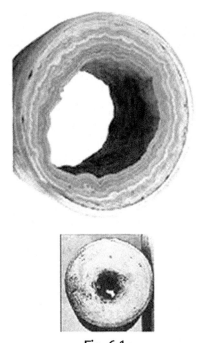

Fig. 6.1

Deposit in steam generator tubes. *Reproduced with permission from Apex Engineering Products Corporation.*

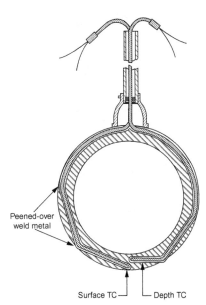

Peened-over
weld metal

Surface TC Depth TC

Fig. 6.2

Chordal thermocouple. *From Fig. 18, P 40–12, Chapter 40: Pressure, Temperature, Quality and Flow Measurement. Steam Its Generation and Use (41st edition). Courtesy of The Babcock & Wilcox Company.*

Fig. 6.3
Formation of scale.

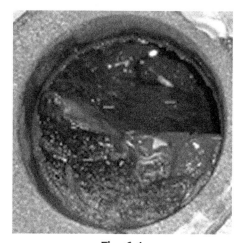

Fig. 6.4
Formation of sludge. *Reproduced with permission from Heatmeister. From http://mechanical-hub.com/sites/hydronics/magnetic-separation-a-concept-that-will-stick/.*

If the precipitates of dissolved salts are soft, loose and slimy they are termed as sludge (Fig. 6.4). Sludge consists of suspended particles that are hardly soluble in cold water, eg, $MgSO_4$, $MgCO_3$, $MgCl_2$, $CaCl_2$, dirt, etc. Sludge enters the steam generator as suspended solids and gets precipitated at comparatively colder section of the steam generator where the flow rate is slow. Sludge deposits can be removed by wire brush or mild acid. In the event sludge gets entrapped in scale; sludge turns hard, dense, and tenacious and may become as problematic as scale to remove [3].

Films of oil, grease, and temporary protective coatings remaining inside a tube or pipe render an insulating effect to heating surfaces and, thus, impair heat transfer, leading to failure of tubes due to overheating (Figs. 6.5 and 6.6) [1,4]. Failure of tubes may also result from thinning of the tube metal surface following excessive corrosion of tube surface areas (Fig. 6.7) [1]. There are also instances when the steam generator tube failed due to hydrogen embrittlement

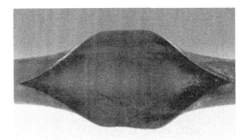

Fig. 6.5

Failure due to overheating. *From Fig. 21, P 42–19, Chapter 42: Water and Steam Chemistry, Deposits and Corrosion. Steam Its Generation and Use (41st edition). Courtesy of The Babcock & Wilcox Company.*

Fig. 6.6

Failure due to overheating. *From Fig. 9–32, P 9–22, Chapter 9: Water Technology. Clean Combustion Technologies, fifth ed., Alstom, Windsor, CT, 2009.*

Fig. 6.7

Rupture due to thinning of pipe. *From Fig. 14, P 42–16, Chapter 42: Water and Steam Chemistry, Deposits and Corrosion. Steam Its Generation and Use, 41st ed. Courtesy of The Babcock & Wilcox Company.*

caused by entrance or absorption of hydrogen into the metal resulting in loss of ductility (Figs. 6.8 and 6.9) [1,4].

In a running steam generator, if the aforementioned failure of steam generator tubes necessitates replacement of more than 10% surface area of tubes, it is imperative to chemically clean the steam generator to maintain the high purity of steam [5].

Fig. 6.8

Tube failure due to hydrogen embrittlement. *Source: From Fig. 18, P 42–17, Chapter 42: Water and Steam Chemistry, Deposits and Corrosion. Steam Its Generation and Use (41st ed.). Courtesy of The Babcock & Wilcox Company.*

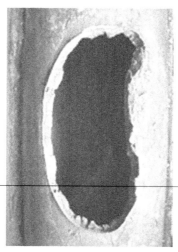

Fig. 6.9

Waterwall failure due to hydrogen embrittlement. *Source: From Fig. 9–19, P 9–17, Chapter 9: Water Technology. Clean Combustion Technologies, fifth ed., Alstom, Windsor, CT, 2009.*

Before undertaking chemical cleaning of deposits from internal surfaces of tubes of old steam generators, it is recommended to cut a representative section of suspected steam generator tube, collect deposits from it, and analyze these deposits in a laboratory. Result of such analysis would facilitate to determine which solvent shall be used to ensure successful removal of deposits.

Most of the internal deposits of a steam generator can be removed by employing either a mineral acid, or an organic acid or a chelant. While use of mineral acids would require the shortest time to complete the chemical cleaning process, using chelating agents may need longer time for completion [5]. From a safety point of view, mineral acids are more potent, have more potential to severely hurt plant and personnel, while organic acids and chelating agents are more eco-friendly. Copper deposits in tubes are generally cleaned by adding a suitable copper complexing agent, such as "thiourea," in the chemical solution [2].

Waste disposal following the chemical cleaning process is an area of concern. Either the plant shall have its own waste disposal plant or wastes should be dispatched through trucks and dumped at a place as accepted by environmental agencies. In either case, disposal of wastes must conform to guidelines laid down by local environmental regulations [4].

Against the above backdrop of the heterogeneous requirement and alternatives, prior to undertaking chemical cleaning of a steam generator, concerned personnel must first decide which solvent to choose.

One mineral acid that is commonly used for chemical cleaning of steam-water circuit is *5% hydrochloric acid* (HCl) *solution* suitably inhibited with Rodine. Inhibitor is used to prevent the acid from aggressively attacking the metal surface. This acid solution is particularly attractive in natural circulation steam generators, wherein complete circulation of solution is difficult to be ensured. Inhibited HCl solution in combination with *0.25% ammonium biflouride*, a.k.a. *ammonium fluoric acid* (NH_4HF_2), becomes very effective at removing silica deposits from steam generator tubes [5]. Advantage of using ammonium biflouride is that it promotes deposit penetration. HCl is also effective for removing calcium and iron deposits. HCl solution, however, has a strong tendency to cause stress corrosion in austenitic regions of the circuit (Figs. 6.10 and 6.11).

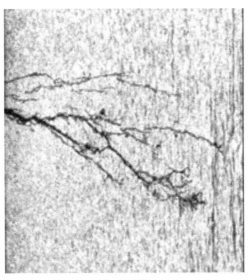

Fig. 6.10
Failure of tube resulting from stress corrosion. *Source: From http://www.corrosionlab.com/Failure-Analysis-Studies/Failure-Analysis-Images/20052.ClSCC. stainless-steel-Pipe/20052. branch-crack.jpg.*

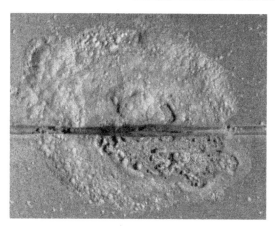

Fig. 6.11

Failure resulting from stress corrosion. *Source: From https://degradationeng.wordpress.com/2014/10/23/ forms-of-corrosion/.*

The inhibitor, used with HCl, remains bonded up to a finite time and must be drained out within this time, regardless if all deposits are removed or not, lest the inhibitor would break down, resulting in acidic attack on steam generator tubes [5]. If this situation arises before all deposits are removed, the cost of the chemical cleaning process would get increased as whole sequence of activities has to be repeated with additional dozing of chemicals for complete removal of deposits. The use of HCl can create problems on heavily deposited steam generators, as the acid often undercuts the deposits and causes sloughing of larger pieces of material [5]. Another drawback of using HCl is that instead of abating oxide fouling rates in boilers operating in two-shift, HCl yields unacceptable corrosion rates [6].

In view of the above, many utilities consciously avoid use of HCl as a means for chemical cleaning of steam generators, instead preferring *3–4% citric acid* (CA—a weak organic acid) $(C_6H_8O_7 - COOH \cdot CH_2 \cdot C(OH)(COOH) \cdot CH_2 \cdot COOH)$ *solution* even though the cost of citric acid is higher than the cost of HCl [6]. Citric acid is an excellent chelating agent. It is capable of removing mill scale from steam generators and evaporators. Citric acid is inhibited by dozing *ammonia* to maintain a pH of 3.5–4.0 [6].

One successful means to reducing oxide fouling rates in steam generators that operate mainly in two-shift operations is to use *inhibited 1–2% hydrofluoric acid* (HF) *solution* as a solvent for chemical cleaning because HF exhibits lower corrosion rates than HCl [6]. This solution is considered to be the fastest cleaning solvent to remove iron and silica deposits [5]. One advantage of using HF is its contact period is less than that of HCl. Waste treatment of HF cleaning is also relatively simple. The procedure involves neutralization of the spent acid with lime to adjust the pH to 6.5–9.0 (Neutralization results in conversion of the acid to calcium fluoride and iron hydroxide.) [5,7]. In Europe, HF has been in use for chemical cleaning for many years. In contrast, use of HF for chemical cleaning of steam generators is not very attractive in North America. Instead they prefer to use HCl in combination with ammonium biflouride; the reaction of these two chemicals yields a certain quantity of

HF that improves the iron removal process and facilitates solubilization of silicates in the deposits [7].

Another commonly used mixture of organic acids in the industry is *hydroxyacetic-formic acid (HAF)*, which is a combination of *2% hydroxyacetic acid*, a.k.a. *glycolic acid* ($HOCH_2COOH/C_2H_4O_3$) and *1% formic acid*, a.k.a. *methanoic acid* ($HCOOH/HCO_2H$). Inhibited HAF with *0.25% ammonium biflouride* is an effective iron removal solvent that exhibits low affinity for other deposit constituents. This solution is especially effective in steam generators having stainless steel components [5]. In supercritical and once-through steam generators, this solvent is used quite often. The advantage of using these acids is, in the event of incomplete flushing, they decompose into gases.

One successful alternative to citric acid is *4–6% ethylene-diamine-tetra-acetic acid or, in short, EDTA*, which is far more efficient than citric acid for scavenging metal ions during chemical cleaning of steam generators. EDTA is the most tolerant solvent and unlike HCl can remain in circuit for longer periods without jeopardizing the health of the tubes [5]. It is widely used to dissolve mill scales. EDTA facilitates solubilization of ferric ions and is able to sequester metal ions in aqueous solution. It is a good chelating agent. After being bound by EDTA, metal ions remain in solution but exhibit diminished reactivity. For pH control, EDTA should be inhibited with *ammonia*. In natural circulation steam generators, tetra-ammonium EDTA is generally used, while diammonium EDTA is preferred in forced-circulation steam generators [5]. The chemical formula of EDTA is $C_{10}H_{16}N_2O_8$, or $(HO_2CCH_2)_2NCH_2 \cdot CH_2N(CH_2CO_2H)_2$.

Two different methods are followed in the industry to carry out chemical cleaning of steam generators—the soaking or static method and the circulation method. Duration required for successful completion of chemical cleaning of steam generators using the circulation method is almost half that required for using the soaking or static method.

In the soaking method, areas to be chemically cleaned are preheated to a preset temperature then filled with hot solvent, kept in filled-up condition for a period of preset hours, as recommended by steam generator manufacturers, and thereafter drained completely. For complete removal of deposits, strength of the acidic solution needs to be higher than that required by actual conditions [1]. Duration of soaking period would depend on type of "acid" used for cleaning. If mineral acid is used, duration of soaking may vary from 4 to 12 h; in case of organic acid, the duration may extend from 12 to 48 h depending on the quantity of material to be cleaned [7]. This method is especially suitable in steam generators with heavy deposits or in steam generators where it is difficult to ensure positive circulation of solvent in all areas to be cleaned economically [1].

In the circulation method, the acid solution is circulated intermittently through the selected circuit by running chemical circulating pumps until the chemical content and the iron and copper content of the cleaning solution reach a constant value. This method is particularly suitable for cleaning once-through steam generators, superheaters, reheaters, and economizers

[1]. Time taken to complete the cleaning process in this method is expected to be 4–6 h if mineral acid is used, while use of organic acid may require 6–24 h based on amount of materials to be cleaned [7].

Areas that are to be included in the chemical cleaning circuit of drum-type steam generators are quite different from areas to be covered in once-through steam generators, as described in the following paragraphs.

6.1.1 Drum-Type Steam Generator

In drum-type steam generators, the chemical cleaning circuit comprises "temporary piping from the discharge of chemical circulating pump to downstream of economizer inlet feed check valve to boiler drum through economizer and from boiler drum through down comers, bypassing steam generator circulating pumps (if applicable), to furnace water walls back to boiler drum, wherefrom through temporary piping back to the temporary chemical dissolving tank" (Fig. 6.12).

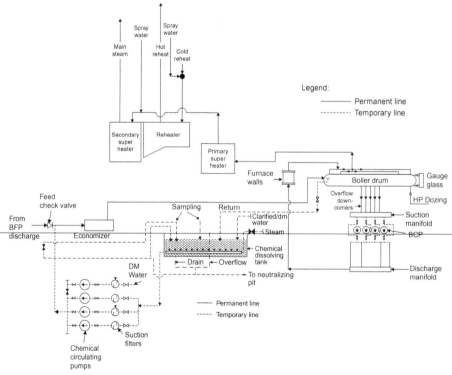

Fig. 6.12
Chemical cleaning of drum-type steam generator.

Superheaters are excluded from this circuit. Each connection from the boiler drum to superheaters shall be plugged prior to filling the superheaters full with ammoniated hydrazine water of pH 10 with 200 mg kg^{-1} concentration of hydrazine in order to prevent ingress of chemicals and/or chemical vapors to superheaters. Reheaters will remain automatically bypassed due to configuration of piping layout.

BFP discharge (BFD) pipe line of drum-type steam generators also is not considered for chemical cleaning as long as alkali flushing of this line is successfully completed. It is more so since the boiler drum is provided with a separate arrangement of chemical dozing to get rid of calcium and magnesium salts contained in feedwater, which are removed by boiler drum water blowdown. Purity of feedwater of a once-through steam generator, in contrast, essentially has to be as good as the required purity of steam.

6.1.2 Once-Through Steam Generator

Because of once-through configuration of these steam generators and of the requirement of high purity feedwater and steam, areas that are to be covered under this configuration are shown in Fig. 6.13. The following four circuits are taken into consideration for chemical cleaning of once-through steam generators:

a. "Temporary piping from the discharge of chemical circulating pumps to BFP discharge (BFD) line starting from downstream of BFP discharge valves up to the upstream of inlet valve of highest pressure (HP) feedwater heater, through the temporary line bypassing all HP feedwater heaters and connecting downstream of the outlet valve of lowest pressure HP feedwater heater, then through the normal BFD line up to the upstream of economizer inlet feed check valve bypassing feed control valve/s (if provided), then through temporary piping back to the chemical dissolving tank."

b. The second circuit takes care of "temporary piping from the discharge of chemical circulating pumps to the downstream of economizer inlet feed check valve, economizer, furnace waterwalls, separator, up to the upstream of the suction valve of steam generator circulating water pumps, then through temporary piping back to the chemical dissolving tank".

c. The third circuit comprises "temporary piping from the discharge of chemical circulating pumps to HP turbine exhaust then through cold reheat line, reheaters, hot reheat line up to IP turbine exhaust, wherefrom through temporary piping back to the chemical dissolving tank".

d. The last circuit starts from "temporary piping from the discharge of chemical circulating pumps to HP turbine exhaust then through HP bypass line, bypassing HP turbine and HP bypass control valves, through the main steam line, superheaters, separator up to the upstream of the suction valve of steam generator circulating water pumps through temporary piping back to the chemical dissolving tank."

Note

Circuits "c" and "d" may be considered for chemical cleaning if recommended by the steam generator manufacturer.

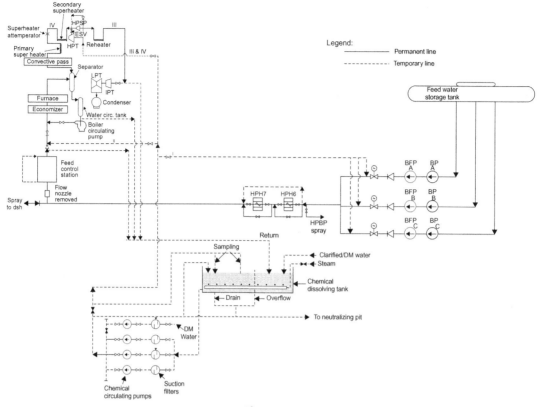

Fig. 6.13

Chemical cleaning of once-through steam generator.

Notes

BFP discharge valves, the inlet valve of HP feedwater heater and the outlet valve of lowest pressure HP feedwater heater, the isolation valves of feed control station, the suction and discharge valves of steam generator circulating water pumps, and HP bypass control valves must remain locked (closed) as long as the chemical cleaning process continues.

6.2 Precautions

The following specific precautionary measures, supplemental to "general precautionary measures" described under Section 2.1.1 in Chapter 2, must be observed before executing chemical cleaning of steam generators:

i.   Acid/alkali solution is injurious to health. When it comes in contact, acid/alkali solution may burn skin; acid fumes may cause injury to eyes. Hence, all probable safety measures shall be exercised by operating personnel while handling acid solution. They must keep on wearing acid/alkali-proof hand gloves, face mask, and protective clothing/apron as well as protective glasses for eyes when working with acids/alkalis and as long as the cleaning process is continued. If contact is made with acids, operating personnel must immediately flush with fresh water.

ii. During the execution of the chemical cleaning process, dirt and debris from various parts of the system get collected into the chemical dissolving tank and may in turn infiltrate and choke the suction filters of the running chemical circulating pumps. Choking of the suction filter could be ascertained by observing a fall in pump discharge pressure. On occurrence of such an event, the standby chemical circulating pump/s shall be put into service; then the running pump is to be stopped; its suction filter is to be cleaned and placed in position. This pump will then act as a standby pump.

iii. Electrical installation shall be at a safe distance from pumps, and care shall be taken to ensure protection against rain.

iv. In order to prevent explosion of hydrogen gas released during the execution of chemical cleaning, all welding work or any other hot work, eg, gas-cutting, and so on, around the steam generator shall be strictly prohibited.

v. Temporary degasifying equipment shall be installed in the drum (applicable to drum-type steam generators only), and gases collected shall be led out through a hose connection to a temporary collector. A special manhole cover is to be provided to accommodate degasifying equipment.

vi. All air vent valves shall be replaced with temporary threaded stubs. Hose pipes from these stubs shall be lead to a temporary collector.

vii. All the safety valves on the boiler drum (applicable to drum-type steam generators only) shall be removed and blank flanged. A small stub shall be welded to each blank flange. Hoses from these stubs shall be lead to a temporary collector.

viii. All the safety valves on superheaters and reheaters (applicable to once-through steam generators only) shall be removed and blank flanged. A small stub shall be welded to each blank flange, from which hoses shall be lead to a temporary collector.

ix. From the temporary collector of all hose pipes described earlier, gases will be lead to the chemical dissolving tank.

Note

In view of probable hazard of chemical solution to plant and personnel, it is essential to engage qualified personnel only, who would select appropriate chemical solvents, lay down the guidelines for cleaning, and be responsible in supervising the chemical cleaning process until the normalization of all circuits.

6.3 Prerequisites

Before undertaking the chemical cleaning process, a thorough inspection of systems under purview is essential to ascertain the following (Table 6.2):

Table 6.2 Areas/items to be checked

S. No.	Areas/Items	Ok (✓)
1	Before erecting the piping systems, protective coating used on internal surfaces as a safeguard against corrosion during transport and storage are removed as far as practicable	
2	Prior to erecting all temporary piping systems, they are thoroughly cleaned mechanically	
3	All piping systems, whether permanent or temporary, are erected and supported properly along with associated valves. Protocol signed jointly by the customer, the engineer, and the supplier (contractor), certifying completion of erection of piping systems, is verified	
4	The availability of protocol jointly signed by the customer, the engineer, the steam generator manufacturer, and the supplier (contractor) is verified, certifying successful completion of "hydraulic test of steam generator (Chapter 9)"	
5	All welded and flanged joints are secured	
6	Completion of "air tightness/leakage test of furnace, ESP and/or bag filter, FGD, the air and flue gas ducts (Chapter 10)" is verified, and relevant protocol jointly signed by the customer, the engineer, the steam generator manufacturer, and the supplier (contractor) is available	
7	Thermal insulation with proper lagging is provided on the furnace and air and flue gas ducts (Chapter 11)	

8	For once-through steam generators, thermal insulation with lagging is provided additionally on the following pipe lines:
	
	   i. main steam, cold reheat, and hot reheat steam lines
	   ii. BFP discharge line up to economizer inlet	

9	The availability of protocol jointly signed by the customer, the engineer, the steam generator manufacturer, and the supplier (contractor) is verified, certifying successful completion of "steam generator initial firing and drying-out of insulation (Chapter 11)"	
10	Steam generator circulating water pumps (BCPs), if provided, are isolated from the cleaning circuit by putting suction and discharge valve of pumps under locked closed condition. A temporary pipe line from the DM water supply line is laid and connected to the normal purge water connection of each steam generator circulating water pump motor cavity for onward discharge to the steam generator. Purge water flow through pump motors shall always be maintained as long as chemicals are inside the steam generator	
11	Normal N_2 blanketing system is made ready in all respects and is operational	
12	The disposal of acidic effluents from chemical cleaning systems directly to plant drainage systems is not permitted. Acidic effluents must be properly neutralized before their disposal in drainage systems conforming to local environmental regulations. Ensure that erection of a neutralization tank located nearby is complete	
13	Flow nozzles, if provided on feedwater line and main steam line for monitoring of flow during normal operation, are removed for carrying out chemical cleaning. Flow nozzle portions on respective piping are replaced with suitable spool pieces	
14	Feed control station control valves are kept closed, keeping their bypass valves fully open	
15	Instruments installed on the permanent piping systems are isolated	
16	Erection of temporary chemical cleaning pumps is complete	
17	Temporary tanks, for dissolving chemicals, along with associated fittings and attachments are erected	
18	Temporary piping fitted with temporary valves is erected	
19	Temporary instruments are in place	
20	Estimated quantity of DM water (Tables 6.5 and 6.6) is available	
21	Estimated quantity of chemicals (Table 6.7) is available	

Table 6.2 Areas/items to be checked—cont'd

S. No.	Areas/Items	Ok (✓)
22	Reagents (Table 6.8) required for analysis of effluent are available	
23	Apparatus required for chemical analysis (Section 6.3.1) are arranged	
24	Safety gadgets (Section 6.3.2) are arranged	
25	Verify that safety tags, eg, "no smoking," "danger," "keep off," and so on (in regional and English languages), are displayed wherever required	
26	Verify also that tanks, pumps, valves, associated equipment, and systems, which are in service or are energized, are provided with proper safety tags	

6.3.1 Typical List of Apparatus

Apparatus that are typically required for chemical analysis during "chemical cleaning of steam generator" are listed as follows. Size and quantity of each of these items would vary from project to project and shall have to be assessed by the chemist associated with cleaning activities.

1. titrating flasks
2. standard flasks
3. polyethylene bottles
4. measuring cylinders
5. pipettes
6. funnels
7. colorimetric cylinders
8. platinum crucible
9. photo colorimeter for determination of phosphate and silica
10. pH meter for determining pH of the solution
11. special pH papers of range 0–12
12. conductivity meter for determining conductivity of the solution
13. electric hot plate

6.3.2 Typical List of Safety Gadgets

Equipment that are generally used for safety of operating personnel are as follows:

1. gum boots of assorted sizes
2. rubber gloves
3. asbestos gloves
4. rubber or polythene aprons
5. hard hats of assorted sizes
6. side-covered safety goggles with plain glass
7. gas masks
8. first aid box
9. safety sign boards
10. safety tags

6.4 Preparatory Arrangements

a. A temporary tank will be required for dissolving chemicals for carrying out the chemical cleaning process. This tank will be an atmospheric tank and will serve as the suction tank for temporary chemical circulating pumps. Chemical solutions supplied by the circulating pumps shall drain back to this tank, thus establishing a continuous circulation. The tank shall be provided with suitable overflow, drain and sampling connections, vent outlets, and so on. Since it would be prudent to use the same tank for carrying out "alkali flushing of preboiler system" (Chapter 3) and "chemical cleaning of steam generator," this tank may be located at a convenient place between the steam generator and the TG hall at an elevated level from the ground.

b. This tank shall also be provided with a bubbler tube for admitting heating steam into the solution. For an extension unit, this steam may be sourced from an existing auxiliary steam system; for a grass root unit, steam may be supplied from a temporary saturated steam-producing steam generator.

   (Alternatively, temperature of clarified water may be raised by continuous recirculation of water, keeping all drain valves closed. This procedure, however, is time consuming and may extend up to 8 h.)

c. Four temporary chemical circulating pumps, each of 50% capacity, with suction filters shall be installed at the ground level just outside the space below the chemical dissolving tank. All four circulating pumps shall remain functional during chemical cleaning of the steam generator. While three of these pumps will be dedicated for circulation of chemicals (this arrangement would ensure uninterrupted continuation of the flushing/cleaning process even in the event of failure of one of the three running chemical circulating pumps); the remaining one pump will be used to backfill and pressurize idle circuits with DM water.

d. To make the pumps suitable for handling acids, the interior of pump casings shall be coated with acid-resistant epoxy paint. The impellers of these pumps may be made of brass.

e. The area surrounding the chemical dissolving tank, circulating pumps should be adequate so that any spillage of chemicals from the system does not damage any permanent equipment.

f. In order to complete the circuit for circulation of flushing fluid, some temporary piping connecting the chemical dissolving tank and the permanent piping systems should be erected and kept ready for putting into service.

g. It is apparent that while carrying out chemical cleaning, some temporary valves, pressure gauges, thermometers, and so on, are required to be installed.

Notes

1. These temporary connections are unit specific and vary from project to project. Hence, layout of temporary piping shall have to be jointly decided at site among the customer, the engineer, the steam generator manufacturer, and the supplier (contractor) responsible for successful completion of the above cleaning activity.

2. Size and quantity of temporary piping materials including elbows, tees, valves, supports, and so on, shall be assessed at the design stage. These requirements, however, may vary based on the exact location of the tanks and pumps at site as well as the location of terminal points of the sources of electric power supply, clarified water and DM water, and the location of effluent discharge and draining points.
   Final quantity of piping materials including associated equipment shall be procured accordingly.
3. To facilitate operation of temporary valves and monitoring of pressure and temperature of circulating fluids, temporary platforms may have to be erected as required.

h. Quantity of chemicals, reagents, various apparatus, and safety gadgets, to be required for the process, should be ascertained beforehand and an adequate quantity of them must be kept available.

i. An adequate quantity of clarified water and DM water must be assessed prior to undertaking the process and shall be made available.

j. The following instruments shall be installed:
   - pressure gauges at the discharge of temporary chemical circulating pumps
   - a thermometer at the temporary chemical dissolving tank

Notes

Chemicals generally required for alkali boil-out are trisodium phosphate, Na_3PO_4, supplemented with disodium phosphate, Na_2PO_4, and/or caustic soda, $NaOH$. Detergent used during this process is "Teepol" or "Lissapol" or "Dodenol."

For executing the chemical cleaning process, chemicals required are inhibitor, eg, Rodine 213, Stannine LTP, and so on; ammonium biflouride and acid considered for chemical cleaning, ie, HCl/HF/HAF/CA/EDTA; and so on.

6.5 Operating Procedure

Depending on the configuration of the chemical cleaning system layout, the system may be split into one or more circuits for the convenience and effectiveness of the cleaning process as already discussed.

While carrying out chemical cleaning of each circuit, the process is sequentially divided into the following activities. This will ensure reducing total time of completing the chemical cleaning process successfully and economizing total requirement of chemicals used:

   i. water flushing of the system
   ii. alkali boil-out of the system
   iii. rinsing after (ii)
   iv. filling the circuit with DM water and dozing acid with inhibitor
   v. soaking with acid solution or circulation of acid solution

vi. rinsing after (v)

vii. first-stage passivation

viii. second-stage or final passivation

ix. inspection

i. Water flushing of the system

Water flushing is required to get rid of dust and other loose materials accumulated in the system and to remove any volatile protective material used during storage.

Admit DM water (alternatively clarified water may be used to economize) to the temporary chemical dissolving tank. Start one temporary chemical circulating pump. Carry out a hydraulic test of the selected circuit, including temporary pipe lines at 1.5 times the maximum operating pressure of the chemical circulating pump to ascertain that there is no leakage from the system. On successful completion of the hydraulic test, start the process that involves circulation of water from the discharge of the pump through the selected circuit and temporary piping back to the tank. When steady flow is established through the circuit, turn on the heating steam to the tank to raise temperature of water to about 353–358 K. Such heating increases effectiveness of the flushing process. Maintain circulation of water. Take samples from the return line to check the turbidity of water at an hourly interval. In the event turbidity is excessively high, drain a portion of the return water and fill the tank with a fresh supply of water. Continue this circulation process until the samples from the return line are noted to be visually clear. Drain the circuit completely.

The above procedure may be followed in each of the circuits mentioned earlier.

ii. Alkali boil-out

On completion of water washing, alkali boil-out of each circuit is taken afoot to get rid of grease and oil from internal surfaces. Alkali boil-out also facilitates in loosening and removing metal oxides, thus assists chemical cleaning to be more effective.

Fill the chemical dissolving tank with a fresh supply of DM water up to about the middle of the tank. Start pumps and maintain steady circulation. Supply steam and raise the temperature of the water to 343–353 K. Start dozing solution of 1000 mg kg^{-1} of trisodium phosphate, Na_3PO_4, and 500 mg kg^{-1} of disodium phosphate, Na_2PO_4, and/or caustic soda, $NaOH$, in the dissolving tank. Feed also detergent in the tank. During feeding of detergent, foaming will occur, hence detergent shall be fed in stages.

(Alternatively, the above chemicals—1000 mg kg^{-1} of Na_3PO_4 and 500 mg kg^{-1} of Na_2PO_4—may be dissolved in buckets and dozed along with detergent directly into the boiler drum.)

As the circuits of drum-type steam generators are quite different from the circuits of once-through steam generators, alkali boil-out of each type of steam generator is discussed separately under the following paragraphs.

6.5.1 Drum-Type Steam Generators

In drum-type steam generators, prior to carrying out alkali boil-out and chemical cleaning, in order to prevent ingress of chemicals and chemical fume from drum to superheaters, suitable plugs will be provided on each connection from the boiler drum to superheaters.

Start chemical circulating pump/s and fill up the steam generator to the normal operating level of the boiler drum. Stop further dozing of alkaline solution.

Light-up the boiler with oil firing. Initial firing shall be continued at a slow rate, complying with the guideline of the cold start-up curve recommended by the steam generator manufacturer. Gradually raise pressure up to 4 MPa and maintain this pressure for about 24 h. Operate blow-down valves every hour for a period of 10–60 s depending on the quality of the effluent. Take samples for analysis every 2 h or so, and note the values of pH, alkalinity, phosphate, and oil. Once the oil content of effluent falls below 5 mg kg^{-1} stop the boil-out process. Box up the steam generator; allow drum pressure to decay gradually. Once pressure reaches 0.3 MPa, the steam generator may be drained to neutralizing pit for further disposal in the sewage system.

iii. Rinsing (drum-type steam generators)
      Admit a fresh supply of DM water into the tank, start pumps, and establish flow through the circuit. Samples of water shall be collected from the effluent and checked for alkalinity. The process of circulation will continue until alkalinity of the effluent gets eliminated and phosphate content falls to about 50 mg kg^{-1}. This completes the rinsing process.
      On completion of rinsing, the following additional preparatory arrangements are to be carried out for chemical cleaning:
      • drum internals to be removed
      • metal orifice plates to be installed in each of the connecting tubes from boiler drum to downcomers
      • nitrogen blanketing arrangement to be provided by installing nitrogen bottles and laying piping from these bottles to drum vent connections with suitable isolating valves

Note

Steam generators in which operating pressure of steam is less than 6 MPa, cleaning by alkali boil-out followed by rinsing, would suffice; it may not be required to carry out further chemical cleaning of the steam generator.

6.5.2 Once-Through Steam Generator

In once-through steam generators, the area to be taken under alkali boil-out is restricted to economizer and furnace waterwalls. The main steam line and superheater will be back-filled with DM water containing 100 mg kg^{-1} concentration of hydrazine and kept pressurized in order to obviate ingress of chemical vapors to the superheater.

Start chemical circulating pump/s and fill up the economizer, furnace water walls, and separator up to a specified level below its center line as recommended by the steam generator manufacturer. Stop further dozing of alkaline solution. Maintain the temperature of chemical solution at 343–353 K as mentioned above. When the circulation is in progress, open blow-down valves for draining of solution from the system in order to maintain the level in the separator as specified. Take samples for analysis every hour or so, and note the values of pH, alkalinity, silica, phosphate, and oil. Once the oil content of effluent falls below 5 mg kg^{-1} stop circulating pumps.

In the event that steam generator manufacturers do not recommend to chemically clean circuits (c) and (d) of Section 6.1, these circuits are to be backwashed with DM water of 100 mg kg^{-1} concentration of hydrazine (N_2H_4). The backwashing process will continue until pH of samples collected from SH/RH drain headers exceeds 9. One of the four chemical circulating pumps may be dedicated to backwashing, while a separate temporary tank is to be installed for DM water storage.

Open vent valves and drain the steam generator.

iii. Rinsing (once-through steam generator)
      Admit a fresh supply of DM water into the tank, start pumps, and establish flow through the circuit. Samples of water shall be collected from the effluent and checked for alkalinity. The process of circulation will continue until alkalinity of the effluent gets eliminated and phosphate content falls to about 10 mg kg^{-1}. This completes the rinsing process.
      On completion of rinsing, the nitrogen blanketing arrangement is to be provided by installing nitrogen bottles and laying piping from these bottles to superheater and reheater vent connections with suitable isolating valves.

iv. Filling the circuit with DM water and dozing acid with inhibitor
     Fill the steam generator and associated system with DM water. Reduce water in the chemical dissolving tank to maintain the level in the tank at around minimum position. Admit steam to this tank, and ensure that the temperature inside the tank is maintained at around 338–368 K, as recommended by the steam generator manufacturer, until the completion of the chemical cleaning process. First, charge the inhibitor in the tank and

circulate the solution; check for the presence of inhibitor in the return line. Next, charge ammonium biflouride in the tank and circulate the solution.

Men should be posted to check whether welding joints in the steam generator and associated system are in order. In the event no leakages are noticed, acid may be charged. Ensure that charging of all chemicals is completed within 60–90 min. Also ensure that concentration of acid does not exceed the prescribed limit; for example, if HCl is used, its concentration shall be 5% or so, while concentration of CA shall be about 3%. The mixture of acid and ammonium biflouride should accelerate the removal of silicate-containing scales.

v. Soaking with acid solution

If the soaking method is adopted, the acidic solution would remain static within the system for a specified period of time. Samples of the effluent shall be collected by opening the steam generator drain valves every 1–2 h as recommended by the steam generator manufacturer, to analyze the concentration of Fe^2, Fe^3, acid strength, silica, oil, and pH. Or,

Circulation of acid solution

In the circulation method, the acid solution is intermittently circulated through the selected circuit by running the chemical circulating pumps until the chemical content and the iron and copper content of the cleaning solution reach a constant value. Take samples from the discharge of the temporary chemical circulating pump/s, every 1–2 h, as recommended by the steam generator manufacturer, to analyze the concentration of Fe^2, Fe^3, acid strength, silica, oil, and pH.

In either of the method, as the steam generator internals remain in contact with the acidic solution, Fe content and silica content in samples will gradually rise, while acid strength would start falling, thereby increasing the pH of the solution. However, at the end of the completion period mentioned previously, both Fe content and acid strength will not change further and will become stable, marking the end of the pickling (acid cleaning) process.

vi. Rinsing

Once pickling is completed, the acid solution is drained to the neutralizing pit following the regulations of local environmental agencies; the system is flushed with a fresh supply of DM water, which remains in continuous circulation until it is free of solvent and soluble iron. Check the pH and iron content of the sample from the return line. As long as the pH remains below 7 and the iron content is above 50 mg kg^{-1} keep on draining flushed water; repeat the filling, circulating, and draining process with fresh DM water. Stop the flushing or rinsing process once pH of about 7 and iron content $<$50 mg kg^{-1} are reached, indicating that traces of all acid and iron have been removed. Check also that conductivity of the return water is at 20 μS cm^{-1} or less. In the event of higher conductivity of the return water, replenish the system with DM water and fill the whole circuit. Maintain circulation for about an hour. Drain water from the system completely under nitrogen blanketing.

It must be strictly followed to *keep nitrogen blanketing "on" at a nitrogen pressure of 20–50 kPa whenever the system is drained.*

Note

Further cleaning of the system with a fresh supply of DM water, acid, and chemicals to flush traces of Fe remaining in the system may be carried out if recommended by the steam generator manufacturer. The requirement of additional circulation of chemicals is project/manufacturer specific and may not be applicable to the cleaning of all steam generators.

Passivation

After rinsing, passivation of acid cleaning surfaces is necessary to preserve the freshly exposed surfaces, which are susceptible to oxidation vis-à-vis corrosion. This is ensured by coating a corrosion-resistant layer of magnetite on the inside surfaces of tubes/pipe lines.

If a mineral acid solvent is used, a boil-out should follow to repassivate the metal surface. If a chelant is used, oxygen injection into the cleaning solution would suffice to provide passivation.

Passivation is done in two stages as discussed as follows.

vii. First stage passivation

Refill the system with DM water. Raise the temperature of the liquid to about 363 K and circulate the hot water. Doze liquid ammonia to raise the pH of the water to 10. Hydrazine is charged then to maintain a uniform concentration of 200 mg kg^{-1}. Circulate the water for a period of 12–24 h as per steam generator manufacturer's recommendation. Stop chemical circulating pumps followed by completely draining the steam generator.

In the case of once-through steam generators, prior to draining of neutralization solution, main steam line, and superheaters may be purged with DM water, dozed with ammonia and 50 mg kg^{-1} hydrazine, through the temporary backwash pump. Ammonia is dozed to elevate pH of DM water to 10. Samples may be collected from the superheater drain header every hour or so to analyze the pH of the effluent. The purging process will continue until the pH of the effluent becomes 9.0–9.5.

Thereafter, the system is allowed to aerate for a specified period as recommended by the steam generator manufacturer. The steam generator may now be boxed up in a dry condition under nitrogen blanketing or may be stored under a wet lay-up condition. All temporary connections may now be removed and permanent piping and equipment should be restored. The steam generator is normalized for regular operation. Suction and discharge valves of the steam generator circulating pumps of both drum-type and once-through may be normalized and kept open. Boiler-feed pumps with associated booster pumps shall also be normalized and brought to operative condition. All bypass connections of feedwater heaters and control valves shall be normalized and associated equipment brought into service.

In drum-type steam generators, prior to proceeding with the second-stage passivation, plugs provided on each connection from the boiler drum to the superheaters, as well as the metal orifice plates installed in each of the connecting tubes from boiler drum to downcomers, shall be removed. Also, drum internals shall be thoroughly cleaned mechanically prior to installing them in respective positions.

viii. Second-stage or final passivation

Second-stage passivation shall be completed within the recommended period from completing the first-stage passivation.

Fill the deaerator feedwater storage tank with DM water. Doze liquid ammonia and hydrazine to this feed tank. While dozing of ammonia will raise the pH of the water to 10, uniform concentration of hydrazine will be maintained at $200\,mg\,kg^{-1}$. Start the boiler feed pump and fill the boiler drum/separator up to its normal operating/ start-up level. Light up the steam generator with oil burners. Firing shall be continued at a slow rate, complying with the guideline of the cold start-up curve recommended by the steam generator manufacturer. The firing rate thereafter may be raised following the start-up curve to bring up the steam generator pressure to about 4 MPa. Maintain this pressure for 12–24 h as recommended. Check samples every hour to ensure that pH does not fall below 10. If required, doze additional ammonia to maintain pH level.

Stop firing and allow pressure to decay to about 0.3 MPa. Open vent valves located on boiler drum (applicable to drum-type steam generators) and superheaters. Monitor fluid temperature, whenever this temperature drops to about 363 K drain the steam generator and allow it to aerate.

With the completion of the second-stage passivation, chemical cleaning of the steam generator may be declared to be complete.

ix. Inspection

On completion of chemical cleaning, it is necessary to inspect the steam generator to verify whether expected results have been achieved. An age-old practice is to randomly cut one or two pieces of samples from waterwalls and superheater for visual inspection. A bluish gray look of the inner surface reveals good cleaning. These cut areas may be replaced again to make the steam generator ready for operation. Instead of cutting, boroscope inspection of the tubes may indicate also how successful the cleaning is.

All surfaces of headers, pipelines, and drums (applicable to the drum-type steam generator) shall be free of any pitting. The magnetite layer on tube surfaces left after second-stage passivation shall be uniform. Presence of loose particulate matter inside drums/separators and headers could be verified by visual inspection. Accumulation of any such matter would call for additional flushing of the system.

6.6 Estimation of DM Water and Chemicals

Prior to chemical cleaning of steam generators, it is essential to estimate the requirement of DM water and chemicals correctly. It is advisable to procure higher quantity of these items than estimated, as any shortfall of any of these items during execution of the process will jeopardize whole sequences of activities and may become detrimental to areas falling within this process. To facilitate estimation of each of above items, it would be prudent to assess the following beforehand.

6.6.1 Water-Holding Capacity

Tables 6.3 (for drum-type steam generators) and 6.4 (for once-through steam generators) identify the areas for which water-holding capacity shall be assessed beforehand. Actual volume of each area is project specific and has to be supplied by the steam generator manufacturer.

Table 6.3 Areas of the drum-type steam generator for which water-holding capacity is assessed

Item	Volume (m^3)
HP feedwater pipe line starting from the downstream of the boiler-feed check valve to the economizer inlet	
Economizer	
Boiler drum	
Downcomers	
Furnace waterwalls	
Temporary piping	
Temporary chemical cleaning tank	
Superheater	

Table 6.4 Areas of the once-through steam generator for which water-holding capacity is assessed

Item	Volume (m^3)
Deaerator feedwater storage tank	
HP feedwater pipe line, starting from downstream of boiler-feed pumps up to economizer inlet	
Economizer	
Furnace waterwalls	
Separator	
Superheater	
Main steam pipe line	
Reheater	
Cold reheat pipe line	
Hot reheat pipe line	
Temporary piping	
Temporary chemical cleaning tank	

6.6.2 Estimation of DM Water Requirement

Areas wherein DM water will be required are shown in Tables 6.5 (for drum-type steam generators) and 6.6 (for once-through steam generators).

Table 6.5 Areas of the drum-type steam generator where DM water will be required

Activity	Areas to Be Considered	Quantity (m^3)
Filling with DM water Water flushing of the system	Superheater HP feedwater pipe line (starting from downstream of the boiler-feed check valve to the economizer inlet), economizer, boiler drum, downcomers, furnace waterwalls, and all temporary pipe lines	
Alkali boil-out	HP feedwater pipe line, economizer, drum, downcomers, and furnace waterwalls	
Rinsing	HP feedwater pipe line, economizer, boiler drum, downcomers, and furnace waterwalls	
Boiler filling and preheating Acid cleaning Rinsing First-stage passivation and neutralization Second-stage passivation	HP feedwater pipe line, economizer, drum, downcomers, and furnace waterwalls	

Table 6.6 Areas of the once-through steam generator where DM water will be required

Activity	Areas to Be Considered	Quantity (m^3)
Water flushing of the system	HP feedwater pipe line, economizer, furnace waterwalls, separator, superheater, main steam pipe line, reheater, cold reheat pipe line, hot reheat pipe line, and all temporary pipe lines	
Alkali boil-out Rinsing	Economizer, furnace waterwalls, and separator	
SH back-filling	Main steam line and superheater; if recommended by the steam generator manufacturer include reheater, cold reheat pipe line, and hot reheat pipe line	
Boiler filling and preheating	HP feedwater pipe line, economizer, furnace waterwalls, and separator	
Acid cleaning Rinsing	HP feedwater pipe line, economizer, furnace waterwalls, and separator; if recommended include separator, main steam pipe line, superheater, reheater, cold reheat pipe line, and hot reheat pipe line	
First-stage passivation and neutralization	HP feedwater pipe line, economizer, furnace waterwalls, and separator, main steam pipe line and superheater; if recommended include reheater, cold reheat pipe line, and hot reheat pipe line	
Second-stage passivation	Deaerator feedwater storage tank, HP feedwater pipe line, economizer, furnace waterwalls, and separator; if recommended include separator, main steam pipe line, superheater, reheater, cold reheat pipe line, and hot reheat pipe line	

6.6.3 Estimation of Chemical Requirement

During execution of each of the activities discussed above, various types of chemicals, applicable to each activity, shall have to be used. Actual amount of these chemicals shall be assessed beforehand and procured accordingly. Table 6.7 summarizes the typical list of such chemicals.

Table 6.7 List of chemicals

Activity	Name of Chemical	Quantity (kg)
Alkali boil-out	Trisodium phosphate, Na_3PO_4, of concentration 1000 mg kg^{-1} Disodium phosphate, Na_2PO_4, of concentration 500 mg kg^{-1} Caustic soda, NaOH Detergent: "Teepol" or "Lissapol" or "Dodenol"	
Backwash with DM water of main steam pipe line, superheater, separator, cold reheat pipe line, reheater, and hot reheat pipe line (if these areas of the once-through steam generator are excluded from chemical cleaning)	Hydrazine, N_2H_4, of concentration 100 mg kg^{-1}	
Chemical cleaning	Inhibitor: Rodine 213, Stannine LTP, and so on Acid: HCl/HF/HAF/CA/EDTA Ammonium biflouride, NH_4HF_2, of concentration 0.25% Nitrogen cylinder	
First-stage passivation	Liquid ammonia to raise pH of water to 10 Hydrazine of concentration 200 mg kg^{-1}	
Purging of main steam line and superheaters of once-through steam generators following first-stage passivation	Liquid ammonia to raise pH of water to 10 Hydrazine of concentration 50 mg kg^{-1}	
Second-stage passivation	Liquid ammonia to raise pH of water to 10 Hydrazine of concentration 200 mg kg^{-1}	

In addition to the above, other chemicals that are required for the analysis of effluent are shown in Table 6.8. Type and quantity of these chemicals are to be assessed and procured by the chemist prior to proceeding with the cleaning/flushing process.

Table 6.8 List of reagents

S. No.	Reagent
1	Reagents for the determination of alkalinity in the degreasing solution
2	Reagents for the determination of acid in the pickling solution
3	Reagents for the determination of Fe^2, Fe^3 ions in the pickling solution
4	Reagents for the determination of SiO_2 content in the pickling solution
5	Reagents for the determination of N_2H_4
6	Any other chemical/reagent as required for complete analysis of effluent coming from various activities associated with the alkali flushing/chemical cleaning process

6.7 Conclusion

The steam generator is now ready to be lit up for executing the next stage of preoperational activity, ie, steam blowing of the MS, CR, HR, and other steam pipe lines.

A joint protocol may be signed by all concerned (Table 6.9).

Table 6.9 Protocol (chemical cleaning of steam generator)

Type of Boiler	Process	Results of Chemical Analysis	Remarks/Inference
Drum	Alkali boil-out	pH, alkalinity, oil, and PO_4 (samples taken from boiler blow-down valve every 2 h)	Once the oil content of effluent falls below 5 mg kg^{-1}, stop the boil-out process
Drum-type steam generator	Rinsing after alkali boil-out	pH and phosphate (samples taken from the effluent)	Once alkalinity gets eliminated and phosphate content falls to about 50 mg kg^{-1}, stop the rinsing process
	Dozing of chemicals	Concentration of chemicals shall conform to the type of chemicals used as given in Section 6.1	
	Pickling	Fe^2, Fe^3, acid strength, silica, oil, and pH (samples from the discharge of pump every 20–30 min)	Fe content and acid strength will become stable, marking the end of the pickling process
	Rinsing after pickling	pH, conductivity, and iron content (samples taken from the return line)	Stop the rinsing process once pH becomes about 7 and iron content is <50 mg kg^{-1}
	First-stage passivation	pH and hydrazine (samples taken from inlet to and outlet from the system every hour)	Circulate the water for 12–24 h as per the manufacturer's recommendation
	Second-stage passivation	pH and hydrazine (samples taken from the inlet to and outlet from the system every hour)	Circulate the water with pH at 10 for 12–24 h
	Inspection		Cut samples from waterwalls; a bluish-gray inner surface reveals good cleaning
Once-through steam generator	Alkali boil-out	pH, alkalinity, silica, phosphate, and oil (samples taken from boiler blow down valve every 2 h)	Once the oil content of effluent falls below 5 mg kg^{-1}, stop the boil-out process
	Rinsing after alkali boil-out	pH and phosphate (samples taken from the effluent)	Once alkalinity gets eliminated and phosphate content falls to about 10 mg kg^{-1}, stop the rinsing process
	Dozing of chemicals	Concentration of chemicals shall conform to type of chemicals used as given in Section 6.1	

Continued

Table 6.9 Protocol (chemical cleaning of steam generator)—cont'd

Type of Boiler	Process	Results of Chemical Analysis	Remarks/Inference
	Pickling	Fe^2, Fe^3, acid strength, silica, oil, and pH (samples from the discharge of pump every 20–30 min)	Fe content and acid strength will become stable marking the end of pickling process
	Rinsing after pickling	pH, conductivity and iron content (samples taken from the return line)	Stop the rinsing process once pH becomes about 7 and iron content is <50 mg kg^{-1}
	First-stage passivation	pH and hydrazine (samples taken from inlet to and outlet from the system every hour)	Purging continues until pH of the effluent becomes 9.0–9.5
	Second-stage passivation	pH and hydrazine (samples taken from inlet to and outlet from the system every hour)	Circulate the water with pH at 10 for 12–24 h
	Inspection		Cut samples from waterwalls; a bluish-gray inner surface reveals good cleaning
Signed by the customer	Signed by the engineer	Signed by the steam generator manufacturer	Signed by the supplier (contractor)

References

[1] S.C. Stultz, J.B. Kitto, Steam Its Generation and Use, 41st ed., The Babcock and Wilcox Company, Barberton, OH, 2005.

[2] GE Power and Water, Water and Process Technologies. Chemical Cleaning of Steam Generator Systems.

[3] GE Power and Water, Water and Process Technologies. Boiler Deposits: Occurrence and Control.

[4] C. Bozzuto, Clean Combustion Technologies, fifth ed., Alstom, Windsor, CT, 2009.

[5] D. Daniels, Boiler chemical cleaning: doing it correctly, Power (January) (2014).

[6] British Electricity Institute, Modern Power Station Practice, Vol. H: Station Commissioning, third ed., Pergamon Press, London, 1992.

[7] K.J. Shields, Chemical cleaning of fossil power station steam generators; past, present and future, in: 14th International Conference on the Properties of Water and Steam in Kyoto.

CHAPTER 7

Flushing of Lube Oil Piping System

Chapter Outline

Thermal Power Plant. http://dx.doi.org/10.1016/B978-0-08-101112-6.00007-1

7.1 Introduction

In 1991, the Electric Power Research Institute (EPRI) reported that bearing failures in power plant rotating equipment cost the electric utility an estimated $200 million per annum [1].

A study by General Electric Company (GE) revealed that failure in the lube oil system is the major contributor to forced outages of a turbine [2]. Reasons attributed to bearing failures are shaft misalignment, vibration, and fluid contamination. While shaft misalignment can be tackled by taking utmost care during erection, which eventually will be able to take care of any vibration problem to a great degree, it is fluid contamination that needs special attention. A study by Franklin Research Center revealed that "contamination in lube oil systems contributed to 54% of bearing-related forced outages" [3].

Particulate contamination may be of various types; of these, the following abrasive type particles are accountable to cause wear-related damages [3]:

  i. weld beads and slag
 ii. work-hardened journal material
iii. metal cuttings, shavings, grindings
 iv. grit from abrasive paper and sand blast material
  v. fly ash and coal dust
 vi. glass and sand

Over and above particulate contaminants, water can exist in solution at varying concentrations depending on the compounding of the lube oil, the temperature, and so forth. For example, lube oil may contain 50 ppm of water at 294 K and 250 ppm at 364 K. In general, under normal operating conditions, dissolved water has no significant adverse effect on lubricating properties of the oil and does not cause significant corrosion in these systems. Yet there are reports that dissolved water has caused bearing damages [4]. The most detrimental effect on lubricating properties, however, is caused by free water (portion of total moisture content in lube oil above its saturation point) [3, 4].

As explained earlier, reliability of prime-movers used in thermal power plants—steam turbines, gas turbines, and diesel engines, greatly depends on the use of lubricating oil free from detrimental contaminants and free water. In new units, the emphasis is on the removal of contaminants introduced during manufacture, storage, field fabrication, and installation. In running units, it is essential to remove contaminants that are generated during operation arising from malfunctions within the lube oil system or contaminants that are introduced during overhaul, or both.

During erection of lube oil piping, precaution should be taken to keep off dirt, dust, and other undesirable debris from the lube oil system as far as practicable. On completion of assembly, the whole lube oil circulation system, including bearing housings, oil tank, oil pipes, coolers, filters, gear boxes, and so on, shall be cleaned manually, and if required chemically, as thoroughly as possible. In cases where pickled (chemically cleaned) lube oil pipe lines are supplied to the plant by the prime mover manufacturer, the entire piping system has to be water washed before normal oil flushing is begun. The purpose of undertaking a water wash is to get rid of the protective coating, which is applied to the inner surface of lube oil pipe lines after completing chemical cleaning of these lines at manufacturer's works.

Particulate contaminants of size 5 μm or smaller can be removed by centrifuging the lube oil. From lube oil of low viscosity with appropriate throughput, centrifuge can also reduce water to less than 0.5%. However, it is recommended by ASTM that prior to putting a lubrication system in service, it must be flushed with an oil to remove these undesirable elements [4].

The flushing process discussed in this chapter is applicable to the lube oil piping system of all prime-movers mentioned earlier. The process entails that the entire lube oil system is flushed exhaustively by circulation of oil of same grade as the normal lube oil (Table 7.1), as recommended by the prime mover manufacturer, by running a starting oil pump or auxiliary oil pump/s and/or by temporary flushing oil pump if that is the requirement of the equipment manufacturer. A successful flush ensures that system piping and components meet acceptance criteria per original manufacturer of prime movers, within a minimum length of time and a minimum effort.

The advantage of using the same grade of circulation oil as the grade of the normal lube oil is that oil left over in the system will not be of inferior quality as that of the new charge of lube oil. Another advantage is the antirust additive in the flushing oil gets absorbed in the metal surfaces of the lube oil circulation system, giving excellent protection against corrosion once the system is cleaned by the oil flushing process.

Table 7.1 Typical analysis of lube oil [5]

Parameter	Unit	Normal Condition (Maximum)
Specific weight at 288 K	$kg\ m^{-3}$	868
Kinematic viscosity at 313 K	$mm^2\ s^{-1}$	41.4–50.6
Flash Point	K	>458
Pour point	K	<252
Total acid number	$mg\ KOH\ g^{-1}$	0.1
FZG, fail load stage	–	>10
Turbine oil stability	h	>10,000
Moisture content	$mg\ kg^{-1}$	<100 [6]

Note: $1\ mg\ kg^{-1} = 1\ ppm = 0.0001\%$.

In order to provide effective lubrication of a lube oil system, the lube oil must possess the following properties (Table 7.2) [5].

Table 7.2 Typical properties of lube oil

S. No.	Property
1	Outstanding thermal and oxidation stability
2	Prevention of sludge formation
3	Low oil degradation
4	High antiwear capability
5	Good load carrying capacity
6	Excellent water separability
7	Resistance to formation of water emulsion
8	Superior rust and corrosion inhibition
9	Low foaming tendency
10	Good air release properties
11	Resistance to chemical degradation to provide excellent equipment protection
12	Reliable operation
13	Reduction of downtime
14	Extended service life

Before going ahead with the discussion on flushing the lube oil piping system, certain equipment and guidelines common to lube oil systems of all prime movers are presented in the following paragraphs for the convenience of readers.

7.1.1 Purpose and Type of Lubrication [6]

Lubrication of prime-movers is required to minimize rotating friction in journal bearings, thrust bearings, and reduction gears, as well as to cool the journal and other bearing surfaces. Heat from hot parts of prime-movers is also transferred from the shaft to bearings. Heat thus transmitted is carried away by lube oil, thereby maintaining bearing temperature at a safe level. Depending on the size and type of prime-movers and also on their age, the lubricating oil systems vary widely and may fall into either one or a combination of "ring oiling" and "circulating (pressure or gravity type)" system.

7.1.1.1 Ring oiling

In small size prime-movers, the oil quantity required for lubrication is little, and the supply system is relatively simple. Bearings in this type of lubrication system comprise a flat ring (item B in Fig. 7.1), of considerable larger inside diameter than the diameter of the journal (item C), rides loosely on the journal, and its lower part dips into the oil reservoir. As the shaft rotates, the ring revolves; as a result, sufficient quantity of oil from the reservoir clings to the ring and is carried to the top of the journal to which it transfers oil by direct contact. Grooves in the bearing permit the oil to travel in axial directions and so provide proper lubrication at bearing areas along the length of the journal. Excess oil drains back to the reservoir beneath the bearing and aids in cooling the journal and bearing. Fresh supply of oil may be introduced to the sump through the inlet opening (item D). In small machines, the surface areas of the reservoir are large enough to radiate this accumulated heat and maintain the oil at a safe temperature. For some machines, it is necessary to cool the oil by passing circulating water through a chamber adjacent to the reservoir or through a coil immersed in the oil.

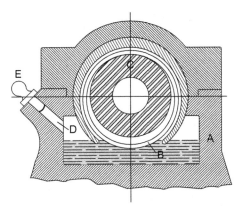

Fig. 7.1
Ring oil lubrication: (A) bearing housing, (B) flat ring, (C) journal, (D) oil inlet, and (E) stopper.
©*2007 http://chestofbooks.com/.*

7.1.1.2 Circulating systems

In larger units, the need for greater oil flow and larger heat dissipation dictate the use of the oil circulating system. Details of this system are discussed under individual prime-movers (Sections 7.2.1–7.2.3).

7.1.2 Oil Coolers [6]

Some form of oil cooling is found in any type of lubrication system. While direct radiation, water jackets, and simple coils are common in small units, separate oil coolers are usual for circulation systems. They serve to remove the heat developed in bearings that in turn maintain low lube oil temperature and retard oxidation of oil. Excessive cooling of the oil, however, is detrimental as it causes precipitation of soluble products of oxidation on oil cooler tubes. The water pressure is maintained less than the oil pressure so that, in case of a leak, no water will get entrapped in the oil, thereby impairing lubricating property. Normally 2 x 100% capacity oil coolers are used in a unit.

7.1.3 Oil Filters

Oil filters are used in all types of lube oil systems. Oil filters in general contain a screen or fiber mat that removes particles from oil by physically trapping them in or on the screen or mesh. Generally a duplex filter (Fig. 7.2) or two filters each of 100% capacity are installed in parallel. During normal operation, one of these filters remains in service, and the other one either is taken under maintenance or kept as standby. In clean oil, pressure drop across the filter is about 10 kPa. Pressure drop will start increasing when the lube oil system is put in normal circulation. A filter is construed to be clogged or choked when pressure drop across it reaches or exceeds 50 kPa. The standby filter is then taken into service; after that, the clogged filter is isolated and the filter element/strainer basket is taken out for cleaning with air blowing or any other means as recommended by the manufacturer. Once the element is cleaned, it is reinstalled inside the filter. This filter is then kept under standby duty.

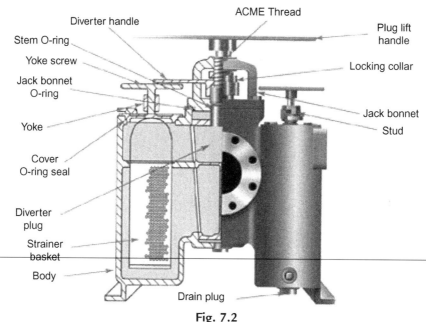

Fig. 7.2
Half cross-section of duplex filter. *Reproduced with permission of Eaton.*

7.1.4 Oil Pumps [6]

Circulation systems are designed generally with a main oil pump for bearing oil supply that is driven by either the main shaft of the prime mover or electric motors. Auxiliary oil pumps are used for lubrication during start-up of the prime-mover bearings and other areas when its speed is not up to the rated value or so. These pumps are usually of the electric motor-driven type. The auxiliary oil pumps are particularly useful during starting and stopping of the prime-mover.

7.1.5 Care and Supervision of a Lubrication System [6]

A careful check on the operating conditions of lubrication systems and proper treatment of oil are the operational means of prolonging oil life and maintaining adequate lubrication. In general, the following basic procedures are recommended:

1. Practice a regular method of purifying oil in service by an oil purification or treatment system. Maintain treatment equipment in good order, for example, by cleaning or changing filters, and cleaning centrifuges and precipitation tanks.
2. Follow the manufacturer's recommendations for periodic cleaning and overhaul, as the frequency of cleaning varies widely in different plants operating under different service conditions. Yearly cleaning is generally good practice for average units.
3. Clean oil strainers regularly, usually daily.
4. Drain the bottom of the oil reservoir every day to remove water and other contaminants.
5. Check on water leakage by inspecting glands of various lube oil valves and oil cooler.
6. Check on oil leakage by inspecting bearing seals and oil piping.
7. Check oil level daily, and keep a record of the type and quantity of make-up oil added. Reservoirs of ring-oiled bearings should be drained monthly, and the oil replaced by new or clean oil. In circulation systems, users of inhibited oils should secure a brand of oil for make-up similar to that in the system.
8. While starting, follow manufacturer's recommendations, but check on whether all lube oil pumps are functioning and whether the correct oil pressure is developed.
9. Keep a daily log of bearing-oil pressure and oil pressure to the governor mechanism.
10. Keep a daily log of water temperatures entering and leaving the cooler. A decrease from normal temperature difference of these indicates sludge or water scale in the cooler.
11. Keep a daily log of oil temperatures entering and leaving the bearings, entering and leaving the cooler, entering and leaving the gear box, if provided.
12. Keep a log of all maintenance items such as overhauls, oil changes, oil additions, filter cleanings, centrifuge cleanings, and so on.

7.2 Description of the Lube Oil System

The lube oil system of different prime-movers are not identical in all respects; hence each system is addressed separately under the following paragraphs.

7.2.1 Lube Oil System of Steam Turbine [6, Section 6.7.2, p. 228]

During normal operation, the main oil pump, mounted on and driven by the turbine shaft, draws oil from the main oil tank (MOT) and discharges it at a high pressure to the bearing lubricating oil system (Fig. 7.3). The lubricating oil pressure is usually maintained constant at about

100–150 kPa, depending upon particular turbine and bearing design. During starting and stopping operations of the turbine, when the rotor and main oil pump are not up to the rated speed, an externally driven auxiliary oil pump is needed to supply the oil.

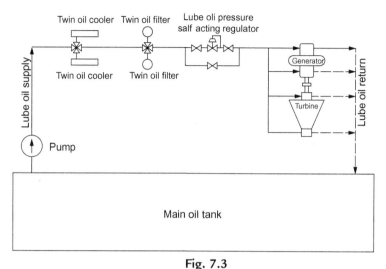

Fig. 7.3

Typical lube oil circulating system of steam turbine. *Source: From Fig. 6.41, P 229. D.K. Sarkar, Thermal Power Plant—Design and Operation, 2015, Elsevier.*

The main oil tank (MOT) is equipped with vapor extractors to keep oil tank atmosphere free of oil vapor, and strainers to arrest suspended impurities. Oil used for the bearing lubricating oil system is cooled in turbine oil coolers in order to control bearing oil/metal temperature. Thereafter, oil passes through the oil filter to supply dirt-free lube oil to bearings. The oil that has passed through the filter is then led to bearings through the lube oil supply header via a lube oil pressure-reducing regulator. From the bearings, lube oil drains back, by gravity, through the lube oil return header to the MOT, where any heavy impurities that may have been picked up by oil during circulation will tend to settle out.

7.2.2 Lube Oil System of Gas Turbine [6, Section 7.5.6, p. 267]

Lubrication and power oil system is provided to meet the lubrication requirement of gas turbine bearings, compressor bearings, generator bearings, thrust bearings, and so on, and depending on equipment design a portion of the fluid may be diverted for use by gas turbine hydraulic control devices (Fig. 7.4).

The lube oil is stored in a tank provided with a heater that maintains the lube oil at the minimum temperature required for operation during prolonged periods of standstill. During operation, the lube oil pump takes suction from the oil tank and circulates lube oil through oil

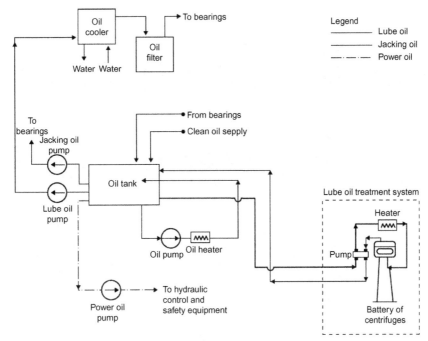

Fig. 7.4

Typical lube oil circulating system of gas turbine. *Source: From Fig. 7.26, P 267. D.K. Sarkar, Thermal Power Plant—Design and Operation, 2015, Elsevier.*

filters to the bearings and other consumers. After lubricating the bearings, oil flows back through various drain lines to the oil tank. The lube oil is also circulated through lube oil coolers in order to maintain oil temperatures within the preset range.

The same oil is also supplied to bearings for lifting the turbine and generator rotors to a small extent during start-up or rotor barring operation. This prevents wear on the bearings and also reduces the starting torque required. The system is named the jacking oil system. A separate pump is used to supply power oil to gas turbine hydraulic control and safety equipment.

The lube oil treatment system is used to clean the oil circulating in the gas turbine lubrication and power oil system. Oil is drawn by a pump from the oil tank, and on completion of desired treatment, oil is returned to the oil tank.

7.2.3 Lube Oil System of Diesel Engine [6, Section 8.5.2, p. 310]

The lubricating oil system of a diesel engine typically includes an external lube oil circuit and an internal lube oil circuit (Fig. 7.5). Primary function of the external lube oil circuit is to filter and cool the circulating oil. It also serves to heat, prime, and scavenge the oil when preparing the diesel engine for starting. An internal lube oil circuit supplies oil to individual units and parts of the diesel engine. Pipe lines of internal circuit are inside the engine.

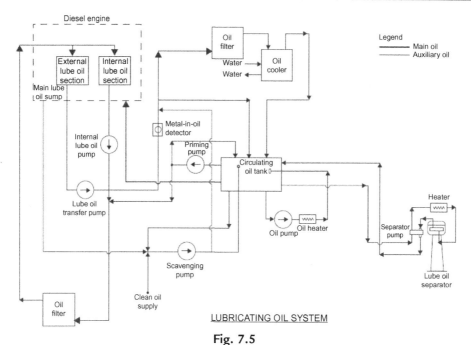

Fig. 7.5

Typical lube oil circulating system of diesel engine. *Source: From Fig. 8.16, P 310. D.K. Sarkar, Thermal Power Plant—Design and Operation, 2015, Elsevier.*

The external lube oil circuit in turn is divided typically into two groups—main lube oil circuit and auxiliary lube oil circuit. The main lube oil is either delivered by road tankers with the help of lube oil unloading pumps or downloaded from drums containing lube oil to the diesel engine main lube oil sump. The lube oil transfer pump delivers lube oil from the lube oil sump through a metal-in-oil detector, oil filter, and oil cooler into a circulating oil tank. When the pressure in the system exceeds the preset maximum permissible limit, a bypass valve diverts part of the lube oil directly into the circulating tank. Lube oil temperature leaving the oil cooler is controlled by adjusting the flow of cooling water through the oil cooler automatically by a thermo-regulator.

Lube oil temperature in the circulating oil tank is maintained with the help of an oil pump provided with oil heater, which cuts in automatically in the event oil temperature drops down to typically 313 K and cuts out when the temperature rises above 323 K.

The lube oil separator is used to clean the oil circulating in the lubrication system. Oil is drawn by the separator pump, either from the system or from the tank, and after separation it is returned to the circulating tank. A temperature of 328 K or more, as recommended, is maintained in this circuit, if required, with the help of an oil heater by putting it in service.

Oil in the internal lube oil circuit is circulated with the help of an engine-driven internal lube oil pump. This pump draws oil from the circulating tank and delivers it to the engine through an oil filter. Prior to starting the diesel engine it is primed with oil by the priming pump.

The internal circuit is exclusively used for piston cooling and lubrication of crosshead and crankpin bearing and lubrication of camshaft main bearings. Oil is drained from the pistons and bearings into the diesel engine sump, from which the scavenge pump takes its suction and delivers oil back to the circulating tank. Separate lube oil pipe lines connect to various lubricating points of major and minor accessories, drives, and the turbo-charger.

7.3 Precautions

Over and above "general precautionary measures" described under Section 2.1.1: Quality Assurance, the following specific precautionary measures must be observed before executing flushing of the lube oil piping system:

i. During manual cleaning of tanks, and so on, it is necessary to ensure that they are well ventilated before any person is allowed to enter them.

ii. Manual cleaning should be followed by a thorough visual inspection to ensure that a good cleaning job has been performed and no extraneous matters, eg, waste cloth, working tools, nuts, bolts, and/or other material, are left inside, before the particular section is boxed up.

iii. During chemical cleaning, it is to be kept in mind that acid/alkali solution is injurious to health. When it comes in contact, acid/alkali solution may burn skin; acid/alkali fumes may cause injury to eyes. Hence, all probable safety measures shall be exercised by operating personnel while handling acid/alkali solution. They must keep on wearing acid/alkali-proof hand-gloves, face mask, and protective clothing/apron as well as protective glasses for eyes when working with acid/alkali and as long as the cleaning process is continued. If contact is made with acid/alkali, operating personnel must immediately flush with fresh water.

iv. The reaction of additives in acidic materials with additives in the flushing oil can form insoluble soaps, which can deposit throughout the system. Hence, compatibility of these additives must be considered [4].

v. Authorized personnel must have proper knowledge of all attributes of the oil flushing process. Precautions must be taken to protect personnel from hot oil spray [4].

vi. Extreme fire and spark protection precautions must be taken. Even lubricants can ignite if heated and sprayed; conventional mineral oils may form explosive mixtures under such conditions [4].

vii. The hose, lance, and any other fittings used must be pressure rated for the full flush pump outlet pressure [4].

viii. Prior to filling the MOT with flushing oil, it must be centrifuged to remove dirt and water content.

ix. When the system operating oil is used for flushing, it should not be used for operation unless it has been tested and shown to be suitable for system operation following flushing. This precaution is necessary because oil-soluble contaminants picked up during the flush may be incompatible with system components or cause foaming, emulsification, reduced oxidation resistance, or a combination thereof [4].

x. Emergency stop-push buttons must be provided in the vicinity of the flushing area for stopping circulating oil pumps as may necessitate.

7.4 Prerequisites

Before carrying out the lube oil flushing process, thorough inspection of the following areas/items must be undertaken (Table 7.3) as a measure of quality assurance:

Table 7.3 Areas/items to be checked

S. No.	Areas/Items	Ok (✓)
1	Before erecting the piping systems, protective coating used on internal surfaces as a safeguard against corrosion during transport and storage are removed as far as practicable	
2	Prior to erecting all temporary piping systems, they are thoroughly cleaned mechanically	
3	If chemical cleaning of the permanent lube oil piping was not carried out at the manufacturer's works, verify that the piping system is chemically cleaned (Section 7.5.2) at the site. Ensure also that chemical cleaning of temporary flushing oil piping is completed at the site. Check protocol of chemical cleaning, jointly signed by the customer, the engineer, the prime-mover manufacturer, and the supplier (contractor), certifying successful completion of chemical cleaning is available (Table 7.9)	
4	In the event that permanent lube oil piping is supplied to the plant as factory pickled, ensure that they are water washed (Section 7.5.3) at the site before erection	
5	All piping systems, whether permanent or temporary, are erected and supported properly along with associated valves. Protocol signed jointly by the customer, the engineer, and the supplier (contractor), certifying completion of erection of piping systems, is verified	
6	Lube oil pipe lines are hydrotested successfully at 1.5 times of their designed pressures. Protocol signed jointly by the customer, the engineer, the prime-mover manufacturer and the supplier (contractor), certifying acceptability of the hydrotest, is verified	
7	All welded and flanged joints are secured	
8	Construction materials—temporary scaffolding, wooden planks, welding rod ends, cotton wastes, and so on, are removed	
9	Instrument/s installed on the permanent piping systems are isolated	
10	Check erection of temporary tank for chemical cleaning is in order	
11	Temporary platforms are provided wherever required	
12	Estimated quantity of DM water (Table 7.5) is available	
13	Estimated quantity of chemicals (Table 7.6) is available	
14	Various reagents (Table 7.7) are available	
15	Estimated quantity of flushing oil (Table 7.8) is available	
16	Materials required for lube oil flushing (Section 7.4.1) are available	
17	Apparatus required for chemical analysis (Section 7.4.2) are arranged	
18	Safety gadgets (Section 7.4.3) are arranged	
19	Portable fire extinguishers are provided in the main oil tank and lube oil flushing area	
20	Verify that "no smoking," "danger," "keep off," and other safety tags (in regional and English languages) at various places around the area of activity are provided	
21	Verify that tanks, pumps, valves, associated equipment and systems, which are in service or are energized, are provided with proper safety tags (Appendix C) so as to obviate any inadvertent operation or to obviate any injury to equipment and personnel	

7.4.1 Typical List of Bill of Materials

Materials that are typically required for flushing of the lube oil piping are listed as follows. Size and quantity of each of these items would vary from project to project and shall have to be assessed by the equipment manufacturer.

1. temporary tank for water flushing
2. temporary tank for chemical cleaning
3. temporary pipe lines
4. spool pieces
5. temporary valves
6. temporary pump for water flushing
7. temporary pump for oil flushing if recommended by the prime-mover manufacturer
8. heating jacket or oil heater
9. temporary fine mesh screens
10. rawhide hammer, rubber mallet, or pneumatic vibrator [4]
11. hot oil-free dry air blowing arrangement
12. pressure-measuring instruments
13. temperature-measuring instruments

7.4.2 Typical List of Apparatus

Apparatus that is typically required for chemical analysis during "chemical cleaning of lube oil pipe lines" and analysis of "effluent" during oil flushing are listed as follows. Size and quantity of each of these items would vary from project to project and shall have to be assessed by the chemist associated with cleaning activities.

1. titrating flasks
2. standard flasks
3. polyethylene bottles
4. measuring cylinders
5. pipettes
6. funnels
7. pH meter for determining pH of the solution
8. special pH papers of range 0–14
9. measuring device for assessing size of suspended particles in flushing oil

7.4.3 Typical List of Safety Gadgets

Equipment that are generally used for safety of operating personnel are as follows:

1. gum boots of assorted sizes
2. rubber gloves

3. asbestos gloves
4. rubber or polythene aprons
5. hard hats of assorted sizes
6. side-covered safety goggles with plain glass
7. gas masks
8. first aid box
9. safety sign boards
10. safety tags

7.5 Preparatory Arrangements

In order that the flushing of the lube oil system can be completed satisfactorily within the shortest possible time, it is necessary to complete the following preparatory arrangements before the actual flushing process is undertaken.

7.5.1 Mechanical Cleaning

It is desirable that the whole circulation system including bearing housings, oil tanks, oil storage tanks, oil pipes, oil coolers, oil filters, gear boxes, oil purifier, generator shaft seal oil system, and so on, have been thoroughly cleaned as far as accessibility permits to get rid of dirt, dust, and foreign materials. For this reason, it may be necessary to chip and wire brush rusty parts. Welding slag on pipe lines shall be removed by grinding or filing as far as practicable. Protective coat of paints is to be removed by puffing, brushing, and so on. Proper cleaning of the entire system will be followed by wiping the system clean using a lint-free cloth. Cotton waste or other fibrous cloth should on no account be used for cleaning. Care must be taken against leaving any thread, or so on, in the system. Cleaning of the above parts will ensure providing clean oil to the remaining parts of the system.

7.5.2 Chemical Cleaning

Permanent lube oil pipe lines, if not pickled already at manufacturer's works, must be chemically cleaned at the site along with temporary pipe lines to make the internal surfaces of piping free from mill scales, which may develop in the manufacturing and fabrication stages; and free from grease, oil, and so on, which are used for preservation and may enter into the system during transport and storage.

The following aspects are to be verified before commencing the chemical cleaning process.

a. A temporary tank with drain connection is required for dissolving chemicals for carrying out the chemical cleaning of lube oil pipe lines adopting the soaking technique. This tank is an atmospheric tank and shall be large enough to accommodate both permanent and temporary lube oil pipe lines.

b. The area surrounding the chemical-dissolving tank should be adequate so that any spillage of chemicals from the system do not damage any permanent equipment.
c. Quantity of chemicals (acids, alkali, and inhibitor), reagents, various apparatus, and safety gadgets, to be required for the process, should be ascertained beforehand and adequate quantity of them must be kept available.
d. Adequate quantity of DM water must be assessed prior to undertaking the process and shall be made available.

7.5.2.1 Chemical-cleaning procedure

Chemical cleaning of lube oil pipe lines shall be carried out in line with the recommendation of the equipment manufacturer. Nevertheless, general guidelines are described below to make the readers/users conversant with the cleaning process using the soaking method.

  i. Fill the temporary tank with DM water up to about the middle of the tank or as adequate enough to submerge lube oil pipe lines. Prepare an acidic solution of 3–4% citric acid inhibited with ammonia to maintain a pH of 3.5–4.0. This solution will ensure removing mill scale, and so on, inside pipe lines.
 ii. Dip permanent and temporary pipe lines inside the solution. Avoid contact between individual section of pipes. Dip also a small piece of pipe with quick lifting arrangement for inspection. For effective cleaning of pipe lines, the soaking process may continue from 12 to 48 h depending on the extent of deposits. After 12 h of soaking, lift the small pipe piece and check the extent of its cleanliness. If it is found to be clean, it may be inferred that the remaining piping in the solution is also clean.
iii. Continue with the soaking process. Check iron content in the effluent at every 30-min interval. When the level of iron content becomes constant, it shows all pipe lines are clean and that is the end of pickling.
 iv. Start supplying fresh DM water to the tank and simultaneously start draining effluent from it. Analyze the sample from the tank drain. Ammonia may be dozed in the tank to raising pH faster. When pH of the sample reaches about 7.0, stop further supply of DM water to the tank. At the same time drain of effluent from the tank is to be also stopped

---

Note

Ensure that effluent is drained to a neutralizing pit fulfilling the local environmental regulation.

---

  v. Raise temperature of DM water in the temporary tank to 343–353 K with steam or an immersion heater. Doze caustic soda, NaOH, in the temporary tank until the pH of the solution becomes 14.0. Pipe lines may remain in this solution from 4 to 6 h for removing grease, oil, and so on, and neutralizing residual acid from inside the pipe lines.
 vi. The above sequence of the cleaning process will provide protection of pipe lines from corrosion following acid cleaning.

vii. After 4 h of soaking, lift the small piece of sample pipe and check whether there is any trace of grease or oil in it. If it is clean, the remaining piping inside the solution is expected to be clean. Otherwise continue with the soaking process until the small piece is free from grease, oil, and so on.

viii. Start supplying fresh DM water into the tank and at the same time start draining the effluent from the tank. The process will continue until the pH of water inside the tank comes down to 7.0 or so.

ix. Collect the sample from the tank drain to inspect whether it is clean. In case the sample is dirty, drain effluent from the tank and replenish with a fresh supply of DM water. Repeat the process until the sample is shown to be clean.

x. Thereafter take out the piping from the tank and dry out them with hot, oil-free dry air.

xi. Reinstall both permanent and temporary lube oil piping along with strainers, valves, fittings, and so on, to proceed with the oil flushing process.

Note

Ensure minimum time gap between completion of air drying and commencement of oil flushing, lest the chemically cleaned surface may get subjected to further corrosion.

---

xii. While the process is in progress, a protocol may be maintained, which shall be jointly signed by the customer, the engineer, the equipment manufacturer, and the supplier (contractor) at the end of the process (Table 7.10).

7.5.3 Water Washing of Factory Pickled (Chemically Cleaned) Lube Oil Pipe Lines

On completion of pickling of lube oil pipe lines at manufacturer's works, some protective coating is applied to freshly exposed surfaces, which are susceptible to corrosion. This protective coating needs to be cleaned by water washing before putting it under oil flushing. Water washing is carried out by circulating water through assembled pipe lines.

Normal steps to be followed for executing the circulation process are

i. remove NRVs from each oil pump discharge

ii. erect temporary water header connecting flushing oil pump and all lube oil pump discharge piping

iii. provide proper supports to temporary piping

iv. remove oil filter elements

v. oil coolers are to be bypassed

vi. shaft-driven oil pump suction and discharge are to be blanked

vii. oil return header to MOT is to be blanked

viii. install temporary water tank at an elevation

ix. install temporary water circulation pump at ground level

x. erect temporary piping from the tank to circulation pump suction

xi. erect temporary piping from the circulation pump discharge to the permanent piping

xii. fill the temporary tank with clarified/DM water

xiii. start the temporary water flushing pump

xiv. replenish the tank with a fresh supply of water when the flushing circuit is getting filled up

xv. maintain circulation of water for 2–3 h

xvi. take water samples every hour

xvii. once the water is declared to be clean, stop the temporary pump and drain the system

xviii. drive out water from the nondrainable areas by blowing hot air through them. This process may take some time

xix. it must be ensured that no trace of moisture is left behind, lest the oil flushing process will prolong to remove residual moisture

xx. remove all temporary piping and connections and normalize the system

7.5.4 Installation of Fine Mesh Screen

At rated speed, the clearance between the journal and the bearing housing of a bearing typically varies from 21.8 to 188.0 μm [3]. However, at low operating speed (while running on turning gear or during start-up/shut-down), this clearance drops down to 1.5–21.8 μm [7]. So, it is usual practice during lube oil flushing to install temporary fine mesh screen of 104 μm (150 mesh) to 154 μm (100 mesh) at the oil inlet to each bearing and gear boxes with provision of periodic withdrawal for inspection and cleaning after stopping oil circulation, if necessary. Temporary fine mesh screen shall also be placed over permanent strainers in the oil tank and elsewhere. These are of great value towards the end of the flushing process. If installed in the early stage, they are likely to become choked frequently and will need frequent inspection and cleaning.

It is recommended to install temporary strainers of 154 μm (100 mesh) to 198 μm (80 mesh) on the suction side of the lubrication oil pumps or temporary oil flushing pumps, or both, if used, and on the discharge side of gravity and pressure systems (Clause 7.4.3 [4]).

7.5.5 Bypassing All Vulnerable Systems

The governing system, control gears, generator shaft seals, and so on, should be bypassed and excluded from the oil flushing circuit so as to avoid damages to surfaces machined with close tolerance by loosened abrasive matters carried during oil flushing. The above areas/items shall be manually cleaned and flushed at the final stage once the rest of the system is thoroughly cleaned out.

During the initial phase of oil flushing, oil is not allowed to enter the bearings and oil coolers; hence, each bearing journal and oil coolers are bypassed with suitable temporary arrangements (jumpers) so that the flushing oil is circulated through the entire system without entering into bearings.

Note

If recommended by the equipment manufacturer, flushing oil may be allowed to pass through bearings right from the beginning of the oil flushing process by providing a paper filter at the oil inlet connection to each bearing.

7.5.6 Heating of the Flushing Oil

To achieve the best result, flushing oil flow must be turbulent (Reynolds number >4000) [8]. Turbulent fluid flow can be attained by adopting relatively high fluid velocity and/or lowering fluid viscosity. For lowering the fluid viscosity, it may be necessary to heat the flushing oil to a temperature of about 343–353 K or as recommended by the equipment manufacturer. This temperature is necessary to maintain fluidity of the flushing oil in the system, to dissolve oil soluble materials, and to aid in loosening adherent particles. For this purpose, if the lubricating oil heater has not been installed, or if it is inadequate, heat may be supplied to the flushing oil in several ways. The best method is to use the purifier oil heater. Another plausible alternative is to pass hot water through the cooler; this can be generated by bubbling low-pressure steam through the water somewhere outside the cooler. The cooler must be vented to the atmosphere to prevent pressure buildup. In the event electric heating is adopted, it must conform to the API 614 recommendation that specifies a maximum watt density of 2.3 W cm^{-2} for mineral-based products. Low pressure steam may also be used; however, the cooler should be checked against the manufacturer's recommendation. Great care must be taken to ensure that not over 34 kPa steam is admitted to the cooler so that the cooler is not damaged and the flushing oil is not overheated (Clause 7.4.5 [4]). Whatever may be the type of heating followed, it must be ensured that overheating of oil is to be avoided lest excessive volatilization will take place.

7.5.7 Erection of Temporary Piping and Valves

In order to complete the circuit for circulation of oil, depending on the pipe layout configuration at the site and if recommended by the equipment manufacturer, some temporary piping with associated temporary valves connecting the MOT and the permanent piping systems may need to be erected and kept ready for putting into service. Size and quantity of temporary piping materials including elbows, tees, valves, supports, and so on, shall be assessed at the design stage.

To allow removal of the temporary piping, valves at the MOT and oil purification unit should be provided.

Sampling points are to be installed for monitoring cleanliness of oil during the progress of oil flushing.

7.5.8 Readiness of Auxiliaries

Prior to undertaking the oil flushing process, the following auxiliary equipment must be kept ready to be put into operation:

    i.   temporary flushing oil pump if recommended by equipment manufacturer
    ii.   starting oil pump
    iii.   auxiliary oil pump
    iv.   emergency lube oil pump
    v.   jacking oil pump
    vi.   permanent lube oil filters
    vii.   lube oil coolers
    viii.   centrifuge
    ix.   main oil tank vapor exhausters
    x.   lube oil heating arrangement in MOT

Notes

1. Shaft driven main oil pump, if provided, shall be removed and replaced with a separate temporary flushing arrangement.
2. Pressure-regulating valve, located on jacking oil header, shall be removed and replaced with a suitable spool piece.

7.6 Operating Procedure

For the convenience and effectiveness of the oil flushing process, the system may be split into one or more circuits based on the layout configuration of the oil flushing system and per the recommendation of equipment manufacturer. Fig. 7.6 depicts a typical lube oil flushing scheme of a steam turbine-generator.

While carrying out oil flushing of each circuit, the process is sequentially divided into the following activities. This will ensure reducing total time of completing the oil flushing process, successfully fulfilling the recommendation of the equipment manufacturer:

    i.   oil charge
    ii.   oil circulation
    iii.   purification of flushing oil during circulation
    iv.   duration of circulation
    v.   inspection and cleaning
    vi.   fresh oil filling and circulation

7.6.1 Oil Charge

Flushing charge must be large enough to ensure that oil circulating pumps do not lose suction and the temporary heater coil, if used for oil heating, is completely immersed. As a rule, a

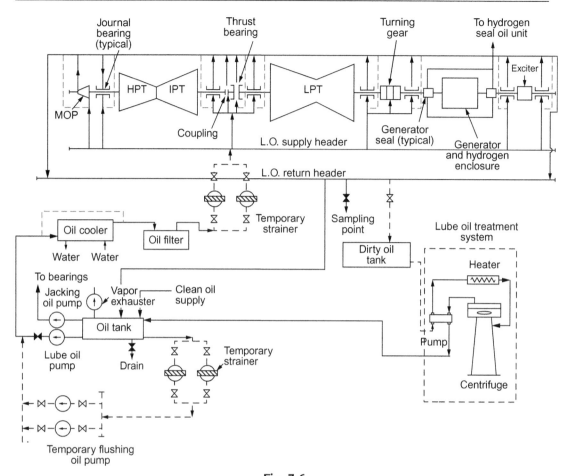

Fig. 7.6
Typical lube oil flushing scheme of steam turbine-generator.

charge of 75–80% of normal requirement of oil of the system is necessary. Nevertheless, if the equipment manufacturer recommends a different quantity of charge, then that recommendation needs to prevail.

Keep vigilance on the overflow line from the dirty compartment of the MOT. Oil should not flow out through this line during the course of the oil charge.

Fresh oil may need to be charged in the MOT in accordance with the progress of the circulation process.

7.6.2 Oil Circulation

The flushing medium must be circulated with the help of the starting oil pump or auxiliary oil pump or temporary flushing oil pump to establish the requisite flow as recommended by the manufacturer. Each lubricating oil pump and auxiliary pump should be used during the flushing period. Put the vapor exhausters and oil heaters of the MOT in service. Raise temperature of oil to 343–353 K, then stop heating the oil. Heating of oil may be recommenced once oil temperature drops to 313–323 K. The circulation shall be generally continuous with stopping from time to time for inspection and cleaning of temporary strainers. Experience reveals that almost all of the foreign matter is collected in the filters or temporary strainers during the first few hours of flushing. During this time, whenever a noticeable increase in pressure drop across the strainers is observed, the strainers should be cleaned and replaced. This may occur as frequently as every 15 min at the beginning (Clause 7.7.2 [4]).

To facilitate dislodging of debris, rust, weld beads that may have adhered to inner surface of piping, particularly at joints and flanges, and other loose materials inside piping, as well as to make removal of debris along with the flushing oil flow easier, it is recommended to pound the oil lines with a rawhide hammer, rubber mallet, or pneumatic vibrator along the direction of the oil flow so that it is just sufficient to initiate some vibration and thermal shock in the pipe section without causing any indentation. The pounding shall continue for a few hours on each section while maintaining as high rate of flow as possible through this section.

In the event any scale is retained in the pipe line even after chemical cleaning, this scale is dislodged by allowing the oil temperature to drop by 28–32 K for 2–3 h during flushing to allow for pipe contraction (Clause 7.7.4 [4]). Repeat the circulation process with proper heating of oil.

During the flushing operation, it may be required to remove the top housing of bearings to permit hosing of journals if it is suspected that damage by abrasive matters might otherwise occur. On completion of oil flushing, it would suffice to examine two or three bearings to determine whether any extraneous matter will have to be removed manually.

7.6.3 Purification of Flushing Oil During Circulation

Circulation of turbine oil removes at a rapid rate dirt, dust, and sludge, which do not get trapped by the temporary wire mesh screen and tend to remain in suspension in the fluid. As a result, they remain in recirculation gathering further matters from the system. The larger particulate matters, however, will get trapped by the temporary wire mesh screens. When the flushing medium gets heavily laden with suspended matters, it is necessary to stop the heating and circulation and transfer the whole oil to a dirty oil storage tank. In this tank the dirty oil is allowed to set for 8–12 h to allow gravity to settle all dirty matters. Thereafter, the same oil charge may be taken back to the MOT through the oil purifier for further heating and resuming circulation.

7.6.4 Duration of Circulation

Flushing times of at least 12 h to as much as several days may be necessary to reach the desired level of cleanliness (Clause 7.7.4 [4]). Circulation and filtration shall continue alternately until the oil charge becomes sufficiently clear with no evidence of contaminants, such as lint, welding beads, or other extraneous matter, on the temporary wire mesh screens and accessible parts of the system. Each lubricating oil cooler should be cleaned separately to provide maximum oil velocity in the cooler to optimize contaminant removal.

Circulation is continued until a strainer screen of 106 μm is cleared. Circulation may be stopped now and flushing oil drained to the MOT. Jumpers provided across bearings and oil coolers may be removed and system normalized. Flushing oil is now allowed to pass through the bearings and circulation and filtration of the flushing oil continued until the final cleanliness is attained, fulfilling the requirement of the equipment supplier.

The size of suspended particulate matter in lube oil must conform to NAS Class 8 (ISO 4406 19/17/14, or as recommended by the equipment manufacturer) specification with moisture content of 100 mg kg^{-1} (Table 7.4) [8]. At the last phase of oil flushing, the governing system, generator shaft seals, and other parts of the system left out of the circulation circuit earlier shall be flushed one by one for a few minutes each so that they are also cleaned. Thereafter, the circulation shall be stopped and the oil transferred to the turbine dirty oil storage tank as soon as practical and safe while the flushing oil is still hot.

Allowing the flushing oil to cool will leave more contamination in the system than if drained while it is still hot. Oil lines should be opened at the lowest points and the oil allowed to drain. Remaining oil should then be removed from sumps, tanks, and coolers (Clause 7.8 [4]).

Note

The flushing oil, after purification, may be reclaimed and used for flushing another installation, after it has been determined that the product is free of contaminants and still contains solvency and rust-inhibiting properties.

Table 7.4 NAS cleanliness level particle count

NAS 1638 (1964)	Based on 100 mL sample				
	5–15 μm	15–25 μm	25–50 μm	50–100 μm	>100 μm
12	1024 k	182 k	32,400	5760	1024
11	512 k	91,200	16,200	2880	512
10	256 k	45,600	8100	1440	256
9	128 k	22,800	4050	720	128
8	64,000	11,400	2025	360	64
7	32,000	5700	1012	180	32
6	16,000	2850	506	90	16
5	8000	1425	253	45	8
4	4000	712	126	22	4
3	2000	356	63	11	2
2	1000	178	32	6	1
1	500	89	16	3	1

7.6.5 Inspection and Cleaning

The turbine oil tank and accessible surfaces should be opened and cleaned manually with lint-free rags to remove all traces of residual contaminants and as much oil as possible. Two or three bearing housings, thrust bearing housing, and barring gear housing shall be opened for inspection and cleaning. If there is evidence of any undesirable element, all the bearing housing must be opened and cleaned manually. Other areas that are to be opened and cleaned manually are the oil injectors in the turbine oil tank, the generator shaft seal housing, and the governing system nozzles, orifices, ports, and so on.

Once all areas and surfaces are satisfactorily cleaned, the lubricating oil system should be again thoroughly inspected for evidence of contamination to determine if it is secure for operation [2].

7.6.6 Fresh Oil Filling and Circulation

After all the accessible parts have been cleaned, inspected, and boxed up, fresh turbine oil of recommended quality shall be taken into the turbine oil tank through the oil purifier. The temporary wire mesh screens shall be left in place for some more time while the fresh oil charge is initially circulated. These screens may be removed once it is declared that the oil is adequately clean.

When the fresh oil charge is circulated, the turbine oil purifier must be put into service and kept in operation continuously, yet with periodic stopping for its bowl cleaning until the oil charge is declared clean and serviceable.

The barring gear shall then be put into service for some time and checked for proper operation.

7.7 Estimation of DM Water, Chemicals, and Flushing Oil

Prior to the chemical cleaning of the lube oil piping and flushing of the lube oil system, it is essential to estimate the requirement of DM water, chemicals, and flushing oil correctly. It is advisable to procure a higher quantity of these items than estimated, as any shortfall of any of these items during the actual execution of processes will jeopardize whole sequences of activities and may become detrimental to areas falling within this process. To facilitate estimation of each of above items, it would be prudent to assess the following beforehand.

7.7.1 Details of Lube Oil Piping

If the permanent lube oil piping is already pickled at manufacturer's works, they may be excluded from further chemical cleaning and only temporary lube oil piping is to be handled. Otherwise both temporary and permanent lube oil piping is to be pickled.

Accordingly, the longest piece of piping will determine the length of the temporary flushing oil tank. Width of this tank will be determined based on the dimension required to put all piping sections side by side under a submerged condition. Actual dimensions of the tank have to be large enough to facilitate maneuver of piping to turn, rotate, or shift inside the tank during the chemical cleaning process (soaking method).

Once length and width are assessed, the depth of the tank should be enough to dip all piping adequately to ensure effective chemical cleaning.

7.7.2 Estimation of DM Water Requirement

From the knowledge of all three dimensions of the temporary flushing oil tank from Section 7.7.1 the required volume of DM water requirement can be finalized.

Processes, in which DM water will be required, are shown in Table 7.5.

Table 7.5 Activities where DM water will be required

Activity	Quantity (m³)
Preparation of acid solution of pH 3.5–4.0 to carry out pickling Rinsing after pickling to raise pH of the solution to 7.0 Preparation of alkali solution of pH 14.0 to carry out degreasing Rinsing after degreasing to lower pH of the solution to 7.0 Supplying fresh DM water to the tank until the drain sample is shown to be clean	

7.7.3 Estimation of Chemical Requirement

During chemical cleaning of the lube oil piping, various types of chemicals (Table 7.6) and reagents (Table 7.7) required for the analysis of effluent are used. The actual amount of these chemicals shall be assessed beforehand and procured accordingly.

Table 7.6 Type of chemicals

Activity	Name of Chemical	Quantity (kg)
Pickling	3–4% Citric acid Inhibitor: liquid ammonia	
Neutralizing Degreasing	Liquid ammonia to raise pH of water to 7 Caustic soda, Na OH, of concentration 500 mg kg^{-1} Detergent: "Teepol" or "Lissapol" or "Dodenol"	

Table 7.7 Type of reagents

S. No.	Reagent
1	Reagents for the determination of acid in the pickling solution
2	Reagents for the determination of Fe^2, Fe^3 ions in the pickling solution
3	Reagents for the determination of alkalinity in the degreasing solution

7.7.4 Estimation of Flushing Oil Requirement

For determining the flushing oil requirement, the "oil holding capacity" of the system shall be ascertained (Table 7.8). Actual volume of each area is project specific and has to be supplied by the equipment manufacturer.

Table 7.8 Areas for which oil holding capacity is to be assessed

Item	Volume (m^3)
Main oil tank (MOT) From MOT to the suction of the flushing oil pump, if provided, starting oil pump, auxiliary oil pump, and emergency lube oil pump From oil pump discharge to the lube oil supply header and various places of the lube oil system through oil filters and oil coolers Oil coolers Oil filters Return line from bearings and other areas to centrifuge From centrifuge to MOT Any other area as recommended	

7.8 Conclusion

The lube oil system is now ready to be put into service for normal operation.

The following two protocols—one during the progress of chemical cleaning of the lube oil piping (Table 7.9) and another on completion of the lube oil flushing (Table 7.10)—are to be jointly signed by all concerned.

Table 7.9 Protocol (chemical cleaning of lube oil piping)

Parameters	Record of Data								
Type of acid used Strength of acid solution									
Half-hourly record of iron content in effluent during the end of pickling									
On completion of pickling pH of fresh water in the tank Type of alkali used Temperature of water prior to degreasing Strength of alkaline solution On completion of degreasing the pH of fresh water in the tank Complete drying out of pipe lines									
Signed by the customer	Signed by the engineer			Signed by the prime-mover manufacturer			Signed by the supplier (contractor)		

The sample should be free of visual contamination and debris for an acceptable level of cleanliness (Table 7.10). Two acceptable samples, obtained at least 8 h apart, are required to verify the cleanliness of the system or portion of the system that is being flushed [4].

Table 7.10 Protocol (flushing of lube oil system)

Areas	Parameters	Inference	
Duplex oil filter	Collection of debris, and so on	Number of particles in 100 mL of sample	
Temporary strainers	Collection of debris, and so on	Particle size, μm	Maximum allowable number of particles
		5–15	64,000
		15–25	11,400
		25–50	2025
		50–100	360
		>100	64
Sampling point from MOT	Analysis of oil	Free water content $\leq$100 mg kg^{-1} [8]	
Signed by the customer	Signed by the engineer	Signed by the prime-mover manufacturer	Signed by the supplier (contractor)

References

[1] EPRI, Fossil Plant News, Spring, 1991.
[2] H.G. Stoll, GE enhancing steam turbine generator reliability-availability-maintainability performance, in: Power Gen Conference, December, 1991.
[3] Pall Corporation, Modernize Turbine Lube Oil Filtration to Improve Reliability and Availability, Pall Industrial Hydraulics Company, 1993.
[4] ASTM D6439–99, Standard Guide for Cleaning, Flushing, and Purification of Steam, Gas, and Hydroelectric Turbine Lubrication Systems, ASTM International, West Conshohocken, PA, 2011.
[5] 4454_TURBINE_OIL_46. http://www.77lubricants.nl/fileadmin/user_upload/77lubricants.nl/productsheets/19_Turbine/product/4454_TURBINE_OIL_46.pdf.
[6] D.K. Sarkar, Thermal Power Plant—Design and Operation, Elsevier, Amsterdam, Netherlands, 2015.
[7] EPRI, Guidelines for maintaining steam turbine lubrication systems, Report No. CS-4555, July 1986.
[8] GE Energy, GE K 110483b: Cleanliness Requirements for Power Plant Installation, Commissioning, and Maintenance, General Electric Company, 2002.

CHAPTER 8

Steam/Air Blowing of Main Steam, Cold Reheat, Hot Reheat and Other Steam Pipe Lines

Chapter Outline

8.1 Introduction

It is evident from Chapter 6: Chemical Cleaning of Steam Generator that superheaters, main steam, cold reheat, hot reheat, and other steam pipe lines of drum-type steam generators are left out from the purview of chemical cleaning. Hence it is essential to blow-out these areas prior to first start-up of a new steam generator to remove rust, mill scales, loose pieces of scales, welding slag, and any other loose material left over from manufacturing, transportation, storing, and erection stages [1,2].

In case of once-through steam generators even if superheaters, main steam, cold reheat, and hot reheat steam pipe lines are subjected to a high standard of the chemical cleaning process, in accordance with the recommendation of manufacturers, some undesired elements or debris will

Thermal Power Plant. http://dx.doi.org/10.1016/B978-0-08-101112-6.00008-3

always remain in these lines due to piping geometry [1,2]. Thus, it is imperative to get rid of the remaining undesired elements from these lines by blowing them out before a new steam generator is started up.

In addition there are other steam pipe lines—auxiliary steam, turbine gland sealing steam [3], atomizing steam, and any other steam pipe lines—that also need to be cleaned by blowing out to remove impurities inside.

In running steam generators, deposits may develop inside superheater tubes and other steam pipe lines due to improperly controlled water treatment (Fig. 3.1) or from process contaminants (Fig. 6.1). Development of deposits may impair heat transfer and result in tube failure from overheating (Figs. 6.5 and 6.6). Excessive formation of deposits may lead to corrosion originating from concentration of boiler water salts. Corrosion of inside surfaces of the above tubes and pipe lines also may take place if the pH of the flowing fluid falls below the recommended limit. There may be cases of encountering undesired elements in old steam generators following replacement of superheater/reheater tubes and/or after carrying out renovation and modernization of steam generators. In order to get rid of different types of undesired elements, the affected tubes and lines shall be blown out to make them healthy.

Blowing out of aforementioned areas is a necessity since steam turbine performance is greatly affected by the presence of any debris in flowing steam—especially those present in main steam and reheat steam systems, which may seriously damage turbine blades and stop and control valves. The extent of damage will depend on the amount of impurities carried with steam. The more the amount of impurities, the worse would be the turbine performance. Impingement of even trace amounts of impurities on high speed turbine blades and valves may cause dents or scars on them. The presence of a higher amount of these undesired elements may result in erosion of blades and valve seats. Occurrence of all these deformations of turbine blades would lead to consequent degradation in turbine performance. The severe effect of impingement of suspended particles on turbine blades would make it even worse and lead to blade failure.

In order to dispose of impurities from flowing steam, an age-old practice is to blow-out steam pipe lines with "steam" as the blowing medium. However, since the 1990s "compressed air" as the blowing medium is becoming attractive to the industry for successful completion of the blowing-out process [4]. Compressed air blowing of steam pipe lines had been tried in all types of steam generators—drum-type steam generators, once-through steam generators, HRSG, and so on, of power plants with ratings of 35–700 MW; the result was noted to be as effective as that of steam blowing [5]. One advantage of compressed air blowing is that it eliminates "thermal cycles associated with saturated steam blowing." Thus, steam generator life gets extended. In addition, it reduces the start-up schedule since compressed air blowing could be commenced once erection of the steam generator, along with associated pressure parts, are complete, leaving out burners, the draft system, dampers, the ash handling system, associated equipment, and so on, erection of which may continue in parallel. On the contrary, while adopting steam blowing, erection of the entire steam generator with all associated auxiliary systems as narrated above must be complete in all respects. Both methods may be cost competitive. While adopting steam blowing, expenses will be incurred to burn fuel oil to raise steam pressure; in addition, DM water, a costly commodity, will be used during steam blowing, which would get wasted to atmosphere. In the event compressed air blowing is utilized, cost has to be incurred for installing or renting high pressure air compressors along with cost of power to run them. If the drive chosen for the air compressor is a diesel engine instead of an electric motor, then cost of power would get replaced with cost of diesel fuel.

Two different methods are followed in the industry while steam is used as the blowing medium—continuous blowing and puffing. While continuous blowing may be adopted in low pressure-low temperature and medium pressure-medium temperature steam generators, puffing is more effective and economic in blowing out of steam pipe lines in high pressure-high temperature steam generators. Current preference, however, is to follow the puffing method in all types of steam generators.

Blowing out of steam pipe lines should be so carried out as to ensure that even if any debris is remaining in the system on completion of blowing out, it would not get disturbed under normal operating conditions. This could be achieved when disturbance effect or momentum ratio or drag of any suspended particulate matter during blowing is greater than the disturbance effect or momentum ratio or drag experienced by the particle under steam generator maximum continuous rating (BMCR). The disturbance factor (K) or momentum ratio of a particle is defined as:

$$K = \left[\frac{Q_{blow}}{Q_{bmcr}}\right]^2 \times \frac{v_{blow}}{v_{bmcr}} \qquad (8.1)$$

where Q_{blow} = mass flow of blowing medium during blowing out (kg s^{-1});

Q_{bmcr} = mass flow of blowing medium at BMCR (kg s^{-1});
v_{blow} = specific volume of blowing medium during blowing out (m^3 kg^{-1});
v_{bmcr} = specific volume of blowing medium at BMCR (m^3 kg^{-1}).

The greater the drag, the more effective is the cleaning. In adopting steam blowing, it is a general practice in the industry to adopt 30–60% higher disturbance effect during blow-out than the disturbance effect achieved during BMCR; that is, the value of K should be about 1.3–1.6 [1]. For air blowing of steam lines, the value of K usually is selected to be 1.2 [5], but could be 1.3 as well.

Example 8.1

Fig. 8.1 shows a simple steam blowing scheme of a main steam (MS) line. The technical specification of the corresponding steam generator and other parameters during steam blowing are as follows:

   i.   Steam generator maximum continuous rating (BMCR) flow: 416.25 kg s^{-1}
   ii.   Superheater outlet (SHO) MS pressure: 17.7 MPa
   iii.   SHO MS temperature: 813 K
   iv.   MS pressure at point 1 during steam blowing: 4 MPa
   v.   MS temperature at point 1 during steam blowing: 683 K
   vi.   Internal diameter (ID) of each MS line: 322 mm
   vii.   ID of temporary pipe line: 450 mm
   viii.   MS pressure at point 2 during steam blowing: 3.18 MPa
   ix.   MS temperature at point 2 during steam blowing: 649 K
   x.   Atmospheric pressure/MS pressure at point 3 during steam blowing: 101.33 kPa

Using the above inputs determine the following:

a. assuming 1.6 disturbance factor at point 1 what is the required quantity of steam flow (Q_{blow}) through each MS line during steam blowing;
b. with calculated Q_{blow} from (a) what is the disturbance factor at point 2;
c. the sonic velocity at point 2.

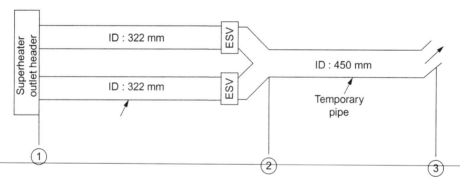

Fig. 8.1
Simple MS piping scheme for steam blowing.

Solution

a. From the steam table corresponding to MS pressure and temperature of 17.7 MPa and 813 K, respectively, the specific volume of MS, v_{bmcr}: 0.018658 m^3 kg^{-1}
During steam blowing corresponding to MS pressure and temperature of 4 MPa and 693 K, respectively, the specific volume of MS, v_{blow}: 0.076019 m^3 kg^{-1}
BMCR flow through each MS pipe line: 416.25/2 = 208.125 kg s^{-1}
Hence, applying Eq. (8.1), the required quantity of steam flowing through each MS pipe line during steam blowing:

$$1.6 = \left[\frac{Q_{blow}}{208.125}\right]^2 \times \frac{0.076019}{0.018658}$$

Example 8.1—cont'd

So,

$$Q_{blow} = 130.42 \, kgs^{-1}$$

Hence, the required quantity of steam flow through each MS line during steam blowing: 130.42 kg s^{-1}.

b. From above, the quantity of steam flowing through point 2 during steam blowing,

$$Q_{blow2} = 2 \times 130.42 = 260.84 \text{ kg s}^{-1}$$

At point 2, corresponding to MS pressure and temperature of 3.18 MPa and 649 K, respectively, the specific volume of MS, v_{blow2}: 0.089454 m^3 kg^{-1}.
So applying Eq. (8.1),

$$K = \left[\frac{Q_{blow}}{Q_{bmcr}}\right]^2 \times \frac{v_{blow}}{v_{bmcr}} = \left[\frac{260.84}{416.25}\right]^2 \times \frac{0.089454}{0.018658} = 1.88$$

Hence, the disturbance factor at point 2: 1.88.

c. The sonic velocity of an ideal gas is given by

$$V_c = 0.3194\sqrt{(k \times g \times P \times v)}$$

where V_c = sonic velocity (m s^{-1});
    k = specific heat ratio = 1.3 (for steam);
    g = acceleration due to gravity (9.81 m s^{-2});
    P = pressure (Pa abs);
    v = specific volume of gas (m^3 kg^{-1}).
At point 2 MS pressure, $P = 3.18$ MPa $= 3.18 \times 10^6 + 101.33 \times 10^3 = 3,281,330$ (Pa abs).

Hence, sonic velocity at point 2, $V_c = 0.3194\sqrt{(1.3 \times 9.81 \times 3,281,330 \times 0.089454)}$
$= 617.97 \, ms^{-1}$.

Note

1000 Pa = 102 kg m^{-2}.

As explained earlier, the prime concern of carrying out steam/air blowing of main steam, and so on, pipe lines is to prevent damage of steam turbine internals; hence, the onus of determining the effectiveness of the steam-line blowing process mainly rests with the turbine manufacturer's representative [6].

Successful completion of the typical blowing-out process may be assessed by the degree of indentation made on target plates by undesired particles in the flowing medium. Target plates, made from bar stocks, are generally located in temporary piping systems downstream of temporary blow-out valves, but upstream of silencer, suitably mounted and supported.

Typical specifications of a target plate are given below:

   i. The target plate is made of aluminum/copper/mild steel (softer material) and alloy steel;
  ii. It shall be suitably machined; its surface must have a mirror finish, and it should be free from any black surface;
 iii. The thickness of the target plate may vary from 3 to 5 mm;
 iv. The length of the target plate must cover 90% of the internal diameter of temporary blow-out piping;
  v. The width may be 25–40 mm depending on the diameter of the temporary piping.

Before declaring the completeness of the blowing-out process, the degree of indentation made on target plates must meet certain minimum requirements. Criteria of making satisfactory assessment of target plates may vary from project to project depending on the guidelines laid down by turbine manufacturers. Two types of criteria that are generally followed in the industry are given below. Two/three successive target plates must satisfy these conditions.

Criteria 1

   i. Number of crest (tip) or cavity (pit) having diameter greater than 0.5 mm (Fig. 8.2) does not exceed 1 in 625 mm^2 at the central area of the target plate;
  ii. Not more than 5 crest or cavity of diameter greater than 0.25 mm on the total surface area is visible to the naked eye;
 iii. Near the rim of the target plate, there must not be any crest or cavity.

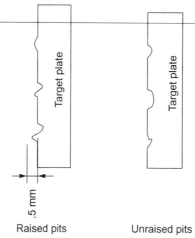

Fig. 8.2
Condition of target plate after blow-out.

Criteria 2

i. Not a single crest (tip) or cavity (pit) of size 1 mm or more should be visible on an area of 3750 mm^2;

ii. No more than 5 crest or cavity visible to the naked eye of any size greater than 0.25 mm on an area of 3750 mm^2;

iii. No raised surface hits, no irregular pockmarks or raised pits, and no embedded material visible to the naked eye on the target.

Note

1. It is to be kept in mind that acceptance criterion of completion of blowing is project specific and the final assessment and acceptance is made by the representative of the turbine manufacturer in consultation with the customer and the engineer.

2. In some projects having a relatively clean steam generator and piping system, as long as the disturbance factor is maintained throughout the blowing process, completion of blowing is ascertained by the number of blows, eg, 25–30 blows for superheater and 40–45 blows for reheater [1]. For record keeping one target plate may be used during the final blow.

Circuits that are included under sequential blowing out of steam pipe lines, either with steam or compressed air as blowing medium, are as follows:

a. Superheater, main steam line up to ESV of HP turbine, through temporary pipe line, exhausting to atmosphere (Figs. 8.3 and 8.4);

b. Superheater, main steam line, HP bypass line, cold reheat line, through temporary pipe line, exhausting to atmosphere (Figs. 8.5 and 8.6);

Note

1. While blowing out circuit b, undesired material may remain lodged in the reheater inlet header and cause blockage of reheater tubes, with resulting overheating and failure of reheater tubing when the unit goes into operation [6]. In order to prevent such blockage of reheater tubes a debris filter, similar to the one shown in Fig. 8.7, may be installed on the cold reheat pipe line inlet to reheater.

2. When blowing out circuit b, it may be noted that while adopting steam blowing the temporary blow-out piping is brought back from the reheater inlet to the turbine house (Fig. 8.5), since it will be difficult to blow-out steam to atmosphere in the steam generator house. This difficulty will not be faced in the steam generator house if air is blown out to atmosphere. Hence, during compressed air blowing of circuit b, temporary blow-out piping is laid near to the reheater (Fig. 8.6). In this case, however, the SHO valve, instead of the temporary blow-out valve, will be used during blowing.

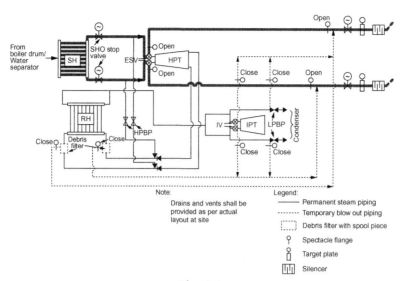

Fig. 8.3

Superheater, main steam line up to ESV of HP turbine, through temporary pipe line, exhausting to atmosphere (steam blowing).

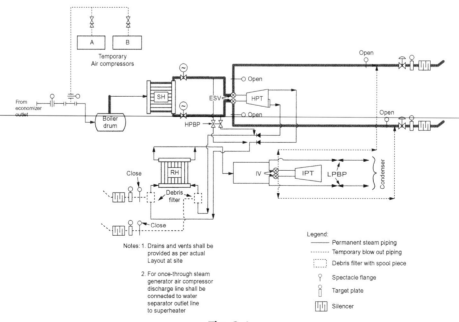

Fig. 8.4

Superheater, main steam line up to ESV of HP turbine, through temporary pipe line, exhausting to atmosphere (compressed air blowing).

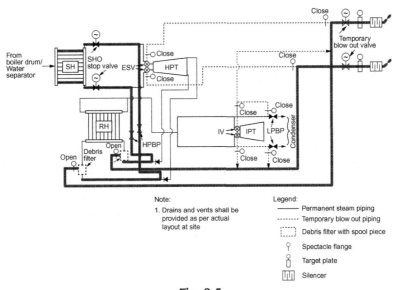

Fig. 8.5
Superheater, main steam line, HP bypass line, cold reheat line, through temporary pipe line, exhausting to atmosphere (steam blowing).

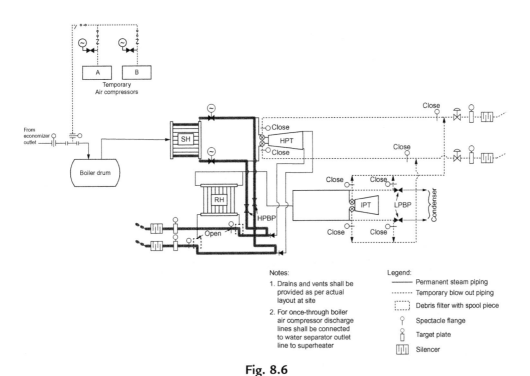

Fig. 8.6
Superheater, main steam line, HP bypass line, cold reheat line, through temporary pipe line, exhausting to atmosphere (compressed air blowing).

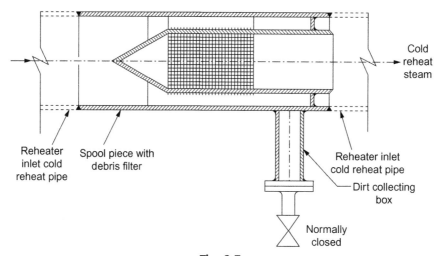

Fig. 8.7
Sectional view of a typical debris filter with spool piece.

c. Superheater, main steam line, HP bypass line, cold reheat line, reheater, hot reheat line up to IV of IP turbine, through temporary pipe line, exhausting to atmosphere (Figs. 8.8 and 8.9);

d. Superheater, main steam line, HP bypass line, cold reheat line, reheater, hot reheat line, LP bypass line, through temporary pipe line, exhausting to atmosphere (Figs. 8.10 and 8.11);

e. Supply from the cold reheat header to the auxiliary steam header, to the BFP drive turbine (if provided), to the deaerator, and then to the turbine gland sealing steam header and through the temporary pipe line, exhausting to atmosphere (Fig. 8.12). While blowing out

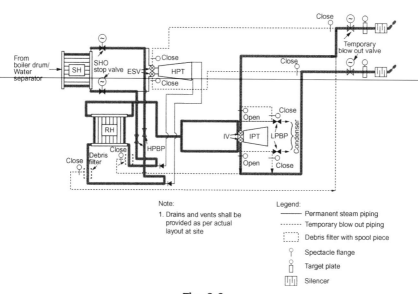

Fig. 8.8
Superheater, main steam line, HP bypass line, cold reheat line, reheater, hot reheat line up to IV of IP turbine, through temporary pipe line, exhausting to atmosphere (steam blowing).

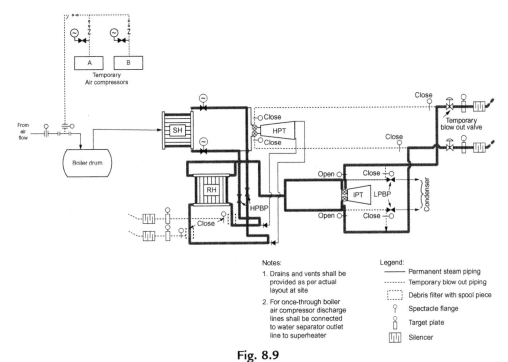

Fig. 8.9

Superheater, main steam line, HP bypass line, cold reheat line, reheater, hot reheat line up to IV of IP turbine, through temporary pipe line, exhausting to atmosphere (compressed air blowing).

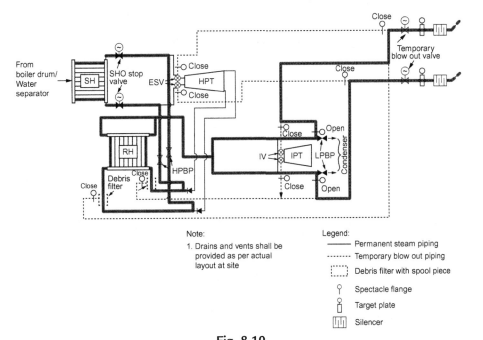

Fig. 8.10

Superheater, main steam line, HP bypass line, cold reheat line, reheater, hot reheat line, LP bypass line, through temporary pipe line, exhausting to atmosphere (steam blowing).

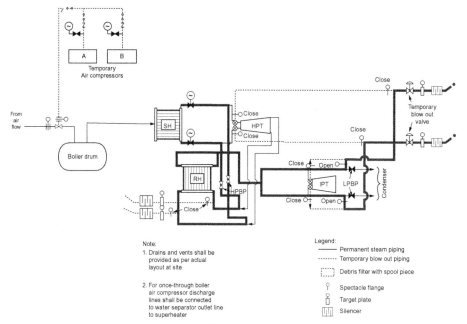

Fig. 8.11

Superheater, main steam line, HP bypass line, cold reheat line, reheater, hot reheat line, LP bypass line, through temporary pipe line, exhausting to atmosphere (compressed air blowing).

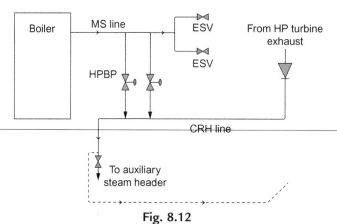

Fig. 8.12

Supply from cold reheat header to auxiliary steam header through temporary pipe line, exhausting to atmosphere (typical) (scheme identical for cold reheat header to BFP drive turbine to deaerator, to turbine gland sealing steam header).

circuit e and other lines as detailed below, installation of target plates is not essential. Visual confirmation of clean exhaust steam would suffice to declare pipe cleanliness.

Once the blowing of the auxiliary steam header (Fig. 11.7) is complete, steam from this header may be used at a later stage to blow the following lines.

- atomizing steam lines
- pulverizer inerting system steam lines
- heating steam lines
- turbine gland sealing line
- any other low-pressure steam lines.

Note

In order to avoid complication of blowing out the turbine extraction steam lines, it is preferable to mechanically clean them before these lines are erected. Nevertheless, if recommended by turbine manufacturers, these lines may be blown out taking steam from the cold reheat line and IP turbine exhaust steam lines.

To facilitate smooth execution of blowing out steam pipe lines within a shorter period of time, it would be prudent to adopt the following sequence of activities.

- Preliminary checking of the blowing-out circuits;
- Steam generator start-up (applicable when blowing medium is steam) or starting of air compressors (applicable when blowing medium is compressed air);
- Temporary heating of steam pipe lines and checking of consequent expansion of piping (applicable when blowing medium is steam);
- Preliminary blowing cycle (at the beginning of operation, carry out some blowing at reduced pressure to check the conditions of the temporary circuit);
- On completion of each daily cycle of blowing-out operations, shut down the steam generator to allow the piping to cool (applicable when blowing medium is steam) or stop the air compressors;
- Start-up of steam generator afresh and heating of temporary pipe lines (applicable when blowing medium is steam) or restarting of air compressors;
- Repeat blowing-out operations of either the previous circuit until this pipe circuit is assessed to be cleaned or start blowing a new circuit;
- Assessment of the condition of target plates and validation of blowing-out results;
- Removal of temporary pipe lines, normalization of permanent pipe lines, and restoration of the plant.

8.2 Description of Main Steam, Cold Reheat and Hot Reheat Steam, and HP-LP Bypass System *(Fig. 8.13)* [7]

As discussed under Section 1.2.1, Chapter 1: General Description of Thermal Power Plants, low-temperature water receives heat from the combustion of fuel in a steam generator to become high-pressure, high-temperature steam. Before leaving the steam generator, steam is passed through superheaters to reach the superheat state and through attemperator, where the main steam temperature is maintained by spray water. Steam thus generated is known as the main steam and is sent from the steam generator SHO to the high-pressure (HP) turbine inlet. The steam then expands through the HP turbine to carry out mechanical work.

The exhaust steam from the HP turbine is known as cold reheat steam, which is sent back to the steam generator in the reheater section, where it is again superheated. At inlet to reheater, attemperator is provided to control the reheat steam temperature to a level equal to or higher than the main steam temperature. The reheated steam, known as hot reheat steam, is directed from the steam generator to the intermediate-pressure (IP) turbine. This steam expands in IP and low-pressure (LP) turbines to carry out additional mechanical work and then exhausts into the condenser, where exhaust steam is condensed and condensate is collected in the hotwell.

A high-pressure/low-pressure (HP/LP) turbine bypass system is used to facilitate fast start-up and to allow quick reloading of the unit following a turbine trip. The HP bypass system takes steam from the main steam line, upstream of the HP turbine inlet emergency stop valves (ESVs), and discharges it to the cold reheat line through a pressure-reducing valve and a desuperheater. Likewise, the LP bypass system takes steam from the hot reheat line ahead of the IP turbine inlet interceptor valves (IVs) and discharges it to the condenser through another pressure-reducing valve and a desuperheater.

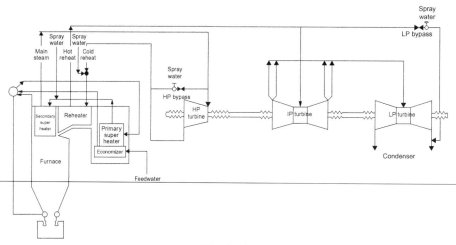

Fig. 8.13

MS, CR, and HR steam and HP/LP bypass systems. *Source: From Fig. 9.15, P 331, Chapter 9: Steam Power Plant Systems, D.K. Sarkar, Thermal Power Plant—Design and Operation, Elsevier, Netherlands, 2015.*

8.3 Precautions

In addition to general precautionary measures discussed under Section 2.1.1 in Chapter 2: Quality Assurance and Quality Control (Applicable to Preoperational Activities), following specific precautionary measures must be observed prior to steam/air blowing:

i. While the blow-out process is underway, all activities including welding work or any other hotwork, eg, gas-cutting, and so on, around the area shall be strictly prohibited;

ii. An audible and a visual alarm must be sounded to clear the area, especially the discharge area, prior to initiating a blow;

iii. Use ear muffs when blowing is in progress;

iv. A proper communicating system shall be established to give the signal to start and stop the blow-out process as well as to coordinate the insertion of target plates;

v. The following precautionary measures are especially applicable to the steam blow-out process:

    a. While the process is underway, dozing of sulfite and phosphate to the boiler drum shall be prohibited to obviate carryover of these chemicals from the boiler drum that may impinge on superheater internal surfaces, resulting in possible erosion and deposits;

    b. The design pressure of temporary pipe lines is much less than the design pressure of the steam generator and permanent steam pipe lines; hence, the set pressure of electromatic relief valves (Section 12.1, Chapter 12: Floating of Steam Generator Safety Valves) to relieve steam from the steam generator must conform with the design pressure of temporary pipe lines;

    c. Permanent and temporary pipe lines are insulated to avoid the risk of personal injury or fire. However, some portion of the temporary piping at the exhaust end purposefully may be left without insulation for ease of insertion and removal of target plates. This area may be provided with additional barriers for providing additional safety to concerned personnel;

    d. Since the steam drum is the most stressed component, drop in saturation temperature, during each steam blowing, must not exceed 40 K. Fig. 8.14 shows the relationship between "drum pressure before commencing steam bowing vs. drum pressure on completion of steam blowing," corresponding to drop in saturation temperature by 30, 35, and 40 K (applicable to drum-type steam generator);

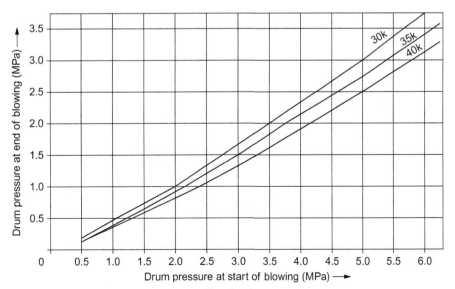

Fig. 8.14

Drum pressure before commencing steam bowing vs. drum pressure on completion of steam blowing.

e. Ensure that under no circumstances during steam blowing, the furnace exit flue gas temperature exceeds 813 K. This would protect the reheater since during the warm-up period and initial sequence of blowing out there would be no flow through the reheater.

vi. During the execution of the blowing-out process, the general noise level of a plant including surrounding areas would get raised substantially much above the tolerable hearing limit of human beings (typically peak noise levels of 90–100 dBA at 10 m from exhaust silencer and 65 dBA at the site boundary may be expected during the highest flow/pressure blows of the blowing medium). This condition may remain in force for 20–30 s during each blow and recur every 90 min (approximately) if the puffing method is adopted. In the event the continuous blowing method is followed, this condition may continue for about 30 min at a stretch.

vii. Consequent to the above, in order to avoid any public commotion resulting from unbearable noise pollution, it would be preferable to carry out blow-out operation starting from early morning, say 06:00 up to about 19:00 on weekdays only, excluding Sundays and public holidays.

8.4 Prerequisites

Blowing out of steam pipe lines is a hazardous process; hence, before any action is taken to carry out the process, a thorough inspection of the entire system is essential to ascertain the following (Table 8.1):

Table 8.1 Areas/items to be checked

Sl. No.	Areas/Items	Ok (√)
1	If steam blowing is adopted, then verify availability of protocols jointly signed by the customer, the engineer, the steam generator manufacturer, and the supplier (contractor) certifying successful completion of the following preoperational activities: i. Hydraulic test of steam generator (Chapter 9) ii. Air-tightness test of the furnace, air, and flue gas paths of the steam generator (Chapter 10) iii. Alkali flushing of the preboiler system (Chapter 3) iv. Steam generator initial firing and drying-out of insulation (Chapter 11) v. Chemical cleaning of the steam generator (Chapter 6) vi. Flushing of the lube oil piping system (Chapter 7)	
2	In the event compressed air blowing is followed, verify availability of protocols jointly signed by the customer, the engineer, the steam generator manufacturer, and the supplier (contractor), certifying successful completion of the hydraulic test of steam generator (Chapter 9)	
3	Protocol jointly signed by the customer, the engineer, the steam turbine and generator manufacturer, and the supplier (contractor), certifying that erection completion of turbine, generator, and associated systems is available	

Continued

Table 8.1 Areas/items to be checked—cont'd

Sl. No.	Areas/Items	Ok (√)
4	Before erecting permanent piping systems, protective coating used on internal surfaces, as a safeguard against corrosion during transport and storage, are removed as far as practicable	
5	Prior to erecting all temporary piping systems, they are thoroughly cleaned mechanically	
6	All relevant pipe lines are hydrotested at 1.5 times the design pressure. Protocol signed jointly by the customer, the engineer, and the supplier (contractor), certifying acceptability of the hydrotest, is verified	
7	In case steam blowing is adopted, ensure that temporary piping systems are provided with proper insulation and expansion devices	
8	Before erecting silencer at the end of the temporary piping, ensure that it is cleaned thoroughly and that internal parts of the silencer are not damaged	
9	All piping systems, whether permanent or temporary, are erected and supported properly along with associated valves. Since the blowing medium from temporary piping systems exhausts to atmosphere, thrust from the exhaust pipe will be transmitted to the pipe supports; hence, verify that the design of these supports are checked taking proper care	
10	All welded and flanged joints are secured	
11	Construction materials—temporary scaffolding, wooden planks, welding rod ends, cotton wastes, and so on, are removed from internals of blowing pipes	
12	The flow nozzle in main steam lines, if provided for monitoring of flow during normal operation, is not installed before blowing out these lines. Flow nozzle portions on respective piping are replaced with suitable spool pieces	
13	Verify that the following instruments installed on permanent piping systems, which must remain in service during blowing out, are available *Steam blowing:* i. drum steam pressure and temperature (applicable to drum-type steam generator) ii. drum level (applicable to drum-type steam generator) iii. separator outlet steam pressure and temperature (applicable to once-through steam generator) iv. separator level (applicable to once-through steam generator) v. SH outlet steam pressure and temperature vi. HP turbine inlet steam pressure and temperature vii. RH inlet steam pressure and temperature viii. RH outlet steam pressure and temperature ix. IP turbine inlet steam pressure and temperature x. feed water flow *Compressed air blowing:* i. air compressor discharge pressure gauge ii. air compressor discharge temperature gauge iii. pressure gauge upstream of temporary blow-out valve iv. SH outlet air pressure and temperature v. HP turbine inlet air pressure and temperature vi. RH inlet air pressure and temperature vii. RH outlet air pressure and temperature viii. IP turbine inlet air pressure and temperature	

Continued

Table 8.1 Areas/items to be checked—cont'd

Sl. No.	Areas/Items	Ok (√)
14	All other instruments installed on permanent piping systems, barring above (step 13), are kept isolated	
15	All associated control loops are operational	
16	Relevant protections, alarms, and interlocks are checked and kept ready	
17	Verify that the LP chemical dosing system is available for dozing of 50 mg kg^{-1} concentration of hydrazine and 15 mg kg^{-1} concentration of ammonia in feedwater (applicable when the blowing medium is steam)	
18	Estimated quantity of DM water (Section 8.7) is available	
19	Estimated quantity of fuel oil (Section 8.8) is available	
20	Required quantity of temporary materials (Section 8.4.1) is available	
21	Safety gadgets (Section 8.4.2) are arranged	
22	Temporary platforms are provided wherever required	
23	Adequate quantity of aluminum/copper/mild steel target plates and alloy steel target plates are arranged	
24	Verify that safety tags, eg, "no smoking," "danger," "keep off," and so on (in regional and English languages), are displayed wherever required	
25	Verify also that fans, compressors, tanks, pumps, valves, and associated equipment and systems, which are in service or are energized, are provided with proper safety tags	

8.4.1 Typical List of Bill of Materials

Materials that are typically required for steam/air blowing of various steam lines are listed below. Size and quantity of each of these items would vary from project to project and shall have to be assessed by concerned personnel associated with steam/air blowing activity.

1. blow-out kit for ESV, IV, HPT, exhaust NRV;
2. protective plugs for HP turbine control valves and IP turbine control valves;
3. blow-through kit and hydraulic blanking kit for HPBP pressure control valve;
4. motor-operated temporary blow-out valve/s for steam blowing; pneumatically operated temporary rotary blow-out valve/s for compressed air blowing;
5. target plates;
6. target plate holders;
7. temporary pipe lines;
8. spool pieces;
9. pressure-measuring instruments;
10. temperature-measuring instruments;
11. DCS to record all parameters and to compute disturbance factor at different points in the blow-out pipe line;
12. stop-watches to monitor time required to complete each blow.

8.4.2 *Typical List of Safety Gadgets*

Equipment that is generally used for safety of operating personnel and the area covered under blowing are as follows:

1. safety shoes of assorted sizes;
2. asbestos gloves;
3. hard hats of assorted sizes;
4. ear muffs;
5. first aid box;
6. sufficient lighting arrangements around temporary system;
7. cordoning of the exhaust area;
8. hooters at various places to send warning sound prior to commencing each blow-out operation;
9. thermal insulation of temporary piping;
10. safety sign boards;
11. safety tags;
12. camera for taking photographs of target plates.

Notes

Ensure qualified personnel only are involved in laying down the blow-out guidelines and in supervising the blow-out process until the normalization of all circuits.

8.5 *Preparatory Arrangements*

Prior to carrying out the blowing-out process, the following preparatory work shall be completed.

a. In order to complete the circuit for blowing out, some temporary piping connecting the permanent piping systems to the atmosphere should be erected and kept ready for putting into service;
b. A cross-section of temporary piping systems shall be at least the same as the permanent piping systems. In case of temporary cold and hot reheat pipe lines, respective cross-sections may have to be made larger than the corresponding cross-section of the permanent pipe lines so as to accommodate increase in volume of fluid flow resulting from reduction in pressure of the flowing medium;
c. The above piping systems are provided with drain valves for discharging any condensate;
d. Remove the internals of the HP turbine inlet ESVs and IP turbine IVs. Replace original covers of ESV and IV with temporary blow-out covers having the arrangement for connecting temporary piping systems per the recommendation of turbine

manufacturers. Seats of ESV and IV shall be protected with suitable protective covers supplied by turbine manufacturers. During blowing out, these temporary piping systems would facilitate cleaning of main steam line/s up to ESVs and cold reheat and hot reheat lines up to IVs;

e. Internals of HP turbine control valves and IP turbine control valves shall be removed. Install protective plugs at the inlet to these control valves and suitable protective covers on valve seats as supplied by turbine manufacturers;

f. Strainers located upstream of turbine shall be removed;

g. Remove the internals of HP turbine exhaust NRVs;

h. It is preferable to remove the internals of HP and LP bypass valves. If this is not permitted by the manufacturer, these control valves are to be kept closed and provided with temporary bypass lines;

i. Two temporary oil-free air compressors, each of 100% capacity, with suction filter and silencers, should be installed at a convenient location in the power house building near the boiler drum (drum-type steam generator) or to the separator (once-through steam generator). This location should not disturb the permanent equipment or piping installed in that area. These compressors would serve the source of blowing medium in the event compressed air blowing is adopted;

j. Air flow capacity and required discharge pressure of these compressors would vary from project to project and shall be decided jointly by the customer, the engineer, and the supplier (contractor);

k. Temporary piping from temporary blow-out air compressors may be connected either to the flange provided on the drum dish end to connect the feedwater inlet line downstream of the economizer (applicable to the drum-type steam generator) (Figs. 8.4, 8.6, 8.9, and 8.11) or at the inlet terminal end of the superheater header where steam coming from the separator will get connected (applicable to the once-through steam generator);

l. Wherever temporary piping systems interfere with passages, ramps, or gangways, suitable temporary crossings over the piping systems shall be provided. These crossings must not interfere with the thermal expansion of the piping systems;

m. Temporary blow-out valves followed by silencers shall be installed on the horizontal portion of temporary piping systems near the exhaust end. Final exit from temporary piping systems will be assembled with an inclination towards the sky. It shall be ensured that debris coming out from the exhaust end would not hit any structure or equipment;

n. For effective completion of blowing out, the speed of opening of temporary blow-out valves shall be as fast as possible;

o. The gate valve is generally used as the blow-out valve during steam blowing, opening time of which shall not exceed 60 s;

p. The temporary blow-out valve for compressed air blowing shall preferably be a pneumatically operated rotary valve (either ball or butterfly type) opening time of which is very fast (less than 5 s);

q. Material of target plates should have requisite effectiveness if it complies with the material specification of turbine blades. The intention is to obtain an impact velocity on target plates at least the same as the impact velocity of any particulate matter in the steam flow hitting the turbine blades. However, material of target plates generally used in the industry during initial stages of blowing is either aluminum or copper or mild steel, much softer than the material of turbine blades; for the final 2/3 blows, it is preferable to use alloy steel target plates;

r. Target plates shall be numbered in compliance with the number of blows to be carried out. This would facilitate proper identification of target plates and assessment of chronological improvement in the blowing-out process after each blow-out;

s. Suitable arrangement along with protection for the insertion and taking out of hot target plates in temporary piping systems shall be provided;

Notes

1. Temporary connections are unit specific and vary from project to project. Hence, layout of temporary piping, installed generally in the turbine house, shall have to be jointly decided at the site among the customer, the engineer, and the supplier (contractor) responsible for successful completion of the above cleaning activity;

2. Size and quantity of temporary piping materials, including elbows, tees, valves, supports, and so on, shall be assessed at the design stage. These requirements, however, may vary based on the exact location of the steam generator at the site as well as the location of terminal points of various steam pipe lines, and the layout of temporary blow-out piping systems. The final quantity of piping materials including associated equipment shall be procured accordingly;

3. To facilitate operation of temporary blow-out valves and monitoring of pressure and temperature of circulating fluids, temporary platforms may have to be erected as required.

---

t. Boiler, condensate, and feedwater systems must be ready for safe and reliable operation;

u. Install spectacle plate with flanges on temporary pipe lines as per the layout and configuration of a particular unit; this would facilitate sequential blowing of one circuit to another with minimum interruption in changeover time. Some typical locations of these spectacle plates with flanges is shown in Fig. 8.3 as well as in some other figures;

v. While the blowing-out process is underway, the turbine must be put on its turning gear to safeguard the turbine against damage due to unexpected leakage of blowing medium that may cause rotation of the turbine;

w. Adequate quantity of DM water along with its transferring facilities must be assessed prior to taking up each blow-out step (applicable if steam blowing is adopted) and shall be made available. For this, all available storage facilities, like the DM water storage tank and including the deaerator feedwater storage tank and condenser hotwell, may be used;

x. The quantity and type of safety gadgets to be required for the process should be ascertained beforehand and must be kept available as long as the blowing-out process continues (a typical list of safety gadgets is furnished under Section 8.4.2);

y. If steam blowing is adopted, a suitable arrangement to collect a sample of blowing steam along with relevant instrumentation for the determination of opacity and Fe content of sample are available;

z. The tripping of the steam generator from "drum high-high level and drum low-low level" shall be bypassed (applicable to the drum-type steam generator);

za. The tripping of the boiler circulating pump from level variation in the separator shall be bypassed (applicable to the once-through steam generator);

zb. The following instruments must be installed and readings recorded in DCS (preferable) during each blow (Tables 8.2 and 8.3). These measurements would facilitate calculation of actual *Disturbance Factor* achieved at various points in a specific blow-out sequence and should be recorded for future reference.

Table 8.2 Log sheet of steam blowing

Parameters	No. of Blows												
	1	2	3	4	5	6	7	8	9	10	11	12	n
Drum steam pressure (applicable to drum-type steam generator)													
Drum steam temperature (applicable to drum-type steam generator)													
Drum level (applicable to drum-type steam generator)													
Separator outlet steam pressure (applicable to once-through steam generator)													
Separator outlet steam temperature (applicable to once-through steam generator)													
Separator outlet level (applicable to once-through steam generator)													
SH outlet steam pressure													
SH outlet steam temperature													
HP turbine inlet steam pressure													
HP turbine inlet steam temperature													
RH outlet steam pressure													

Continued

Table 8.2 Log sheet of steam blowing–cont'd

Parameters	No. of Blows												
	1	2	3	4	5	6	7	8	9	10	11	12	*n*
RH outlet steam temperature													
IP turbine inlet steam pressure													
IP turbine inlet steam temperature													
Steam pressure at point 2 (Fig. 8.1)													
Steam temperature at point 2													
Steam pressure at point 3 (Fig. 8.1)													
Steam temperature at point 3													
Feedwater flow													

Table 8.3 Log sheet of compressed air blowing

Parameters	No. of Blows												
	1	2	3	4	5	6	7	8	9	10	11	12	*n*
Drum outlet air pressure (applicable to drum-type steam generator)													
Drum outlet air temperature (applicable to drum-type steam generator)													
Separator outlet air pressure (applicable to once-through steam generator)													
Separator outlet air temperature (applicable to once-through steam generator)													
SH outlet air pressure													
SH outlet air temperature													
HP turbine inlet air pressure													

Continued

Table 8.3 Log sheet of compressed air blowing—cont'd

Parameters	No. of Blows												
	1	2	3	4	5	6	7	8	9	10	11	12	n
HP turbine inlet air temperature													
RH outlet air pressure													
RH outlet air temperature													
IP turbine inlet air pressure													
IP turbine inlet air temperature													
Air pressure at point 2 (Fig. 8.1)													
Air temperature at point 2													
Air pressure at point 3 (Fig. 8.1)													
Air temperature at point 3													

8.6 Operating Procedure

Procedure to blow-out steam pipe lines is addressed under the following paragraphs. Since the procedure for blowing with steam as the blowing medium is different from the procedure for blowing with compressed air, these two methods are addressed separately for convenience of understanding.

When adopting steam blowing, two separate methods—continuous blowing and puffing, are followed in the industry. Procedures of these two methods are not same, so they are treated separately here.

8.6.1 Steam Blowing

Prior to steam blowing, feedwater quality at the economizer inlet must conform to the requirement as given in Table 8.4.

To carry out steam blowing, initial firing of the steam generator may be continued at a slow rate in conformity with the steam generator manufacturer's recommended cold start-up curve.

Ensure that the temporary blow-out valve is closed. Close the SHO stop valve/s along with one of the associate bypass equalizing valves, keeping another bypass equalizing valve/s to remain open. Light-up the steam generator following the cold start-up procedure. Gradually raise the steam generator pressure to about 1.5 MPa. The main steam line/s, ESVs at HP turbine inlet,

Table 8.4 Quality of feedwater

Contaminants	Unit	Recommended Value
Hydrazine (as N_2H_4)	$\mu g\ kg^{-1}$	≥ 200
Iron (as Fe)	$\mu g\ kg^{-1}$	≤ 50
Oxygen (as DO)	$\mu g\ kg^{-1}$	≤ 10
Silica (as SiO_2)	$\mu g\ kg^{-1}$	≤ 30
Copper (as Cu)	$\mu g\ kg^{-1}$	≤ 20
pH	–	9.3–9.6
Specific conductivity, at 298 K	$\mu S\ cm^{-1}$	5.5–11
Conductivity (after cation exchanger), at 298 K	$\mu S\ cm^{-1}$	≤ 0.5

and temporary pipe lines thereafter may be heated at a slow rate by gradually opening the closed bypass equalizing valve. During this heating stage, adequate care should be taken to ensure free expansion of both permanent and temporary pipe lines at frequent intervals. On reaching the required temperature, SHO stop valve/s may be opened fully. Close the associate bypass equalizing valves.

8.6.1.1 Continuous blowing method

In the continuous blowing method, steam velocity is the driving force for removal of debris. In this method, firing in the steam generator is not disturbed and blowing pressure is maintained constant until the completion of blowing. Hence oil firing alone is unable to sustain the firing, and it becomes essential to put at least two pulverizers in service.

Advantages of the continuous blowing method are:

i. Less time is required for completion of blowing (in a typical case it may take 6–7 days);
ii. The steam generator is lit up only once at the beginning of steam blowing;
iii. Consequently, thermal stress developed in tube material is less;
iv. Since the pressure is maintained, the constant consequent reaction force on the temporary pipe line/s gets reduced;
v. On completion of steam blowing, normalization time of the system is less;
vi. Less time is required from final light-up to synchronization to commercial loading due to availability of the pulverized coal-firing system, ash handling system, and so on.

Disadvantages of the continuous blowing method are:

i. The DM water make-up to the system during steam blowing is exorbitant;
ii. Pulverizer, coal feeder, associated equipment and piping system, instrumentation and control, and so on, for at least the lowest two tiers of coal burners, must be made ready in all respects;

iii. Erection of the bottom ash hopper, economizer hopper, air heater hopper, and ESP hoppers must be complete;

iv. The coal handling plant has to be kept ready;

v. Ash evacuation and removal systems along with associated equipment must be made available prior to carrying out the blowing process.

The procedure to be followed during continuous steam blowing is described below:

As explained earlier, continuous blowing is adopted in low pressure-low temperature and medium pressure-medium temperature boilers (that are not provided with reheater and HP-LP bypass valves). As such, only circuit a (Fig. 8.3) falls within the purview of continuous blow. Miscellaneous steam supply lines, circuit e (Fig. 8.12), also follow this method. For circuit e, the target plate is not required. Visual confirmation of clean exhaust would suffice.

Steps generally followed to blow-out circuit a are described below:

i. Increase the firing rate and raise the steam generator pressure to 2.5–3.0 MPa;

ii. The temporary blow-out valve shall be opened gradually to verify the integrity of the temporary pipe line along with its supports and to ensure unobstructed expansion during heating;

iii. On successful verification of step (ii), the steam generator firing rate shall be stepped up to raise the steam generator pressure to 4.0 MPa;

iv. During steam blowing, the drum-water level will be subject to extreme fluctuations. When the steam flow approaches the "blow-out" flow, as the temporary blow-out valve is opened fast, the drum-water level will rise rapidly and may disappear beyond the visible limit of the drum-level gauge glass (in a drum-type steam generator) with consequent risk of carryover of water into superheaters. Hence, before commencing each blow, ensure that the boiler drum level is maintained at a low level.

As the blowing out continues, the drum-water level will reappear and may drop out of sight. Feedwater flow must be established as soon as the water level drops back in sight to prevent an excessively low drum-water level, which in turn will prevent either exposure of riser tubes to furnace heat with consequent burn-out of riser tubes in natural circulation steam generators or suction loss of the boiler circulating pumps in a forced circulation steam generator [6];

v. Open the temporary blow-out valve gradually to commence blowing. Raise the steam generator firing rate simultaneously to prevent pressure from falling. Slowly raise rate of steam flow in accordance with the rate of rise in pressure and temperature as per the start-up curve in order to attain a disturbance factor of 1.3–1.6;

vi. At the beginning of the blowing out operation, debris coming out from the exhaust end will be visible. Continue blowing out the circuit until the exhaust debris become invisible to the naked eye. Close the temporary blow-out valve and cut down the firing rate following the steam generator shutdown curve;

vii. Once the temporary pipe line is sufficiently cooled, insert a target plate. Open the temporary blow-out valve. Slowly raise steam generator pressure to 4.0 MPa along with corresponding temperature and rate of steam flow commensurate with a disturbance factor of 1.3–1.6. Continue blowing under this condition for about 30 min. Close the temporary blow-out valve and cut down the firing rate. Once the temporary line is cooled, take out the target plate. Observe the number of marks (both "crest" and "cavity") on the target plate.

viii. Continue repeating step (vii) until the indentation on the successive target plates remain unchanged and by and large complies with the acceptable limit explained earlier under Section 8.1.

Steps (vi) and (vii) may constitute 30–40 blows depending on the extent of debris remaining in the system. A fresh target plate shall be used for each blow. The target plate shall be withdrawn for inspection as soon as it cools off after a blow. Arrangements may be made to photograph target plates so that impingement marks may be clearly visible.

The process may continue for two or more weeks. Hence, every day at 19:00 h after closing the temporary blow-out valve, firing should be stopped and the steam generator boxed up in preparation for the next day. This procedure has an added advantage in that it results in heating and cooling of pipe lines which are effective in loosening and removing pipe scales and other foreign materials remaining in the piping system.

8.6.1.2 Puffing method

In the puffing method, thermal shock is the driving force. This method may be adopted to blow-out steam pipe lines of circuits a, b, c, and d (Figs. 8.3, 8.5, 8.8, and 8.10) described earlier.

Advantages of the puffing method are:

i. Pulverized coal-firing system is not required;
ii. Eventually, it is not essential to make ready coal handling plant, ash evacuation, and removal system;
iii. Erection of the bottom ash hopper, economizer hopper, airheater hoppers, and ESP hoppers is not essential. In place of hoppers, the bottom of the furnace and the economizer, air heater, and ESP may be temporarily kept sealed with steel plates;

Disadvantages of the puffing method are:

i. Light-up and shutdown have to be repeated;
ii. Tube materials are subjected to more thermal stresses;
iii. Supports experience repeated sudden loading;
iv. More time is required for complete steam blowing due to stagewise blowing (typically 15–20 days);

v. On completion of the blowing system, more normalization time is required;

vi. Time required from first light-up of boiler to synchronization to commercial loading is more.

Steps to be followed during the puffing method are:

i. Light-up the steam generator following the steam generator manufacturer's recommended cold start-up procedure. For preliminary checking, gradually raise steam pressure to 4 MPa at either boiler drum (drum-type steam generator) outlet or water separator (once-through steam generator) outlet;

ii. Maintain the water level in the boiler drum or separator near to the lowest visible limit in the gauge glass prior to opening the temporary blow-out valve. This will prevent carryover of water to the superheater due to a rise in the boiler drum level following a drop in boiler pressure during blowing;

iii. Kill the steam generator fire and rapidly open the temporary blow-out valve;

iv. As explained under Section 8.3 v (d), a drop in saturation temperature during each blow must not exceed 40 K to prevent thermal stress. Hence, when the steam pressure drops down to 1.9 MPa close the temporary blow-out valve;

Note

Since the boiler needs to be re-lighted up, the time interval between successive blows will be about 90 min. Furthermore, as the blow-out process has to be stopped at 19:00 h every day it will be difficult to carry out more than 6 blows per day.

v. Light-up the steam generator and gradually raise steam pressure as recommended by the steam generator manufacturer to about 4–6 MPa while blowing main steam and HP bypass pipe lines and to about 5–7 MPa for blowing cold reheat and hot reheat pipe lines. This would ensure a disturbance factor of the order of 1.3–1.6 MPa during blow-out.

Note

Actual pressure, however, is to be finalized by the customer, the engineer, and the supplier (contractor) at the design stage while developing a project-specific steam blowing procedure;

vi. Maintain the lowest water level in the boiler drum or separator;

vii. Kill the steam generator fire and rapidly open the temporary blow-out valve;

viii. In order to restrict drop in saturation temperature during blowing to 40 K, close the temporary blow-out valve once the steam pressure drops down to 1.9–3.1 MPa (main steam and HP bypass lines) or to 2.5–3.7 MPa (cold reheat and hot reheat lines);

ix. At the beginning of the blowing-out operation, debris coming out from the exhaust end will be visible;

x. Repeat steps v to viii for executing the next blow-out of pipe lines;

xi. Continue blowing out the circuit until the exhaust debris become invisible to the naked eye;

xii. Once the temporary pipe line is sufficiently cooled, insert the first target plate, which is of softer material (aluminum, copper, or mild steel);

xiii. Repeat steps v to viii for the next blow-out;

xiv. Once the temporary line is cooled, take out the target plate. Observe the number of marks (both "crest" and "cavity") on the target plate;

xv. Insert a new target plate of softer material;

xvi. Continue repeating steps xiii, xiv, and xv until the indentation on the successive target plates remains unchanged;

xvii. For the final 2/3 blows, use target plates of alloy steel;

xviii. Verify that indentation on all the plates are identical and complies with the acceptable limit explained earlier (Section 8.1);

xix. This circuit may now be declared sufficiently clean.

The above steps may be followed for circuits a, b, c, and d mentioned earlier under Section 8.1.

Note

Every day at 19:00 h after closing the temporary blow-out valve, firing should be stopped and the steam generator boxed up in preparation for the next day.

8.6.2 Compressed Air Blowing

Compressed air blowing is commenced on completion of hydraulic test of the steam generator. During the process of compressed air blowing, erection of steam generator auxiliaries, like burners, fans, air preheaters, air and flue gas ducts, coal-firing systems, and so on, may continue in parallel.

Quality of compressed air to be used during blow-out must conform to the following specification:

- must be clean and dry;
- must be practically oil free (oil content should be less than 1 mg kg^{-1});
- size of suspended particulates in air should not be larger than 40 µm;
- temperature of air delivered by the compressor should be about 20 K higher than ambient temperature.

Discharge capacity and pressure of air compressors shall be decided during the design stage jointly by the customer, the engineer, and the supplier (contractor). In certain projects, working pressure of about 4 MPa would suffice to ensure successful compressed air blowing, in large steam generators (500 MW and above), working pressure may have to be raised to 6 MPa or higher as recommended.

Prior to starting compressed air blowing, verify that temporary piping from the temporary blow-out air compressors is connected to either the feedwater inlet line to drum (applicable to the drum-type steam generator) (Figs. 8.4, 8.6, 8.9, and 8.11) or steam inlet line to superheater downstream of water separator (applicable to the once-through steam generator).

The sequence of activities can be as follows:

i. Close the temporary blow-out valve. Open the SHO stop valve/s, keeping associate equalizing bypass valves closed;
ii. Start one of the temporary blow-out air compressors. Raise the air pressure to about 0.2–0.4 MPa as recommended by steam generator manufacturers. Open all drain valves in the selected circuit to impart free blow through them to the atmosphere for removing any liquid and/or loose material collected at low points. Likewise open all vent valves in the selected circuit to clean them by imparting free blow to the atmosphere. Thereafter close all drain and vent valves;
iii. Verify that the selected circuit is free from any leakage;
iv. Raise the air pressure at the SHO to 2 MPa. Open the temporary blow-out valve (valve opening time must be less than 5 s to promptly establish the desired flow conditions and disturbance factor);
v. Continue blowing the steam pipe line until air pressure at the SHO drops to 0.7 MPa, below which velocity in the line would not be adequate to remove any debris. Close temporary blow-out valve;
vi. On completion of the above preliminary blow, verify the integrity of permanent and temporary piping systems along with their supports and anchors;
vii. Raise air pressure at the SHO to about 4 MPa. Open the temporary blow-out valve. Continue blowing the steam pipe line. Close the temporary blow-out valve when air pressure at the SHO drops to around 0.7 MPa;
viii. Air pressure thereafter may be raised to 6 MPa or to the maximum pressure as recommended by steam generator manufacturers; then the temporary blow-out valve may be opened. Once air pressure decays to 0.7 MPa, close the temporary blow-out valve;
ix. At the beginning of the blowing-out operation, debris coming out from the exhaust end will be visible (Fig. 8.15);
x. Continue blowing out the circuit until clear air comes out through the exhaust (Fig. 8.16);

   xi.   Insert the target plate of softer material (aluminum, copper, or mild steel) downstream of the temporary blow-out valve, but upstream of the silencer;

  xii.   Repeat steps vii or viii for the next blow-out;

 xiii.   Take out the target plate. Observe the number of marks (both "crest" and "cavity") on the target plate;

  xiv.   Insert the new target plate of softer material;

   xv.   Continue repeating steps xi, xii, and xiii as often as necessary until the indentation on the successive target plates remain unchanged;

  xvi.   For final 2/3 blows, use target plates of alloy steel;

 xvii.   Verify that indentation on all these plates are identical and complies with the acceptable criteria explained earlier under Section 8.1;

xviii.   This circuit may now be declared to be sufficiently clean.

The above steps may be repeated for effective cleaning of circuits a, b, c, and d (Figs. 8.4, 8.6, 8.9, and 8.11) described beforehand.

Note

Every day at 19:00 h after closing the temporary blow-out valve, the process of blowing should be stopped and steam generator boxed up in preparation for the next day.

Fig. 8.15

Debris coming out through the exhaust at the beginning of blow-out operation. *Source: From http:// www.omnicompressedair.com/plant-air-blows.*

Fig. 8.16
Clear air coming out through the exhaust. *Source: From http://www.omnicompressedair.com/plant-air-blows.*

8.7 Estimation of DM Water

During steam blowing of a steam generator, there would be a large demand of DM water. Hence, prior to performing each steam blow, adequate stock of DM water must be built up. As such, it is essential to estimate the requirement of DM water correctly. It is advisable to procure a higher quantity of DM water than estimated, as any shortfall of this item during actual execution of the process will jeopardize whole sequence of activities and may become detrimental to areas falling within this process.

A typical example for determining DM water during steam blowing is given below for the benefit of readers.

Example 8.2

A steam generator is capable of generating 225 kg s^{-1} of steam at pressure and temperature of 15.8 MPa and 813 K, respectively, under BMCR condition. Determine the quantity of DM water to be kept in stock before carrying out each stage of blowing out of the superheater and main steam pipe lines at 5 MPa pressure and 723 K temperature of steam adopting the continuous blowing method. The disturbance factor to be maintained during steam blowing is 1.5.

Solution
Table 8.5 may be prepared based on the given data:

Example 8.2—cont'd

Table 8.5 Parameters of steam

Parameter	SHO	During Blowing
Pressure (MPa)	15.8	5.0
Temperature (K)	813	723
Specific volume (m^3 kg^{-1})	0.021206	0.063234
Flow (BMCR) (kg s^{-1})	225	–

Taking the help of Eq. (8.1), quantity of steam required during steam blowing is

$$Q_{blow} = \left\{ \sqrt{1.5 \times \frac{0.021206}{0.063234}} \right\} \times 225 = 160 \text{ kg s}^{-1}$$

Considering the duration of each blow to be of 30 min and keeping a margin of 50% on the quantity of blow-out steam, stock of DM water is required to be maintained before commencing each blow,

$$Q = 160 \times 30 \times 60 \times 1.5 = 43,200 \text{ kg}$$

So the quantity of DM water to be kept in stock before carrying out each steam blow: 43,200 kg.

8.8 Estimation of Fuel Oil

During steam blowing, all the load carrying fuel oil burners would be operated at their maximum capacity (30% BMCR) for about 6 h per day. Hence, in order to estimate total fuel oil requirement it is to be taken into account that duration of steam blowing for each circuit may continue for a period of 15–20 days depending on the condition of pipe internals.

A typical example pertaining to blowing out of superheater and main steam pipe line is addressed below for the convenience of readers to calculate the fuel oil requirement during steam blowing. Same approach may be followed for calculating fuel oil requirement during blow-out of remaining circuits.

Example 8.3

If the efficiency of steam generator of Example 8.2 is 86.9% at 30% BMCR load, which can be attained by burning fuel oil having GCV of 41,000 kJ kg^{-1}, estimate the requirement of fuel oil during steam blowing.
For complete removal of debris from superheater and main steam line steam blowing will continue for 14 days and load carrying fuel oil burners will operate at 30% BMCR capacity for about 6 h per day.

Continued

Example 8.3—cont'd

Pressure and temperature of main steam at SHO and pressure and temperature of feedwater at economizer inlet along with other salient parameters during 30% BMCR load are summarized in Table 8.6.

Table 8.6 Pressure and temperature of main steam and feedwater

Parameters	SHO	ECO IN
Pressure (MPa)	15.20	15.72
Temperature (K)	748	455
Sp. enthalpy (kJ kg^{-1})	3233.90	778.86
Flow (kg s^{-1})	67.5	67.5[a]
Efficiency (%)	86.9	

[a]Neglecting blow down or any leakage from the boiler.

Solution

From given data quantity of fuel oil to cater 30% BMCR steam flow,

$$Q_{FO} = \frac{67.5 \times (3233.90 - 778.86)}{41,000} \times \frac{1}{0.869} = 4.65\,kg\,s^{-1}$$

Hence for completion of steam blowing of superheater and main steam line total requirement of fuel oil will be

$$Q_{FOT} = 4.65 \times 6 \times 3600 \times 14 = 1,406,160\,kg$$

8.9 Conclusion

When the target plate is showing only a little improvement on completion of consecutive blowing out of pipe lines, the blowing-out process should be considered complete.

Following blowing out of various steam pipe lines, floating of steam generator safety valves (Chapter 12) may be undertaken. Once the floating process is successfully completed, the steam generator along with various steam piping systems may be declared to be ready for further use after normalizing the permanent piping systems.

Table 8.7 depicts a typical protocol, which must be signed jointly by all concerned.

Table 8.7 Protocol (steam/air blowing of main steam, cold reheat, hot reheat, and other steam pipe lines)

Circuit Under Blowing	Status of Aluminum/ Copper/Mild Steel Target Plate	Status of Alloy Steel Target Plate		
"a"	Fulfills Criteria 1/Criteria 2 Section 8.1: Introduction	Shows no further improvement		
"b"	Fulfills Criteria 1/Criteria 2 Section 8.1: Introduction	Shows no further improvement		
"c"	Fulfills Criteria 1/Criteria 2 Section 8.1: Introduction	Shows no further improvement		
"d"	Fulfills Criteria 1/Criteria 2 Section 8.1: Introduction	Shows no further improvement		
Steam generator is ready for floating of safety valves				
Signed by the customer	Signed by the engineer	Signed by the steam generator manufacturer	Signed by the steam turbine manufacturer	Signed by the supplier (contractor)

References

[1] British Electricity Institute, Modern Power Station Practice, Volume H—Station Commissioning, third ed., Pergamon Press, London, 1991.
[2] CEGB Design Memorandum 068/25, Steam Purging, February 1970.
[3] GE Power Systems, Cleaning and Blowing Steam Seal Lines, October 2003.
[4] S.C. Stultz, J.B. Kitto (Eds.), STEAM Its Generation and Use, 41st ed., The Babcock and Wilcox Company, Barberton, OH, 2005.
[5] GE Power Systems, Experience with compressed air cleaning of main steam piping, in: The American Power Conference, Chicago, Illinois, April 23–25, 1990.
[6] C. Bozzuto, Clean Combustion Technologies, fifth ed., Alstom, Windsor, CT, 2009.
[7] D.K. Sarkar, Thermal Power Plant—Design and Operation, Elsevier, Netherlands, 2015.

PART 2

Activities That Make Critical Equipment Ready to Put Them in Service

CHAPTER 9

Hydraulic Test of Steam Generator

Chapter Outline

9.1 Introduction

More than 110 years back on March 10, 1905, a major explosion in a steam generator took place in the Grover Shoe Factory in Brockton, Massachusetts. This catastrophic failure prompted all concerned responsible for maintaining safety of plant and personnel to formulate the steam generator safety rules and regulations and made it mandatory to implement them, thereby ensuring safety, security, healthiness, welfare, and so on, of plant personnel and equipment.

Taking cue from this practice, Boiler Inspectorate of many countries of the world, who employ steam generators, laid down stringent rules and regulations, which must be implemented in the associated country for controlling the design of steam generator and for selecting materials during the manufacturing and erection stages, and also during the operation of steam generator. Any violation of these rules and regulations, if not followed wherever applicable, can cause associated plant authorities to land in legal trouble.

Thermal Power Plant. http://dx.doi.org/10.1016/B978-0-08-101112-6.00009-5

Preparation of steam generator safety rules and regulations is an onerous task that requires many toilsome hours; hence, there are countries that instead of developing their own rules and regulations adopt the rules and regulations already in vogue in technically advanced countries or organizations—ASME boiler and pressure vessel code (ASME BPVC), ISO 16528-1-Boilers and Pressure Vessels-Part 1-Performance Requirements, and so on.

As a part of fulfilling requirements of "steam generator safety rules and regulations," it is mandatory to carry out hydraulic test or in short "hydrotest" of pressure parts of new steam generators on completion of their erection as well as pressure parts of running steam generators after undertaking any repair in pressure parts or following renovation and modernization of steam generators. It is also statutory under the jurisdiction of regulatory authorities to undertake a hydrotest of pressure parts annually to verify any degradation of pressure part material and to ensure continuous safety of steam generators.

The hydrotest should be conducted under the supervision of and witnessed by an authorized representative of "boiler inspectorate" or "regulatory authorities" who should represent a state or municipality having jurisdiction, or the "insurance company" that indemnifies financial protection against future loss of the steam generator resulting from any hazard.

Pressure parts of drum-type steam generators and of once-through steam generators are not identical, barring the reheat circuit, which is common for both.

In drum-type steam generators pressure parts constitute piping from the economizer inlet header, economizer, economizer outlet header, and interconnecting piping up to the boiler drum, downcomers, waterwall bottom ring headers, waterwalls, intermediate headers, saturated steam piping from boiler drum up to the superheater inlet headers, superheated steam piping up to the back-pass junction headers, steam-cooled walls, superheaters, and superheated steam piping up to the superheater outlet.

Pressure parts of once-through steam generators typically comprise piping from the economizer inlet, economizer, economizer outlet interconnecting piping, furnace ring header, lower waterwalls, intermediate headers, upper waterwalls, waterwall outlet headers, steam separator, saturated steam piping from steam separator up to superheater inlet headers, superheated steam piping up to back-pass junction headers, steam-cooled walls, back-pass outlet headers up to the superheater outlet, water collection tank, and start-up system.

Reheat circuit of large steam generators covers cold reheat piping starting from downstream of the HP (high pressure) turbine exhaust NRVs (non-return valves) up to the reheater inlet header, HP bypass downstream piping, reheater, hot reheat piping starting from the reheater outlet header up to the upstream of the IP (intermediate pressure) turbine inlet interceptor valve, LP (low pressure) bypass upstream piping.

It is general practice in the industry to utilize "water" as the test medium during the hydrotest of steam generators since water possesses the special property of incompressibility. Advantage of "incompressibility" is that after completely filling up the pressure parts of large steam generators it would require hardly 0.20% or so of additional water to raise pressure from atmospheric to about 25 MPa. Thus, a test could be conducted within a short period of time with less effort. Furthermore, in the event of any failure in pressure parts, water pressure could be relieved almost immediately, thus obviating further damage within the system. In addition, water is readily available in nature, it is nontoxic, and it does not pose any environmental concern.

Note

In the case of pressurized systems, other than steam generators, where pressure parts cannot be completely filled with water, the test medium during the pressure test could be "air" or "nitrogen" or any nonflammable or nontoxic gas.

The hydrotest of the steam generator is conducted prior to carrying out successful completion of "air-tightness/leakage test of the furnace, air, and flue gas ducts of the steam generator" (Chapter 10). It is advised to ensure that the time interval between conducting the "hydrotest" and "chemical cleaning of the steam generator" (Chapter 6) is as short as possible lest wet surface of pressure parts following the hydrotest lead to corrosion-related problems.

9.2 Objective

In large steam generators there are about tens of thousand tons of steel tubes, pipes, drums, or steam separators, other vessels that contain the high pressure steam as is evident from the earlier discussion. To build an integrated vessel, the steam generator, utilizing the above items, more than 60,000 welding joints need to be carried out at the manufacturer's works and also on site. During the manufacturing stage and also during erection of steam generators, there may remain some omissions and voids in these joints overlooked by visual inspection. Hence one of the objectives of the hydraulic test or hydrotest is to confirm the integrity of these welding joints and fittings.

The hydrotest is a nondestructive endurance test to finding out any weak link within the pressure parts of steam generators that would get reflected by eventual failure of weak links during the test. Once conducted successfully, this test would ensure the safety, reliability, and leak tightness of pressure parts, thereby ensuring also safe, trouble-free, reliable operation of steam generators. During the manufacturing stage, a steam generator gets subjected to various tests to ensuring its integrity. Hence, the hydrotest may be construed as the final test to ensuring completeness, compliance of rules and regulations, and safety of steam generators prior to generating steam.

9.3 Precautions

Pressure of water during the hydrotest of large steam generators is raised to such a high level that water coming out from even a pinhole leakage may act as a sharp machete and slash any being or thing coming across the water flow path. Hence extreme precautionary measures must be taken while conducting the hydrotest.

Over and above observing "general precautionary measures" discussed under Section 2.1.1: Quality Assurance, the following "specific precautionary measures" shall have to be observed during the hydrotest:

i. Authorized personnel must have proper knowledge of all attributes of the hydrotest and should be capable of taking emergency actions in the event of an accident or of preventing likely occurrence of an accident;

ii. Availability of safe access to the test area must be ensured for inspection and repair;

iii. While the hydrotest is underway, all activities around the area shall be strictly prohibited;

iv. The temperature of the water for filling pressure parts shall be within 293–323 K [1].

Note

Contrary to the above guidelines, another practice is exercised in the industry of maintaining temperature of filling water within 7–15 K above the drum external surface temperature.

v. It is advised to use DM water having specific conductivity of <0.2 μS cm^{-1} for pressure parts filling;

vi. Dozing of ammonia and hydrazine to DM water is desirable to maintaining pH of 9–10.0 and hydrazine concentration of 100–200 mg kg^{-1};

vii. Availability of adequate plant drainage systems must be ensured for safe discharge of a large quantity of water on completion of successful hydrotest of the steam generator; and

viii. While raising the pressure for the first time, it shall be strictly prohibited to enter the enclosure of the test circuit as the test pressure exceeds 2.5 MPa.

9.4 Prerequisites

Before conducting the hydrotest of the steam generator, a thorough inspection of systems under purview is essential to ascertain the following (Table 9.1):

9.4.1 Typical List of Bill of Materials

Materials that are typically required for a hydrotest of steam generators are listed below. Size, type, and quantity of each of these items would vary from project to project and shall have to be assessed by concerned personnel associated with the hydrotest.

1. Temporary piping for chemical dozing;
2. temporary tank;
3. temporary chemical dozing pump;
4. floodlights;
5. handlamps;
6. Sky climbers;

Table 9.1 QA checklist

Sl. No.	Areas/Items	Ok (√)
1.	Erection completion protocol for the following equipment signed jointly by the customer, the engineer, the steam generator manufacturer, and the supplier (contractor) is available: i. economizer; ii. steam generator drum/separator; iii. downcomers (drum-type steam generators); iv. waterwalls; v. boiler water circulating pumps, if provided; vi. pendant superheater; vii. primary and secondary superheaters; viii. reheaters; ix. interconnecting piping; x. safety valves on boiler drum (drum-type steam generators); xi. safety valves on superheater outlets; xii. safety valves on reheater inlet header and outlet header; xiii. reheater inlet isolating devices located on cold reheat pipe lines downstream of HP turbine exhaust NRVs; xiv. reheater outlet isolating devices on hot reheat pipe lines located upstream of HP turbine inlet IVs; xv. all drain and vent valves of pressure parts; xvi. continuous blow-down tank; xvii. steam generator DM water filling pump; xviii. test pressure gauges on boiler drum (drum-type steam generators), superheater outlet, and reheater outlet; xix. permanent platforms, staircases, and handrails at various elevations around the furnace, turbine house, and so on; xx. all temporary fittings and blind flanges shall be suitable to withstand hydrotest pressure and should be free from any defect, lest their failure may interrupt the test process; xxi. verify that all welding joints are complete, stress relieved, and radiographed (X-rayed); and xxii. any other items as required.	
2.	Visually inspect all welding joints, as far as practicable, to verify their integrity.	
3.	Pressure transmitters, pressure switches, gauge glasses, or control components, which should not be subjected to hydrotest, must be isolated.	
4.	Construction materials—temporary scaffolding, wooden planks, welding rod ends, cotton wastes, and so on, are removed from the internals of the furnace, buckstays, and piping system.	
5.	Calibrated pressure gauges are available on drum, superheater outlet header, reheater outlet header.	
6.	Verify that the temporary high pressure positive displacement pump (Fig. 9.1 or Fig. 9.2) is available and connected to the steam generator.	

Continued

Table 9.1 QA checklist—cont'd

Sl. No.	Areas/Items	Ok (√)
7.	Temporary chemical injection pump is available.	
8.	Temporary tank for the preparation of chemical solution is available.	
9.	DM plant is operational.	
10.	Adequate quantity of DM water storage is available.	
11.	Instrument air compressors including instrument air system are operational.	
12.	Temporary platforms and staircases are provided wherever required.	
13.	Sky Climbers with suitable scaffolding arrangement are provided inside the furnace.	
14.	Manholes and peepholes on the furnace and the second pass are kept open for inspection.	
15.	Verify that suitable warning signs are displayed to restrict entrance of unauthorized personnel inside the cordoned area.	
16.	Fire-fighting system is ready in all respects around the test area, and fire tender is available.	
17.	Required quantity of DM water is available (Section 9.8.1).	
18.	Required quantity of chemicals are available (Section 9.8.2).	
19.	Ensure that materials required for hydrotest (Section 9.4.1) are available.	
20.	Check that necessary safety gadgets (Section 9.4.2) are arranged.	
21.	Verify that "no smoking," "danger," "keep off," and other safety tags (in regional and English languages) are displayed at various places around the test area.	
22.	Verify also that any equipment and systems, which are in service or are energized, are provided with proper safety tags (Appendix C).	

Fig. 9.1
Portable motor operated high pressure positive displacement pump.
Source: From http://www.directindustry.com/industrial-manufacturer/hydraulic-pump-61099-_9.html.

Fig. 9.2

Portable hand operated high pressure positive displacement pump. *Source: From http://www. rhinopumps.co.za/.*

7. calibrated pressure gauges;
8. gagging devices; and
9. stop watches to monitor waiting time during testing.

9.4.2 Typical List of Safety Gadgets

Equipment that are generally used for the safety of operating personnel are as follows:

1. safety shoes of assorted sizes;
2. hand gloves;
3. hardhats of assorted sizes;
4. side-covered safety goggles with plain glass;
5. safety belts;
6. first aid box;
7. fire tender;
8. ambulance;
9. safety sign boards; and
10. safety tags.

9.5 Preparatory Arrangements

Prior to going ahead with conducting the hydrotest of a steam generator, certain valves (Table 9.2) are to be kept closed, which will vary from project to project. A list of these valves, applicable to a specific steam generator, is to be developed by concerned personnel before undertaking the hydrotest.

Table 9.2 Typical list of valves to be kept closed

Sl. No.	Description
1.	Superheater outlet stop valve/s
2.	Superheater outlet stop valve bypass equalizing valves
3.	Superheater outlet header vent valve/s
4.	Superheater attemperator spray water inlet drain valve/s
5.	Final superheater outlet header drain valve
6.	Final superheater outlet header vent valve
7.	Final superheater inlet header drain valve
8.	Final superheater inlet header vent valve
9.	Primary superheater outlet header drain valve
10.	Primary superheater outlet header vent valve
11.	Primary superheater inlet header drain valve
12.	Primary superheater inlet header vent valve
13.	Pendant superheater outlet header drain valve
14.	Pendant superheater outlet header vent valve
15.	Pendant superheater inlet header drain valve
16.	Pendant superheater inlet header vent valve
17.	Superheater attemperator spray water inlet isolating valves
18.	Isolating valve of electromatic relief valve
19.	Isolating device located on cold reheat line downstream of HP turbine exhaust NRV
20.	Isolating device located on hot reheat line upstream of IP turbine inlet IV
21.	Reheater outlet header drain valve
22.	Reheater outlet header vent valve
23.	Reheater inlet header drain valve
24.	Reheater inlet header vent valve
26.	Reheater attemperator spray water inlet isolating valves
27.	Sootblower steam supply valve
28.	Economizer feedwater inlet header drain valve
29.	Economizer feed water inlet header vent valve
30.	Economizer recirculation valve
31.	Steam generator drum drain valve/s (applicable to drum-type steam generators)
32.	Waterwall bottom header drain valves
33.	Instrument root valves of pressure transmitters and pressure gauges (barring root valve of pressure gauges on steam generator drum, superheater, and reheater), gauge glasses
34.	Sampling line root valves
35.	Continuous blow-down root valve
36.	Phosphate dozing line inlet isolating valve to steam generator drum (applicable to drum-type steam generators)
37.	HP bypass valves
38.	LP bypass valves

In addition to closure of the above valves, the following must also be fulfilled:

i. In the event safety valves on the boiler drum (applicable to drum-type steam generators) and superheater headers are already installed, special hydrostatic plugs or gagging devices

(Fig. 9.3) are to be fitted on safety valve seats in order to protect the valves from damage arising from overpressure;

or

In case safety valves are not installed, test plugs or blind flanges may be installed in their places;

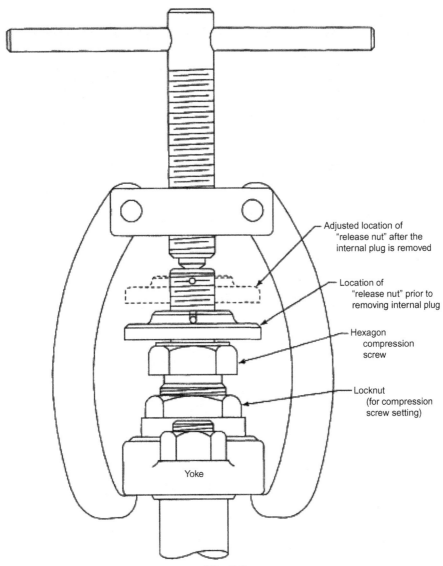

Adjusted location of "release nut" after the internal plug is removed

Location of "release nut" prior to removing internal plug

Hexagon compression screw

Locknut (for compression screw setting)

Yoke

Fig. 9.3

Gagging device. *Source: From Fig. C4.1–4, P48, 2011a Section VII, ASME B&PVC VII, Boiler and Pressure Vessel Code, Part VII, Recommended Guidelines for the Care of Power Boilers, 2010.*

Note

At the option of an authorized representative of boiler inspectorate or regulatory authorities, the hydrostatic test may be conducted at a pressure slightly lower than the set value of the lowest pressure safety valve. This test arrangement would facilitate avoiding the necessity of installing gagging devices in safety valves. This would also avoid isolation of pressure transmitters/gauges, gauge glasses, and so on, during a hydrotest.

---

ii. Prior to filling water in the superheater and main steam lines up to the superheater outlet stop valves, locking pins must be inserted to spring hangers to lock the main steam line spring support.

9.6 Filling of Steam Generator

Prior to filling large steam generators with water, it is desirable to perform the air-tightness test of the pressure circuit in order to identify the following:

i. any omission and void in welding joints;
ii. face of any valve that is not completely secured with pipe lines; and
iii. X-ray plugs remaining unwelded in the circuit by oversight, and so on.

A successful air-tightness test will cut down consumption of DM water during the hydrotest as well as the duration of the test. The air-tightness test is conducted using oil-free dry air from a station instrument air system and pressurizing the test circuit to about 0.2 MPa.

Initial filling of steam generators is undertaken with DM water, properly dozed with ammonia and hydrazine [Section 9.3, v and vi], by employing either a permanent or a portable water filling pump (Fig. 9.4). It is to be kept in mind that in order to ensure complete filling of pressure circuits with water, it is imperative to drive out trapped air from the test circuit so as to avoid any sudden pressure surges or rapid pressure increase during subsequent raising of pressure in the test circuit. Furthermore, entrapped air will drastically increase the time taken to reach the test pressure.

Since the pressure circuit of drum-type steam generators is different from the pressure circuit of once-through steam generators, the step-by-step procedure for filling of each of the types with water is described separately below.

9.6.1 Drum-Type Steam Generators

i. Before filling the steam generator completely, special care must be observed to flush out any debris remaining in the water filling pipe line. This is ensured by filling the steam generator up to a certain level in waterwalls, and then draining out water by opening the waterwall bottom header drain valves;

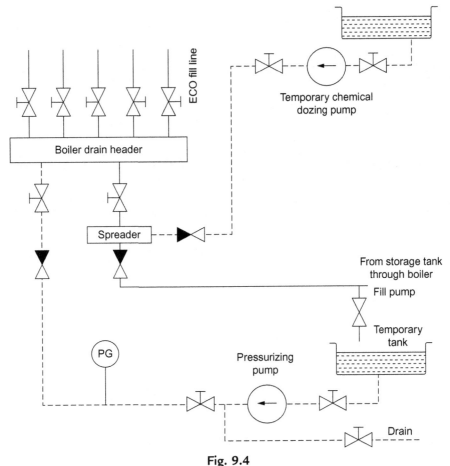

Fig. 9.4
Scheme for steam generator water filling, chemical dozing, and pressurizing.

ii.   On completion of draining, close the waterwall bottom header drain valves;

iii.   Filling of pressure circuits is repeated by keeping drum manhole doors open. This will facilitate driving out air from the pressure circuits;

iv.   Start introducing water through the economizer. When the economizer, drum interconnecting pipe, and downcomers are filled with water, open the waterwall bottom header drain valves to flush out any loose debris remaining in the circuit;

v.   Check the quality of the sample water from the drain lines. If the effluent is noted to be free from any debris, close water wall bottom header drain valves;

vi.   Continue filling through the economizer, downcomers, and waterwalls. When the water level draws close to the drum manholes, stop further water-filling operations. Close drum manholes and secure them tightly;

vii.   Ensure air evacuation from the remaining portion of the test circuit by following either of the methods given below:
  1.   In case of small steam generators, ensure that all drain valves are closed; open all vent valves.
  2.   When filling large steam generators, close all drain and vent valves. Connect a vacuum pump to one of the drum vent valves and open it. Raise vacuum in the remaining air space in the pressure circuits to about 80–90 kPa.
viii.   Start filling the steam generator at a very slow rate to obviate any airlock in pressure circuits. Carry out the following sequence of operation:
  1.   In small steam generators, after ensuring that a full flow of water is coming through a vent valve, close the valve. Continue with this closing process until all vent valves are closed. Stop the steam generator water-filling operation.
  2.   When the pressure circuits including superheaters of large steam generators are completely filled-up, stop the vacuum pump. Simultaneously stop further filling of the steam generator.
 ix.   In continuation to the above step, raise the pressure in the steam generator to the rated pressure of the filling pump (permanent/portable), but not less than about 700 kPa; and
  x.   Maintain the above pressure and verify the expansion of the steam generator and various clearances around it.

9.6.2 Once-Through Steam Generators

  i.   Before filling the steam generator completely, special care must be observed to flush out any debris remaining in the water filling pipe line. This is ensured by filling the steam generator up to lower waterwalls and then draining out water by opening the water wall bottom header drain valves;
 ii.   On completion of draining, close the waterwall bottom header drain valves;
iii.   Start introducing water through the economizer. When the economizer, lower waterwalls, and upper waterwalls are filled with water, open the waterwall bottom header drain valves to flush out any loose debris remaining in the circuit;
 iv.   Check the quality of the sample water from the drain lines. If the effluent is noted to be free from any debris, close the waterwall bottom header drain valves;
  v.   Ensure that all drain valves are closed. Open all vent valves for evacuating air from the test circuit while water filling is continuing;
 vi.   Start filling the steam generator through the economizer, lower waterwalls, and the rest of the pressure circuit at a very slow rate to obviate any airlock in the test circuit;
vii.   When full flow of water comes through a vent valve, close the valve. Continue with this closing process until all vent valves are closed. Stop the steam generator water-filling operation;

viii. Raise pressure in the steam generator to the rated pressure of the filling pump (permanent/portable), but not less than about 700 kPa; and

ix. Maintain the above pressure and verify expansion of the steam generator and various clearances around it.

9.7 Operating Procedure

For successful completion of the hydrotest of steam generators, the test circuit is split into two parts; in the first circuit the complete steam generator, barring reheat section, is put under test pressure; the second circuit covers testing of the reheat section.

Gradually raise pressure (typically at the rate of 350 kPa per minute) in the test circuit to 2.5 MPa with the help of a temporary high pressure positive displacement pump. Stop further raising of pressure and inspect the test circuit thoroughly to ensure integrity of the test circuit. On confirmation of secured integrity, raise pressure at the rate of typically 1 MPa per minute or as recommended by the steam generator manufacturer. After reaching every 10 MPa rise in test pressure, hold the pressure for about 10 min and check the integrity of the test circuit. Continue with this process until the design pressure is reached. Stop the pump and close its discharge valve and maintain this pressure. Visually inspect the drum or steam separator and pressure parts thoroughly for any signs of wetting of the surfaces or water leakages from the test circuit. Any leak detected at this stage must be rectified first prior to raising pressure further. Depending on the size and capacity of steam generators, the whole process of pressure raising and inspection may take a few hours.

On completing repair of all leaks, open the discharge valve of the temporary pump and start the pump. Once the pressure within the test circuit reaches at about 80% of the test pressure, reduce the rate of pressure raising to about 0.1 MPa per minute or as recommended by the steam generator manufacturer. Continue with pressurizing the test circuit gradually to full test pressure (drum-type steam generator: 1.5 times the design pressure of the drum; once-through steam generator: 1.5 times the maximum working pressure) under proper control at all times so that it never exceeds 6% of the required test pressure [1].

Note

In the event the steam generator drum is tested at the manufacturer's works at 1.5 times the design pressure and a proper statutory certificate signed by the boiler inspectorate or regulatory authorities is available, the hydrotest of the steam generator on site may be carried out at 1.25 times the design pressure in consultation with the authorized representative of the boiler inspectorate or regulatory authorities [1].

Maintain the test pressure for 30 min. Relieve the pressure to the maximum allowable working pressure; while maintaining this pressure, thoroughly inspect the drum or steam separator and pressure parts. In the absence of further leakage of water from the test circuit or undue deflection or distortion of its parts for at least 10 consecutive minutes, the test may be considered to be in order and no further testing is required by raising pressure beyond the design pressure.

Notes

1. Should any part of the boiler show undue deflection or indication of permanent set during the progress of the test, the pressure shall be released immediately when such indications are observed. The working pressure for the part shall be 40% of the test pressure applied when the point of permanent set was reached. This procedure shall apply to any steam generator at any test.
2. At the first hydraulic test of a steam generator prior to the issue of an original certificate, deflection measurements shall be made before, during, and after the test of each furnace length, fire-box, and flat end or other plates [1].

Based on the number of leakages to be attended, duration of successful hydrotesting of steam generators starting from initial pressure raising, then intermediate inspection, rectification of leakage areas, further pressure raising, and final inspection, may take about a week.

9.8 Estimation of DM Water and Chemicals

Prior to conducting the hydrotest of steam generators, it is essential to estimate the exact requirement of DM water and chemicals correctly. It is advisable to procure a higher quantity of DM water and chemicals than estimated, as any shortfall during actual execution of the test will jeopardize a whole sequence of activities.

9.8.1 Estimation of DM Water

To facilitate estimation of DM water, it would be prudent to assess the water-holding capacity of the pressure parts beforehand (Table 9.3: drum-type steam generators and Table 9.4: once-through steam generators). Actual volume of each area is project specific and has to be supplied by the manufacturer.

Table 9.3 Water-holding capacity of drum-type steam generators

Item	Volume (m^3)
HP feedwater pipe line starting from downstream of boiler feed check valve to economizer inlet	
Economizer	
Boiler drum	
Downcomers	
Furnace waterwalls	
Superheaters	
Interconnecting piping	
Reheaters	
Cold reheat and hot reheat pipe lines	

Table 9.4 Water-holding capacity of once-through steam generators

Item	Volume (m³)
HP feedwater pipe line starting from downstream of boiler feed check valve to economizer inlet Economizer Furnace waterwalls Superheaters Steam separator Interconnecting piping Reheater Cold reheat and hot reheat pipe lines	

9.8.2 Estimation of Chemical Requirement

Prior to filling the steam generator with DM water, hydrazine and ammonia are to be dozed to water to maintain pH and residual hydrazine [Section 9.3, v and vi]. Based on the quantity of DM water to be used during the hydrotest, the actual amount of these chemicals shall be assessed beforehand by the station chemist and procured accordingly.

9.9 Conclusion

The hydrotest of steam generators is witnessed by the authorized representative of the boiler inspectorate or regulatory authorities. If the test is observed to be satisfactory in all respects, he approves and certifies the test.

On completion of the hydrotest, slowly release the pressure through a drain valve so as not to endanger personnel or damage equipment. When the pressure inside the test circuit drops to about 200 kPa, fully open all drain and vent valves. The pressure circuit must be drained completely to prevent freezing of water during extreme cold weather and also to minimize corrosion of the metal surfaces. Special care shall have to be taken for the nondrainable portion of the test circuit by putting it under wet preservation by either dozing 400 mg kg^{-1} of hydrazine and 100 mg kg^{-1} of sodium sulfite or dozing of chemicals per the recommendation of the steam generator manufacturer.

Remove all hydrostatic plugs/gagging devices from safety valve seats or blind flanges installed in lieu of safety valves [Section 9.5, i]; install safety valves at their respective positions.

The steam generator is inspected further by the authorized representative of the boiler inspectorate or regulatory authorities to find out whether piping arrangements, safety valves, pressure transmitters, gauges, controls, and other equipment on the unit are installed fulfilling the requirement of "steam generator safety rules and regulations" and/or other jurisdictional requirements.

Note

On successful fulfillment of all statutory requirements, a typical certificate issued by the authorized representative of the boiler inspectorate or regulatory authorities delineates the following:

Inspected the steam generator after completion of erection of pressure parts (both drainable and nondrainable) and hydrotested to test pressure and found satisfactory.
Steam generator is cleared for applying refractory, insulation, and lagging only.

A protocol (Table 9.5) jointly signed by the customer, the engineer, the steam generator manufacturer, and the supplier (contractor) may be issued thereafter.

Table 9.5 Protocol of hydrotest of steam generator

Circuit	Test Method	Remarks
Economizer, boiler drum, downcomers, waterwalls, superheaters, interconnecting piping	Raising of test pressure up to 1.5 times the drum design pressure (for a drum-type steam generator; typically about 30 MPa) or 1.5 times the maximum working pressure of the test circuit of a once-through steam generator (for a supercritical steam generator, typically 40 MPa or more).	The test circuit is hydrotested successfully without any noticeable deformation or leakage
Cold reheat piping, reheater, hot reheat piping, HP bypass downstream piping, LP bypass upstream piping	1.5 times the maximum working pressure of the circuit (typically about 10 MPa for a supercritical steam generator)	The test circuit is hydrotested successfully without any noticeable deformation or leakage
Signed by the customer	Signed by the engineer	Signed by the steam generator manufacturer	Signed by the supplier (contractor)

References

[1] Central Boilers Board, Indian Boiler Regulations (IBR), New Delhi, 1950.
[2] ASME B&PVC VII, Boiler and Pressure Vessel Code, Part VII, Recommended Guidelines for the Care of Power Boilers, 2010.

CHAPTER 10

Air-Tightness/Leakage Test of the Furnace, Air, and Flue Gas Ducts of a Steam Generator

10.1 Introduction

In a steam power plant, atmospheric air is fed into the furnace of a steam generator for combustion of fossil fuel (coal, fuel oil, and fuel gas), generating heat released by burning of fuel. Hence, it is essential that fuel is burnt completely in the furnace to release as much heat energy as possible under all operating conditions. As a consequence of combustion, flue gases are produced, which after giving up part of their heat by radiation to waterwalls, leave the furnace at a safe temperature (below the initial deformation temperature of ash) and enter the convective zone. Heat remaining in flue gases is absorbed in different heat transfer surfaces (superheater, reheater, and economizer) to heat the working fluid (water and steam) and in an air heater to heat the ambient air before air enters into the furnace. After passing over the heating surfaces at various zones, the flue gases get cooled and discharged to the atmosphere through a stack [1].

For sustaining a healthy combustion process in a furnace, the furnace needs to receive a steady flow of air and at the same time remove flue gases from the furnace without any interruption. When only a stack is used for the supply of air and removal of flue gases, the system is called a *natural draft* system. When the stack is augmented with forced-draft (FD) fans to supply "secondary air" to the furnace, and/or induced-draft (ID) fans to remove flue gases from the furnace, the system is called a *mechanical draft* system. Small steam generators use natural draft, while large steam generators need mechanical draft to move large volumes of air and gases against flow resistances. A furnace utilizing an FD fan only is subjected to "positive draft" (pressurized), while one using an ID fan only is subjected to "negative draft" (subatmospheric pressure). Normally large pulverized coal-fired steam generators utilize *balanced draft*, employing both FD fans and ID fans.

Fig. 10.1 depicts the air and flue gas flow path of a typical large pulverized coal-fired balanced draft steam generator. It is evident from this figure that in pulverized coal-fired steam generators, in addition to FD fans, there is another set of fans named primary air (PA) fans, which is used for drying and transporting pulverized coal from pulverizers to coal burners [1].

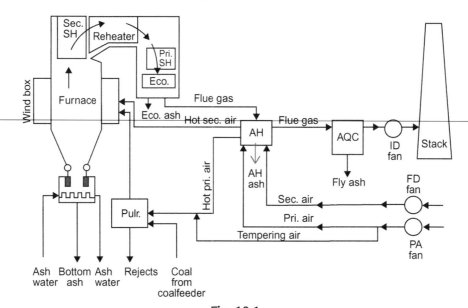

Fig. 10.1

Air and flue gas path. *Source: From Fig. 2.9, P 54. D.K. Sarkar, Thermal Power Plant – Design and Operation, 2015, Elsevier; Amsterdam, Netherlands.*

The most important step in designing a coal-fired unit is to properly size the furnace [2]. The furnace is a chamber in which the chemical energy of fuel is converted into the heat energy of flue gas as the fuel is burned in the furnace space. In order to reduce heat loss from the furnace so as to ensure that waterwalls and convective heat transfer surfaces receive maximum heat of combustion, the furnace chamber is provided with insulation. Insulation is also provided on all hot surfaces at temperatures above 333 K for personal protection. The outer surface of insulation is lagged with all-welded sheet metal.

Different ducts are used as vehicles for transporting air from the discharge of FD fans to the furnace and evacuating flue gases from the furnace to the stack through air heaters, electrostatic precipitators (ESPs) and/or bag filters, flue gas desulfurization units (FGDs) (if provided), and ID fans. These ducts, that is, flue gas ducts, hot air ducts, including outer casings of ESPs and/or bag filters are fabricated from sheet metals of all-welded construction.

It is evident from the above that welding joints on furnace and air and flue gas ducts including ESPs, bag filters are distributed over vast surface area. Thus, there is likelihood that there will be some incomplete welding joints, cracks, and blow holes in welding (Fig. 10.2), burn-through welding joints (Fig. 10.3), welding joints that are left out by oversight, and so on.

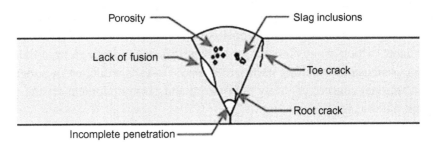

Fig. 10.2

Defects in welding. *Source: From http://www.olympus-ims.com/en/ndt-tutorials/flaw-detection/weld-overview/. Used with permission from Olympus Scientific Solutions.*

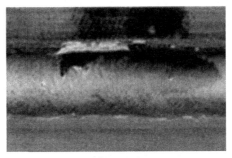

Fig. 10.3

Burn-through welding. *Source: From https://www.millerwelds.com/resources/article-library/the-most-common-mig-weld-defects-on-aluminum-and-steel-and-how-to-avoid-them. Images owned by Miller Electric Mfg. Co.*

These gaps in welding joints including leaks in flange joints of ducting must be detected and rectified before applying thermal insulation to aforementioned vast surface areas and prior to carrying out alkali boil-out of the steam generator as a prequel to chemical cleaning of the steam generator (Chapter 6). It is strongly advised that this leakage test must not be construed as the test to qualify the physical strength of sheet metal surfaces.

Over and above the fossil fuel–fired steam generator, there is another type known as the heat-recovery steam generator (HRSG), which is a heat exchanger that recovers heat from hot exhaust gases, from either a gas turbine or a diesel engine or a blast furnace, and in turn produces steam. Although configuration of HRSG is simpler than configuration of a fossil fuel–fired steam generator, the casing of the HRSG could be quite large. This casing is fabricated also with welded construction. Hence, any gap in welding joints must be detected and rectified before applying thermal insulation on the casing.

10.2 Objective

The objective of conducting an air-tightness/leakage test of the furnace as well as the air and flue gas ducts of a steam generator, ESPs, bag filters, and FGDs is to ensure the following:

i.   to arrest leakage of flue gases from a "positive draft" furnace and flue gas ducts as well as from air ducts (which always remain pressurized) since any leakage from the furnace will affect combustion, resulting in loss of energy vis-à-vis reduction in boiler efficiency with an additional concern of safety of personal and plant equipment arising from leaked-out hot flue gases/air;
ii. to obviate ingress of atmospheric air to a "negative or balanced draft" furnace and flue gas ducts. Ingress of cold atmospheric air in the furnace zone will affect combustion. Any air leakage from the roof or ducts will not participate in combustion yet will reduce the flue gas temperature impairing required heat transfer, causing boiler efficiency to fall; and reduce back-end temperature of air-heater affecting performance of ESP and ash handling system. Moreover the loading on ID fans will enhance, which in turn will raise plant auxiliary power consumption. Reduction in back-end temperature of the air heater also may cause back-end corrosion.

10.3 Precautions

In addition to "general precautionary measures" described under Section 2.1.1: Quality Assurance, the following specific precautionary measures must be observed prior to undertaking an air-tightness/leakage test of furnace, air, and flue gas ducts of a steam generator:

i.   while an air-tightness/leakage test is underway, all activities around the area shall be strictly prohibited;

ii. strict vigilance shall be maintained both near the equipment and the control room to observe the performance of FD and ID fans, especially the rising trend of bearing temperature and bearing vibration as well as motor loading. Any abnormality observed shall be reported to the proper authority for taking corrective action.

10.4 Prerequisites

Before undertaking the air-tightness/leakage test, a thorough inspection of systems under purview is essential to ascertain the following (Table 10.1):

Table 10.1 Areas/items to be checked

Sl. No.	Areas/Items	Ok ($\sqrt{}$)
1	Erection completion protocol of the following equipment signed jointly by the customer, the engineer, the steam generator manufacturer, and the supplier (contractor) is available: i. steam generator including outer casing of furnace and roof ii. stack iii. ID fans with associated lube oil systems iv. FD fans with associated lube oil systems v. PA fans with associated lube oil systems vi. air preheaters with associated lube oil systems vii. ESPs and/or bag filters viii. FGDs (if provided) ix. flue gas duct from economizer outlet up to air preheater inlet including associated dampers x. flue gas duct from air preheater outlet up to ESP and/or bag filter including FGD inlet and associated dampers xi. flue gas duct from FGD outlet to ID fan inlet and from ID fan outlet up to stack including associated dampers xii. FD fan suction and discharge ducts up to air preheater inlet including associated dampers xiii. secondary air duct from air preheater outlet up to windbox including associated dampers xiv. PA fan suction and discharge ducts up to air preheater inlet including associated dampers xv. hot and cold primary air ducts from air preheater outlet up to pulverizer inlets including associated dampers xvi. both coal and oil burners xvii. bottom ash hoppers xviii. water-impounded seal-trough of bottom ash hoppers	

Continued

Table 10.1 Areas/items to be checked—cont'd

Sl. No.	Areas/Items	Ok (√)
	xix. economizer and air preheater hoppers xx. ESP and/or bag filter hoppers xxi. permanent platforms and staircases with handrails at various elevations around the furnace, ESPs, and/or bag filters xxii. any other items as required	
	Note It is advised that soot blowers are not erected prior to completing successful air-tightness/leakage test.	
2	Trial run and commissioning of ID, FD, and PA fans are completed	
3	Construction materials—temporary scaffolding, wooden planks, welding rod ends, cotton wastes, and so on, are removed from the internals of furnace, ESPs, and/or bag filters, FGDs, air and flue gas ducts	
4	Pressure-measuring instruments are available on furnace and on ducts	
5	Sonic-type leakage detection instruments are available	
6	Stop watches are available	
7	Relevant protections, alarms, and interlocks of all fans are checked and kept ready for the protection of running fans	
8	Temporary platforms and staircases are provided wherever required	
9	All required materials (Section 10.4.1) are arranged	
10	All safety gadgets (Section 10.4.2) are arranged	
11	Verify that "no smoking," "danger," "keep off," and other safety tags (in regional and English languages) are displayed at various places around the test area	
12	Verify also that any equipment and systems, which are in service or are energized, are provided with proper safety tags (Appendix C)	

10.4.1 Typical List of Bill of Materials

Materials that are typically required for an air-tightness/leakage test are listed below. Size and quantity of each of these items would vary from project to project and shall have to be assessed by concerned personnel associated with the test.

1. pressure-measuring instruments;
2. sonic leakage detection instruments with microphone;
3. flame torches;
4. stop watches to monitor time required to note drop-in pressure;
5. kerosene;
6. soap solution;
7. smoke bomb.

10.4.2 Typical List of Safety Gadgets

Equipment that is generally used for safety of operating personnel are as follows:

1. hand gloves;
2. hardhats of assorted sizes;
3. side-covered safety goggles with plain glass;
4. safety belts;
5. first aid box;
6. safety sign boards;
7. safety tags.

10.5 Preparatory Arrangements

Before completing the air-tightness/leakage test satisfactorily, it is imperative to complete the following preparatory arrangements. These arrangements are discussed separately for fossil fuel–fired steam generators and HRSGs.

10.5.1 Fossil Fuel–Fired Steam Generator

Depending on the configuration of a steam generator on site, the whole flow path may be split into two or more circuits as decided by all concerned associated with the test for the convenience of executing the test. Some typical circuits are listed below:

   i. flue gas duct from ESP and/or bag filter outlet to ID fan inlet;
  ii. flue gas duct from air preheater outlet to ESP and/or bag filter inlet;
 iii. ESP and/or bag filter;
 iv. furnace;
  v. secondary air duct from FD fan outlet to air preheater inlet;
 vi. secondary air duct from air preheater outlet to windbox;
vii. cold primary air duct from PA fan outlet to air preheater inlet and pulverizer inlet;
viii. hot primary air duct from air preheater outlet to pulverizer inlet.

Typical arrangements to be provided for an air-leakage/tightness test of fossil fuel–fired steam generators are discussed below:

a. Openings of soot blowers are closed with blank plates;
b. Bottom of the furnace is sealed by filling the bottom ash hopper and water flow seal trough with water and allowing the water to overflow;
c. Bottom of economizer hoppers, air heater hoppers, ESP, and/or bag filter hoppers is sealed with blank plates;

d. Coal burner inlet gates are closed tightly;

e. Isolation valve of oil burners is kept shut;

f. Peepholes and manholes are sealed tightly;

g. Instrument tapping points, barring furnace pressure-measuring instrument, are kept plugged;

h. Section of the air and flue gas ducts to be tested is isolated by closing or opening associated dampers;

i. Temporary blower/s, for supplying pressurized air to various circuits, is installed at a convenient location prior to conducting an air-tightness/leakage test of some of the above circuits.

Note

Some circuits may be tested by operating FD or PA fans.

10.5.2 Heat-Recovery Steam Generator

Typical arrangements that are to be provided for an air-leakage/tightness test of an HRSG are given below:

a. Guillotine damper at "exhaust gas" inlet to HRSG is tightly closed;

b. Exhaust duct from HRSG to stack is blanked with a seal plate;

c. All instrument tapping points are plugged;

d. Manholes are sealed tightly.

10.6 Operating Procedure

The first step towards conducting an air-tightness/leakage test is to visually inspect the completeness of all welding joints and flange joints by authorized personnel by roving around the enclosures of the furnace, ESP/bag filter, and various ducts. Voids and omissions in welding and flange joints could be detected during this inspection and repaired accordingly. In the majority of occasions this step will identify the leakage problem substantially.

On completion of satisfactory visual inspection, depending on the size and type of steam generators, various methods that are adopted in the industry for detection of air ingress or leakage are as follows:

a. smoke bomb test;

b. kerosene test;

c. soap solution test;

d. noise detection test;

e.   flame torch test;

f.   pressure drop test.

Each of the above tests is described below separately.

10.6.1 Smoke Bomb Test

A smoke bomb is made of potassium chlorate ($KClO_3$) or potassium nitrate (KNO_3), also known as "saltpeter," as the oxidizer, sugar (sucrose or dextrin) as the fuel, sodium bicarbonate (baking soda) to moderate the rate of chemical reaction, and a powdered organic dye for emitting colored smoke [3] (Fig. 10.4). Materials are contained in a canister provided with a pull-ring igniter (Fig. 10.5). Once the ring is pulled, ignition takes place, emitting about 1000 m^3 of thick smoke that may last for 100–150 s. The heat of reaction evaporates the organic dye generating colored smoke. The canister is biodegradable and the smoke is eco-friendly; hence, these do not pose any health problem.

A smoke bomb test is particularly convenient for the detection of leakage in HRSGs, the physical configuration and boundary arrangement of which is much simpler than the configuration and boundary arrangement of fossil fuel–fired steam generators. After pulling the

Fig. 10.4
Colored smoke emitting from a smoke bomb canister. *Source: From https://www.youtube.com/watch? v=bF2UFg-UwMY. Reproduced with permission from Global Pyrotechnics.*

Fig. 10.5

Smoke bomb/grenade canister with pull-ring igniter. *Source: From http://ammotechph.com/index.php? route=product/product&product_id=50.*

ring igniter of a smoke bomb canister, it is thrown into the enclosure of an HRSG. Colored smoke will eventually fill the enclosure. Any opening in the enclosure can be easily detected by observing colored smoke sneaking out of the opening.

10.6.2 Kerosene Test

This test is useful for detecting leakage from a small section of a circuit, for example, a duct between ID fan discharge and stack inlet. Kerosene is applied around a welding joint from inside the duct with the help of a swab. Kerosene will ooze out even through small openings or voids in welding joints.

10.6.3 Soap Solution Test

In this method, the selected circuit is first isolated and sealed from both ends, then pressurized to a level as recommended by the steam generator manufacturer. Typically secondary air circuits and flue gas circuits are pressurized to 1–1.5 kPa, while primary air circuits are pressurized to about 5 kPa. Once the selected circuit is pressurized, soap solution is applied around welding and flange joints. In the event of any leakage, bubbling of soap solution could be detected on the defective area.

10.6.4 Noise Detection Test

This test is very convenient for detecting air-tightness/leakage of large enclosures like the furnace, ESP, and so on, along with secondary air and flue gas ducts by running FD fan/s.

Prior to starting FD fan/s, the following dampers are kept fully open:

- air preheater flue gas inlet and outlet dampers;
- air preheater secondary air inlet and outlet dampers;
- ESP/bag filter inlet and outlet dampers;
- ID fan inlet and outlet dampers;
- FD fan outlet damper.

Air preheater primary air inlet and outlet dampers are kept closed.

Run one FD fan, regulate air flow to create "positive draft" in the complete circuit to about (+) 1.5 kPa or to a pressure as recommended by the steam generator manufacturer by modulating the inlet vane (centrifugal fan) or blade pitch (axial flow fan) of the fan. Maintain steady condition inside the furnace and of all associated equipment; scan all welding and flange joints of the furnace, roof, ducts, and so on, with the help of a microphone pick-up of a sonic detector. Any leakage in joints will create noise, which will be detected by the microphone pick-up indicating existence of a leak.

10.6.5 Flame Torch Test

Procedure to conduct this test is similar to the noise detection test; however, instead of positive draft the complete circuit, starting from furnace, ESP, and so on, along with secondary air and flue gas ducts, is subject to a "negative draft" by running ID fan/s.

Before starting ID fans, the following dampers are kept fully open:

- Air preheater flue gas inlet and outlet dampers;
- Air preheater secondary air inlet and outlet dampers;
- ESP/bag filter inlet and outlet dampers;
- ID fan inlet and outlet dampers.

The following dampers are kept fully closed:

- FD fan outlet damper;
- air preheater primary air inlet and outlet dampers.

Start the ID fan, create a negative draft of (−) 1.5 kPa or a draft as recommended by the steam generator manufacturer in the circuit by modulation of the fan inlet vane (centrifugal fan) or blade pitch (axial flow fan) or by variation of the speed of the fan. Traverse flame torches around all joints. In the event of even a minor leakage, the flame envelope will be sucked in; profuse leakage may even extinguish the flame.

10.6.6 Pressure Drop Test

This test is conducted in the similar manner as the soap solution test or noise detection test by running FD fan/s. Prior to starting FD fans, the following dampers are kept fully open:

- air preheater flue gas inlet and outlet dampers;
- air preheater secondary air inlet and outlet dampers;
- ESP/bag filter inlet and outlet dampers;
- FD fan outlet damper.

The following dampers are kept fully closed:

- ID fan inlet and outlet dampers;
- air preheater primary air inlet and outlet dampers.

Start the FD fan and raise and maintain "positive draft" of about 3.74 kPa in the circuit under consideration. Observe the drop-in pressure. If the pressure drop does not exceed 1.25 kPa in 10 min, the test is considered to be satisfactory [4]. In the event the pressure drop exceeds the above limit, find out the leakage employing the soap solution or noise detection test.

10.7 Conclusion

Whatever may be the method employed for an air-tightness/leakage test, any leakage observed during the test must be marked with chalk or paint. A list containing defects and leakages may be prepared; accordingly, these defects and leakages are to be repaired or sealed. If this list is short further test may not be required. However, if too many leakages or voids are listed, then the adopted test must be repeated until all leakages are completely eliminated.

After attending all identified leakages satisfactorily, air-tightness/leakage test of the furnace, ESP and/or bag filter, and various air and flue gas ducts may be declared to be complete in all respects. A protocol certifying completeness of test (Table 10.2) may be signed jointly by the owner, the engineer, the steam generator manufacturer, and the supplier (contractor).

The furnace and various air and flue gas ducts may now be released for applying thermal insulation with suitable lagging over them.

Table 10.2 Protocol of air-tightness/leakage test

Circuit		Test Method	Remarks
Flue gas duct from ESP and/or bag filter outlet to ID fan inlet			
Flue gas duct from air preheater outlet to ESP and/or bag filter inlet			
ESP and/or bag filter			
Furnace			
Secondary air duct from FD fan outlet to air preheater inlet			
Secondary air duct from air preheater outlet to windbox			
Cold primary air duct from PA fan outlet to air preheater inlet and pulverizer inlet			
Hot primary air duct from air preheater outlet to pulverizer inlet			
Furnace, ESP, and so on, along with secondary air and flue gas ducts all together		Soap solution test/noise detection test/flame torch test/pressure drop test	
Signed by the customer	Signed by the engineer	Signed by the steam generator manufacturer	Signed by the supplier (contractor)

References

[1] D.K. Sarkar, Thermal Power Plant—Design and Operation, Elsevier, Amsterdam, Netherlands, 2015.

[2] C. Bozzuto (Ed.), Clean Combustion Technologies, fifth ed., Alstom, Windsor, CT, 2009.

[3] A.M. Helmenstine, How to make a smoke bomb. http://chemistry.about.com/od/demonstrationsexperiments/ss/smokebomb.htm.

[4] S.C. Stultz, J.B. Kitto (Eds.), Steam: Its Generation and Use, 41st ed., The Babcock and Wilcox Company, Barberton, OH, 2005.

CHAPTER 11

Steam Generator Initial Firing and Drying Out of Insulation

Chapter Outline

11.1 Introduction

All hot surfaces at temperatures above 333 K must be insulated for personal protection. The thickness of insulation is chosen to ensure a maximum external surface temperature of 328 K for metal surfaces and 333 K for nonmetallic surfaces. There are four types of thermal

Thermal Power Plant. http://dx.doi.org/10.1016/B978-0-08-101112-6.00011-3

insulation materials used in the industry: granular, fibrous, cellular, and reflective. Granular materials, such as calcium silicate, contain air entrained in the matrix. Fibrous materials, such as mineral wool, contain air between fibers. Cellular materials, for example, cellular glass and foamed plastics, contain small air or gas cells sealed or partly sealed from each other. Reflective insulation materials consist of numerous layers of spaced thin-sheet material of low emissivity, such as aluminum foil, stainless steel foil, and so on. In practice, a combination of two or more of the above four types are in use [1].

Heat from a hot surface is transferred or lost to the surrounding area by radiation and convection. Insulation is applied to the steam generator (SG) enclosure in order to restrict the heat loss through it to a minimum, yet at the same time must also keep the cost of insulation to a minimum. At present, water- and steam-cooled tubes from SGs are provided along the high temperature zones of the inner surface of the enclosure. These tubes are joined by thin membrane bars and the whole structure acts as a gas-tight membrane wall through which flue gases are unable to escape. Insulation and lagging are provided over these membrane walls (Fig. 11.1). Because of membrane construction, no refractory is applied on these tubes. In certain construction of SGs, castable and plastic refractory lining is provided between water- and steam-cooled tubes (that are not joined as membrane) and the furnace casing for sealing (Fig. 11.2) [2].

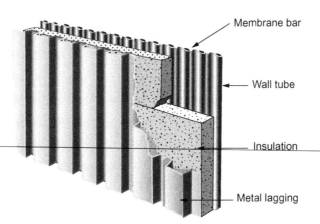

Fig. 11.1

Membrane wall with insulation and lagging. *Source: From Fig. 1, P 23–2, Chapter 23: Boiler Enclosure, Casing and Insulation, Stultz SC, Kitto JB, editors. Steam Its Generation and Use (41st Edition), 2005. Courtesy of The Babcock & Wilcox Company.*

Initial firing of SGs is accomplished on successful completion of alkali flushing of the preboiler system (Chapter 3), the flushing of the fuel oil piping system (Chapter 4), hydraulic testing of the SG (Chapter 9), and air-tightness/leakage testing of the furnace, and air and flue gas ducts of the SG (Chapter 10). SG initial firing and drying out of insulation is followed by chemical cleaning of the SG (Chapter 6).

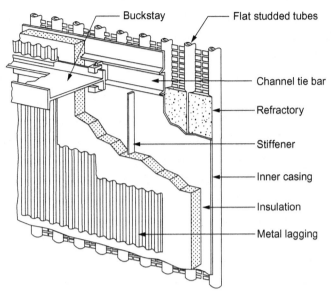

Buckstay — Flat studded tubes

— Channel tie bar

— Refractory

— Stiffener

— Inner casing

— Insulation

— Metal lagging

Fig. 11.2

Refractory lining between water- and steam-cooled tubes and the furnace. *Source: From Fig. 3, P 23–2, Chapter 23: Boiler Enclosure, Casing and Insulation, Stultz SC, Kitto JB, editors, Steam Its Generation and Use (41st Edition), 2005. Courtesy of The Babcock & Wilcox Company.*

11.2 Objective

Before going into normal operation of SGs for the first time, it is imperative to dry out insulating and refractory materials (if provided) by firing SGs with fuel oil. In addition, during initial firing of SGs, combined performance of the following equipment and furnace stability are critically observed:

    i. fuel oil firing equipment;
    ii. induced draft (ID) fans;
    iii. forced draft (FD) fans;
    iv. air preheaters;
    v. steam coil air heater, if provided;
    vi. gas recirculation fans, if provided;
    vii. flame scanner cooling air fans;
    viii. flame scanners;
    ix. boiler circulating water pumps (if provided);
    x. feedwater supply system;
    xi. boiler mountings (Section 11.2.1), and so on.

11.2.1 Boiler Mountings

Equipment that are directly attached to, or within, the SG are generally called boiler mountings. They are essential for safety, economics, and convenience. These mountings include water-level gauges, safety or relief valves, drain and blow-down valves, vent valves, water and steam sample connections, stop-check valves, soot blowers, and so on. Accessories for measuring boiler-operating conditions include pressure gauges, water-level gauges, thermometers, thermocouples, water and steam flow meters, alarms, and so on. There are also combustion control equipment and measuring devices. Interlock and protection devices protect the boiler from abnormal operating conditions such as low drum water level, high steam pressure, high steam temperature, and other off-normal conditions [3].

11.3 Precautions

Prior to initiating firing of SG for the first time, it is essential to verify the fitness of the SG along with all associated auxiliary equipment, for example, ID and FD fans, air preheaters, electrostatic precipitators (ESP)s and/or bag filters, flue gas desulfurization (FGD) equipment (if provided), reduction of nitrogen oxides (DeNOx) equipment (if provided), and so on. It is advised that this preoperational activity be planned so as to enable carrying out chemical cleaning of the SG within 3–4 weeks from the initial firing lest it will become essential to preserve the pressure parts of the SG by either dozing 400 mg kg^{-1} of hydrazine and 100 mg kg^{-1} of sodium sulfite (or per the dozing of chemicals recommended by the SG manufacturer) or blanketing the SG with nitrogen to prevent oxygen corrosion of the inside surfaces of pressure parts.

While executing this preoperational activity, since the SG will be fired for the first time, firing equipment shall be properly tested and initial firing continued at a very slow rate in accordance with the manufacturer's recommended procedure of drying out the SG, which may continue for 24–48 h for complete drying out of insulation and refractory.

Note

It may be kept in mind that the rate of increasing firing is restricted by the superheater (and reheater) metal temperatures and also temperature differential across the boiler drum shell (applicable to a drum-type SG) to obviate overheating or excessive thermal stress of metal surfaces before sufficient steam flow is established through the SG and temperatures are stabilized.

During initial firing of the SG, all activities around the area shall be strictly prohibited.

In addition to the above "specific precautionary measures," "general precautionary measures" as discussed under Section 2.1.1: Quality Assurance shall also be observed.

11.4 Prerequisites

First firing of SGs is a very sensitive activity; hence, before any action is taken to carry out this activity, it is essential to ascertain the following (Table 11.1):

Table 11.1 Prerequisites

Sl. No.	Areas/Items	Ok (√)
1	Protocol jointly signed by the customer, the engineer, and the supplier (contractor) certifying successful completion of the following is available: a. "Alkali flushing of preboiler system" and restoration of permanent piping system b. "Flushing of fuel oil piping system" and restoration of permanent piping system	
2	Protocol jointly signed by the customer, the engineer, the steam generator manufacturer, and the supplier (contractor) certifying successful completion of following preoperational activities is available: a. "Hydraulic test of steam generator" b. "Air-tightness/leakage test of the furnace and air and flue gas ducts of the steam generator"	
3	Erection completion certificate of the following equipment and subsystems is available; and these subsystems are operational: i. Intercommunication (intercom) system ii. Air-conditioning plant iii. Uninterrupted power supply (UPS) system iv. Clarified water system v. DM water system vi. Auxiliary cooling water (ACW) system vii. Closed-circuit cooling water (CCCW) system viii. Instrument and service air system ix. Fire water system x. Auxiliary steam system xi. Make-up water system xii. High-pressure (HP) chemical dozing system xiii. Fuel oil system and atomizing air/steam system xiv. Flame scanner cooling air fan system xv. Fuel oil burners with associated igniters xvi. BMS for oil firing xvii. Alarm annunciation system and DCS for boiler light-up xviii. ESP and/or bag filter xix. FGD xx. DeNOx equipment, if provided xxi. Boiler-feed pumps xxii. Condensate extraction pumps	

Continued

Table 11.1 Prerequisites—cont'd

Sl. No.	Areas/Items	Ok (√)
4	Verify that precommissioning checks of the following systems have been successfully completed and they are operational: i. Steam generator draft system ii. Fuel oil system iii. Feedwater system iv. Condensate system v. Make-up water system	
5	Trial run of the following equipment is successful: i. ID fans ii. FD fans iii. Air preheaters (regenerative type) iv. Flame scanner cooling air fans v. Boiler water circulating pumps (if provided) vi. Boiler-feed pumps vii. Condensate extraction pumps	
6	Furnace, air preheaters, economizer, ESPs, FGD equipment, DeNOx equipment, ID, FD and primary air (PA) fans, flue gas ducts, air ducts, and other passages of a steam generator are inspected and observed to be free from men and foreign material, for example, wooden planks, welding rod ends, scaffolding, cotton wastes, debris, and so on, then boxed up	
7	Verify that the steam generator is in a satisfactory state for firing	
8	All remote-operated control vanes/blade pitch control/speed control, dampers, valves, and so on, are operational	
9	Interlock and protection of ID fans, FD fans, air preheaters, boiler circulating water pumps (if provided), boiler-feed pumps, condensate extraction pumps, and other applicable equipment and systems have been checked	
10	Boiler mountings are in place and are serviceable	
11	Boiler drum pressure gauge and level transmitters are installed (applicable to drum-type steam generator)	
12	Boiler drum gauge glass drains are operable, and water column is in perfect order (applicable to drum-type steam generator)	
13	Check that boiler drum manhole covers are installed and secured properly (applicable to drum-type steam generator)	
14	Verify that safety valves are placed on drum (applicable to drum-type steam generator), superheater and reheater headers; gagging devices from these valves are removed	

Table 11.1 Prerequisites—cont'd

Sl. No.	Areas/Items	Ok (√)
15	The following valves have been checked for proper operation: i. Boiler drum blow-down valves, emergency drain valve, vent valves (applicable to drum-type steam generator) ii. Waterwall bottom header drain valves iii. Continuous blow-down tank-level control valve	
16	Pressure, temperature, and draft measuring instruments for indication and recording, as follows, are calibrated, installed, ready, and serviceable: i. Fuel oil pressure ii. Fuel oil temperature iii. Fuel oil flow iv. Atomizing air/steam pressure v. Furnace draft vi. Furnace temperature vii. Windbox to furnace differential pressure viii. Air path pressures and temperatures ix. Flue gas path pressures and temperatures x. Air flow xi. Boiler drum pressure (applicable to drum-type steam generator) xii. Boiler drum level (applicable to drum-type steam generator) xiii. Boiler drum metal temperature (applicable to drum-type steam generator) xiv. Superheater and reheater metal temperature xv. Superheater outlet pressure xvi. Feedwater pressure at economizer inlet xvii. Feedwater temperature at economizer inlet xviii. Deaerator feedwater storage tank level	
17	Verify that hangers on main steam piping and reheat steam piping are in place, properly adjusted, and free to move	
18	Ensure that soot blowers of the steam generator and air preheater are ready for operation and are retracted	
19	Inspect the operation of furnace temperature probes. Verify that these probes are retracted and cooling air is supplied to them	
20	Verify that fuel oil guns are installed and coupled to the fuel oil supply line; instrument air is available to oil gun advance-retract cylinder	
21	Ensure that bill of materials as listed under Section 11.4.1 below are arranged	
22	Ensure also that safety gadgets as listed under Section 11.4.2 below have been arranged	
23	Verify that "no smoking," "danger," "keep off," and other safety tags (in regional and English languages) are displayed at various places around the test area	
24	Verify also that any equipment and systems, which are in service or are energized, are provided with proper safety tags (Appendix C)	

11.4.1 Typical List of Bill of Materials

Materials that are typically required for initial firing of SGs are listed below. Type and quantity of each of these items would vary from project to project and shall have to be assessed by concerned personnel associated with this preoperational activity.

1. demineralized (DM) water;
2. fuel oil;
3. chemicals for dozing;
4. applicable instruments (Section 11.4 (16));
5. furnace expansion measuring arrangement;
6. tongue tester for measuring motor current at the equipment as well as at respective switchgears.

11.4.2 Typical List of Safety Gadgets

Equipment generally used for safety of operating personnel are as follows and have been arranged:

1. safety shoes of assorted sizes;
2. hand gloves;
3. hardhats of assorted sizes;
4. side-covered safety goggles with plain glass;
5. first aid box;
6. safety sign boards;
7. safety tags.

11.5 Brief Description of Various Systems

It is apparent from Section 11.4 (1) and (2) that initial firing of the SG must be preceded by making certain systems ready and operational. Some of these systems are discussed already under previous chapters (Table 11.2); remaining chapters are discussed below for the convenience of readers.

Table 11.2 List of systems discussed in previous chapters

Sl. No.	Name of the System	Section No.	Name of the Chapter
1	Condensate system, feedwater system, and feedwater heater drains system	3.2	Alkali flushing of preboiler system
2	Fuel oil system	4.2	Flushing of fuel oil piping system
3	Main steam, cold reheat and hot reheat steam, and high pressure & low pressure (HPLP) bypass system	7.2	Steam blowing of MS, CRH, HRH, and other steam pipe lines
4	Steam generator draft system	10.1	Air-tightness/leakage test of the furnace, air, and flue gas ducts of the steam generator

11.5.1 Clarified Water System *(Fig. 11.3)* *[1]*

All natural/raw water contains impurities. Hence, prior to using this water in industrial application, it must be clarified to ensure efficient and reliable operation. Principal mechanisms of clarification of raw water are coagulation, flocculation, and sedimentation. *Coagulation* is the process of destabilizing suspended particles in raw water by neutralizing their charge. It is carried out in a rapid mixing tank to ensure complete and uniform dispersion of the coagulant throughout the entire mass of water. Chemicals, such as alum (aluminum sulfate), ferrous sulfate, ferric sulfate, ferric chloride, activated silica, organic polyelectrolyte, and so on, are dozed to ensure the desired effect of the process of coagulation. *Flocculation* is the process of agglomerating destabilized particles and is carried out by gentle stirring of water in the flocculation tank. Once the water is coagulated and flocculated, the next step is liquid-solid separation, followed by settling of solid particles at the bottom of a tank. This process is known as *sedimentation*.

In a clarified water reservoir, sodium hypochlorite solution is dosed to prevent any microbiological growth.

The clarified water is used as circulating (condenser cooling) water, service water, drinking water, and so on, and as an input to the demineralized (DM) water plant.

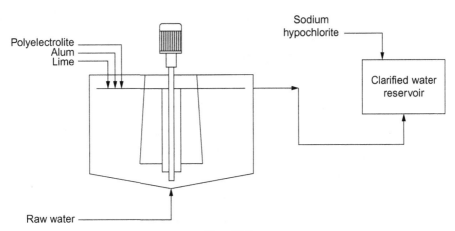

Fig. 11.3

Clariflocculator with clarified water reservoir. *Source: From Fig. 9.3, P 319, Chapter 9: Steam Power Plant Systems, Dipak K Sarkar, Thermal Power Plant — Design and Operation, Elsevier, 2015.*

11.5.2 DM Water System *(Fig. 11.4)* *[1]*

Raw water contains many minerals as cations (positively charged ions) and anions (negatively charged ions) in diverse concentrations, along with silica (SiO_2) and free carbon dioxide (CO_2). Cations in raw water are usually calcium, magnesium, barium, aluminum, iron, sodium, and potassium. Anions generally are chlorides, sulfates, nitrates, carbonates, and bicarbonates.

In an ion-exchange process, that is, displacement of one ion by another, cations are replaced with hydrogen ions in a cation exchanger and anions are replaced with hydroxide ions in an anion exchanger. Resins generally used in an ion-exchange process are weak acid cation resin, strong acid cation resin, weak-base anion resin, and strong-base anion resin.

In the pressure sand filter (PSF), turbidity of incoming water to DM plant is reduced first followed by the activated carbon filter (ACF), where suspended impurities and organic matters are entrapped and "free chlorine" and "free iron" are removed. This water is then passed through either a strong acid cation (SAC) exchanger or a weak acid cation (WAC) exchanger or both. The effluent of the cation exchanger enters the degasser tower to remove CO_2. The outgoing "decationized and decarbonated" water is treated further in either a strong base anion (SBA) exchanger or a weak base anion (WBA) exchanger or both and mixed-bed (MB) unit. DM water coming out of the treatment plant is stored in a DM water storage tank.

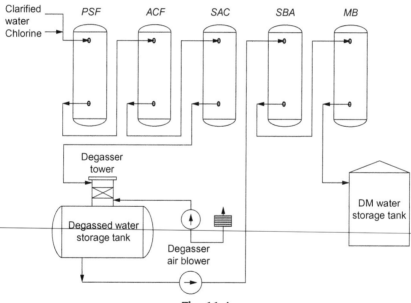

Fig. 11.4

Demineralized (DM) water plant. *Source: From Fig. 9.4, P 321, Chapter 9: Steam Power Plant Systems, Dipak K Sarkar, Thermal Power Plant — Design and Operation, Elsevier, 2015.*

The DM water system primarily meets the heat-cycle make-up water demand to replenish losses from the blow-down of the boiler drum (drum-type SG), leakages from valve glands and flange joints, steam venting, safety valve popping, and so on.

11.5.3 Auxiliary Cooling Water (ACW) System [1]

The ACW system uses clarified water and is used in turbine lube oil coolers, heat exchangers of the closed-cycle cooling water (CCCW) system, as makeup to the ash water system, and so on.

11.5.4 CCCW System (Fig. 11.5) [1]

The CCCW system supplies DM cooling water through heat exchangers to various coolers of the SG package, turbo-generator (TG) package, compressors, ash-handling system, sample coolers, and so on. Hot water return from these coolers then flows through heat exchangers. In these exchangers, CCCW is cooled by rejecting heat to the ACW.

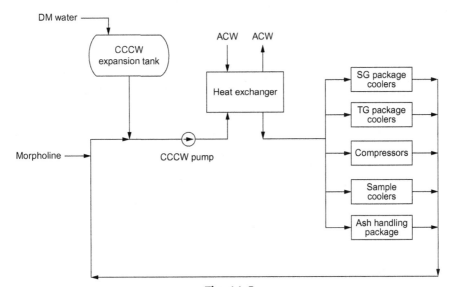

Fig. 11.5

Closed-cycle cooling water (CCCW) system. *Source: From Fig. 9.12, P 328, Chapter 9: Steam Power Plant Systems, Dipak K Sarkar, Thermal Power Plant – Design and Operation, Elsevier, 2015.*

11.5.5 Compressed Air System (Fig. 11.6) [1]

The compressed air system is capable of supplying all station air requirements, either intermittently or continuously, at a pressure of 700–800 kPa. The system comprises two sections: the "instrument air" (IA) for control of pneumatically operated instruments and drives and the "service air" (SA), which provides compressed air for general house cleaning, pneumatic tools, and other miscellaneous purposes.

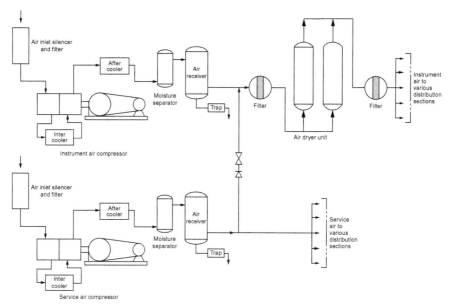

Fig. 11.6

Compressed air system. *Source: From Fig. 9.13, P 328, Chapter 9: Steam Power Plant Systems, Dipak K Sarkar, Thermal Power Plant – Design and Operation, Elsevier, 2015.*

The IA essentially has to be oil-free, dust-free, and dry to prevent condensation of moisture. The system uses compressors with associated suction filters, silencers, intercoolers, aftercoolers, and moisture separators to convert ambient air into high-pressure IA. The discharge from compressors is sent to air receivers for storage and distribution to IA loads. A drain trap is provided at the bottom of each air receiver to drain any accumulated moisture in the receiver and also to remove other impurities. Downstream of air receivers, an air-drying plant is provided to remove any entrained moisture. The air dryer unit is provided with pre- and afterfilters to remove dust, dirt, and so on.

The SA system uses identical compressors as those of the IA system. The discharge from compressors is sent to SA receivers for storage and distribution to SA loads. The discharge header of the SA system is interconnected with the discharge header of the IA system to meet a sudden inrush demand of IA. During this period, the supply to the SA system is temporarily suspended.

11.5.6 Fire Water System [1]

The fire water system uses firewater pumps to provide raw water at 1.0 MPa pressure to the hydrant system and spray systems. The hydrant system feeds pressurized water to a number of hydrant valves located throughout the entire power plant. In the event of a fire, fire hoses are coupled to hydrant valves, and jets of water are then directed towards the fire.

The spray system could be either automatic or manually operated. The automatic spray system detects fire with frangible bulb-type detectors. When the surrounding temperature exceeds the rated temperature of detectors, the frangible bulb collapses, opening the water supply deluge valves (pneumatically operated); thus, water starts coming out through nozzles toward the hot zone. This system protects generator transformers, station transformers, unit auxiliary transformers, CW system transformers, coal-handling plant transformers, and so on.

In a manually operated spray system, fire is detected either visually (at the turbine lube oil tank area or near outdoor LT transformers) or by the use of heat detectors (around indoor LT transformers). Upon detection of fire, the operator would manually open the valve to put the spray system into service.

11.5.7 Auxiliary Steam System (Fig. 11.7) [1]

The auxiliary steam system provides steam at usually 1.6 MPa pressure and 493–503 K temperatures to various plant auxiliaries during all modes of plant operation, including cold start-up and low-load operation. The system uses steam from the extraction steam system during normal operation. However, during the cold start-up of a unit, the auxiliary steam header receives supply from the main steam line, then switches over to the cold reheat line.

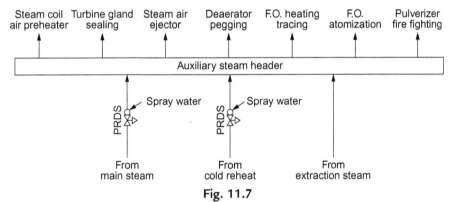

Fig. 11.7

Auxiliary steam system. *Source: From Fig. 9.20, P 336, Chapter 9: Steam Power Plant Systems, Dipak K Sarkar, Thermal Power Plant – Design and Operation, Elsevier, 2015.*

11.5.8 Make-Up Water System (Fig. 3.5) [1]

The level in the condenser hotwell is maintained by adding make-up water to the condenser from the condensate water storage tank as required. Normally, in the event of a low level in the hotwell, condensate water will flow from the storage tank to the hotwell by the static head. If this supply fails to restore the level, an emergency supply through a pump will be cut in.

11.5.9 HP Chemical Dozing System [1]

In the drum-type SG calcium and magnesium salts in the feedwater get deposited on the boiler heat transfer surfaces and form a hard, tightly adhering scale. The HP dozing system uses trisodium phosphate (Na_3PO_4) and sometimes caustic soda ($NaOH$) that are injected into the boiler drum to remove these salts and to maintain the pH in the boiler water. In the presence of phosphate ions (PO_4) and at high temperature and pH, these calcium and magnesium salts become soft, loosely adhering scales that are removed by boiler water blow-down.

11.5.10 Burner Management System (BMS) [1]

BMS standards are formulated by NFPA to eliminate or minimize boiler-furnace explosion/ implosion hazards. The BMS permits starting of equipment in sequence when a set of preset conditions are satisfied. When the equipment is started, the system will continuously monitor safe operating conditions determined in advance, warn operating personnel, and remove equipment from service in sequence again when the preset safe conditions are not met in practice.

The system supervises overall furnace conditions, monitors all critical parameters of fuel-firing system, and cuts out all fuel input to the furnace whenever dangerous conditions occur.

11.5.11 Uninterrupted Power Supply (UPS) System [1]

The UPS system furnishes a reliable source of 240 VAC power supply to equipment vital for plant operation and emergency shut-down. This system feeds the following loads:

- BMS;
- coordinated control system;
- TG electro-hydraulic control system;
- turbine supervisory instruments;
- computers;
- public address and plant communication system;
- fire protection system.

11.6 Preparatory Arrangements

Prior to undertaking initial firing of the SG, ensure that the following preparatory works are completed.

    i. Open the following dampers (Fig. 10.1):
        1. air preheater secondary air inlet and outlet dampers;
        2. air preheater flue gas inlet and outlet dampers;
        3. ESP and/or bag filter flue gas inlet and outlet dampers;

   ii. Close the following dampers (Fig. 10.1):
1. air preheater primary air inlet and outlet dampers;
2. overfire dampers;
3. pulverizer hot air shut-off gates;
4. pulverizer hot air control dampers.

   iii. Verify smooth travel of secondary air dampers/air registers in the windbox;

   iv. Check necessary arrangement is provided for measuring SG expansion following light-up;

   v. Close SG manholes and peepholes;

   vi. Close ash hopper doors;

   vii. Verify cooling water, if provided, is supplied to the SG manholes and ash hopper doors;

   viii. Fill the bottom ash hopper and ash hopper seal trough. Ensure continuous makeup to these is overflowing;

   ix. Open the following valves:
1. boiler drum vent valves;
2. start-up and other vent valves on superheater and superheater headers;
3. start-up and other vent valves on reheater and reheater headers;
4. steam pressure gauge isolation valves;
5. boiler drum chemical inlet valve;
6. continuous blow-down line isolation valve;
7. continuous blow-down tank vent valve;
8. isolation valves of continuous blow-down tank-level control valve;
9. economizer recirculation valve;
10. economizer feedwater inlet valve;
11. low load feed control station isolation valves;
12. auxiliary steam supply valve to air preheaters;
13. isolating valves on atomizing air/steam supply and fuel oil supply to each burner;
14. fuel oil header short recirculation valve, if provided;
15. auxiliary steam pressure control valve pressure-sensing isolating valve.

   x. Close the following valves:
1. nitrogen blanketing valves to boiler drum, superheater, and reheater;
2. isolating valves of electromatic relief valves on superheater and reheater;
3. superheater and reheater attemperator spray water supply block and control valves;
4. boiler, regular-level gauge glass and level transmitter isolation valves and blow-down valves;
5. sample line isolation valves of feedwater, drum water, saturated steam, and superheated steam;
6. waterwall bottom drain header drain valves to intermittent blow-down tank;
7. economizer inlet header drain valve;
8. full-load feed control station isolation valves;

9. soot blowing system supply valves;

10. fuel oil header trip valve.

xi. Remove regular-level gauge glasses from boiler drum and replace them with temporary tubular gauge glasses, to prevent alkaline water from destroying the mica and damaging the regular gauge glasses;

xii. Energize illumination lights of drum gauge glasses;

xiii. Supply cooling air to each flame scanner;

xiv. Place the atomizing steam pressure control valve in service and operate from the control room;

xv. Place the fuel oil flow control valve in service and operate from the control room;

xvi. Fill the SG through the economizer, furnace walls, and, since water level increases with rise in temperature, up to 25 mm above the lowest visibility limit of the boiler drum level gauge glass (applicable to the drum-type SG) or steady level in steam separator (applicable to the once-through SG) at a slow rate as recommended by the SG manufacturer.

Note

Some of the activities discussed above may not be applicable to once-through SGs and may be ignored.

11.7 Operating Procedure

i. Start boiler water circulating pump/s, if provided;

ii. Start both air preheaters (regenerative type);

iii. Place steam coil air heater in service;

iv. Start both ID fans;

v. Verify that the outlet damper of ID fans opens fully;

vi. Start both FD fans;

vii. Verify that the outlet damper of FD fans opens fully;

viii. Start the flame scanner cooling air fan;

ix. Adjust the ID fan control vane/variable blade pitch, ID fan speed (if provided), FD fan control vane/variable blade pitch, windbox secondary air dampers to ensure that at least 30% of total air flow (purge air flow) is maintained through the furnace.

x. Ensure also that the furnace negative draft is about $(-)\,0.10$ to $(-)\,0.15$ kPa;

xi. Verify that none of the SG [master fuel relay (MFR), known as 86 MF relay in the industry] trip conditions are present (Section 11.7.1);

xii. Prior to admitting fuel to an unfired boiler, it is essential to ensure that the furnace is free from gaseous or suspended combustible matters by purging the furnace (Section 11.7.2) with purge air flow (ix above);

xiii. Open the fuel oil trip valve located on the fuel oil supply line near to the burners;

xiv. Open the atomizing air/steam shut-off valve;

xv. Start one or more fuel oil burners as recommended by the SG manufacturer;

xvi. Regulate oil flow to burners to raise the SG water temperature slowly (typically 2 K/min) as recommended by the SG manufacturer;

xvii. The SG water temperature should be such as to appear as a light vapor at open vents;

xviii. Put the furnace temperature probe in service;

xix. Ensure that the furnace exit gas temperature does not exceed 813 K (large subcritical SG) or 838–883 K (as applicable to supercritical SG);

xx. Continue with this warming-up process per the recommendation of the SG manufacturer and observe the following:

1. free and uniform expansion of the furnace;
2. there is no binding or fouling of the furnace structure, pipe work, and SG itself.

Note

Furnace expansion movement must be recorded for comparison with furnace expansion during future start-up.

---

xxi. After maintaining the fire at this rate or raising the firing rate if required for a recommended period (typically about 24–48 h), it may be construed that drying out of insulation is complete;

xxii. Ensure that during drying out of insulation, fluid is not pressurized. Any generation of steam will be vented to atmosphere;

xxiii. Safety limits must be carefully observed for each running equipment all through this activity;

xxiv. A log sheet is to be maintained for noting hourly readings of all salient parameters during initial firing of the SG (Table 11.3).

Table 11.3 Log sheet

Sl. No.	Time	Fuel Oil			Atom. Air/ Steam Press. (MPa)	Furnace Draft (kPa)	Wb- Furnace, Δp (kPa)	Furnace Temp. (K)
		Press. (MPa)	Temp. (K)	Flow (kg/s)				
Air Flow (kg/s)	Feedwater at Eco. Inlet		Boiler Drum			SH Metal Temp. (K)	RH Metal Temp. (K)	SHO Press. (MPa)
	Press. (MPa)	Temp. (K)	Level (mmwc)	Press. (MPa)	Metal Temp. (K)			
Signed by the customer			Signed by the engineer			Signed by the steam generator manufacturer		

11.7.1 Master Fuel Relay Trip (MFT) Conditions [4]

To ensure basic furnace protection of a multiple burner SG, there are certain minimum-required interlocks that should be provided without any lapse. Furthermore, in order to establish integrated operation of the boiler-turbine-generator, these minimum-required interlocks are supplemented by some additional interlocks for the protection of the furnace. Occurrence of any one of these conditions, as follows, will trip automatically the MFR:

1. loss of all ID fans;
2. loss of all FD fans;
3. total air flow decreases below purge rate air flow by 5% of design full-load air flow (reason being whenever total air flow falls below purge rate air flow, removal of combustibles and products of combustion from the furnace gets impaired);
4. furnace pressure in excess of recommended operating pressure (in the event of excessive increase in furnace pressure, MFR has to be tripped so as to obviate furnace explosion);
5. furnace negative pressure in excess of recommended operating negative pressure (in case of excessive fall in furnace negative pressure, MFR should be tripped in order to avoid furnace implosion);
6. boiler drum level (applicable to drum boiler) in excess of recommended operating level (excessive increase in boiler drum water level may interfere with the operation of the internal devices in the boiler drum, which separate moisture from the steam, resulting in carryover of water into the superheater or steam turbine causing mechanical damage);
7. boiler drum level (applicable to drum boiler) falling below the recommended operating level (excessive fall in boiler drum water level may uncover boiler riser wall tubes and expose them to furnace heat without adequate water cooling, resulting in riser tube burnout unless action is taken either to restore the supply of feedwater or to kill the fire, that is, trip the MFR);
8. igniter fuel trip;
9. first pulverizer burners fail to ignite;
10. last pulverizer in service tripped;
11. all fuel inputs to furnace are shut off;
12. loss of all flame (admission of fuel in the furnace at any time without ensuring proper combustion is a potential hazard; hence, fuel input to the furnace must be immediately cut out in case of loss of flame, that is, loss of fire);
13. partial loss of flame that results in a hazardous accumulation of unburned fuel;
14. loss of energy (electric and/or pneumatic) supply for combustion control, burner control, or interlock systems;

15. all sources of HT power (typically 6.6 kV) supply to major auxiliaries tripped;
16. in a reheat boiler, if the following conditions occur simultaneously:
    a. one or both the ESV and IV of the steam turbine are closed;
    b. HP bypass valves closed;
    c. any load-carrying oil burner or coal mill in service.
17. Manual trip.

11.7.2 Furnace Purge [4]

Verify that the following purge permissives are satisfied:

1. Establish AC and DC power supply to BMS and associated equipment;
2. Ensure MFR is in the tripped condition;
3. Verify that none of the MFR tripped conditions (Section 11.7.1) are present;
4. Verify an open-flow path from the inlet of FD fans, through the furnace, to ID fans and the stack;
5. Place all burner air registers in the purge position or ensure all auxiliary air damper positions maintain required differential pressure between the windbox and the furnace as recommended by the SG manufacturer;
6. Ensure all fuel oil burner headers, the igniter header, individual burner, and individual igniter shut-off valve are closed;
7. Verify that all PA fans are not running;
8. Check all flame scanners show "no flame" condition;
9. For pulverized coal-fired SGs, ensure that the mass air flow through the furnace is not greater than 40% of design full-load mass air flow.

When all above conditions are satisfied, "Purge Ready/Push to Purge" indication comes ON. Initiate the command for "Purge Start." Such action will maintain purge air flow for a period of at least 5 min. On completion of the above time period, the "Purge Complete" indication switches ON. The furnace may now be treated to be free from any combustible matter and is ready for fuel firing. Reset MFR.

11.8 Conclusion

On completion of the drying out of insulation, a protocol (Table 11.4) jointly signed by the customer, the engineer, and the steam generator manufacturer may be issued.

Table 11.4 Protocol of drying out of insulation

Activity	Remark	Recommendation
Initial firing of steam generator	Satisfactory	The pressure of the steam generator may be raised at a slow rate to proceed with the next preoperational activity, "Chemical Cleaning of Steam Generator (Chapter 6)"
Drying out of insulation of steam generator	Complete	
Signed by the customer	Signed by the engineer	Signed by the steam generator manufacturer

References

[1] D.K. Sarkar, Thermal Power Plant—Design and Operation, Elsevier, Amsterdam, Netherlands, 2015.
[2] S.C. Stultz, J.B. Kitto (Eds.), Steam: Its Generation and Use, 41st ed., The Babcock and Wilcox Company, Barberton, OH, 2005.
[3] Central Boilers Board, Indian Boiler Regulations (IBR)., Central Boilers Board, New Delhi, 1950.
[4] National Fire Protection Association, USA (NFPA) 85 Boiler and Combustion Systems Hazards Code, National Fire Protection Association, Quincy, MA, 2004.

Floating of Steam Generator Safety Valves

12.1 Introduction

A safety valve is a spring-loaded, pressure-actuated device (Fig. 12.1) mounted on a pressure vessel or a pipe line for protection of the equipment against overpressurization, which may cause severe damage to the pressure vessel or the pipe line. This valve when activated by a gas or vapor pressure above a preset safe working level, opens and allows the gas or vapor to escape to atmosphere until its pressure is reduced to a pressure equal to or lower than the predetermined level. Fig. 12.2 depicts two different designs of spring-loaded safety valves.

To ensure continued safe operation of steam generators as well as to protect against overpressure conditions, spring-loaded safety valves must be installed on all steam generators to meet statutory requirement of "boiler inspectorate" of the country of their manufacturers as well as the country where they would be installed. Any violation of this requirement may result to those involved in criminal or civil penalties under the law. Overpressure protection of the steam generator along with associated steam piping, that is, main steam, cold reheat, and hot reheat pipe lines, is ensured by mounting spring-loaded safety valves on the following:

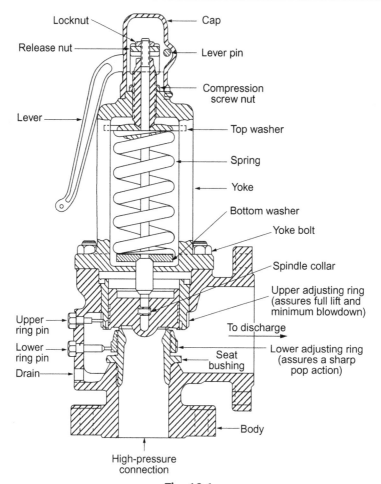

Fig. 12.1

Spring-loaded safety valve. *Source: From Fig. 1, P25–1, Chapter 25: Boiler Auxiliaries, S. C. Stultz and J. B. Kitto, STEAM Its Generation and Use (41st Edition). Courtesy of The Babcock & Wilcox Company.*

i.   boiler drum (drum-type steam generator)/steam separator outlet (once-through steam generator);

ii.   main steam pipe lines [at superheater outlet (SHO)];

iii.   cold reheat pipe lines (at reheater inlet);

iv.   hot reheat pipe lines (at reheater outlet).

In addition to spring-loaded safety valves, electromatic relief valves (ERVs) (Fig. 12.3) are also installed on SHO main steam pipe lines and reheater outlet hot reheat pipe lines. The ERV is operated by a pressure-sensitive element, which automatically relieves pressure within very close limits. ERV does not replace the spring-loaded safety valve but acts as a supplementary operating valve. It is set to operate at a lower pressure than the relieving pressure of spring-loaded safety valves, thereby reducing probability of popping of spring-loaded safety valves substantially.

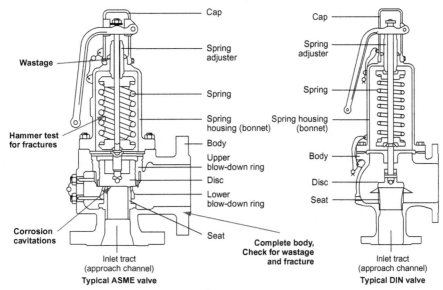

Fig. 12.2

Two designs of spring-loaded safety valves. *From http://marinesurveypractice.blogspot.in/2013/01/boiler-survey.html.*

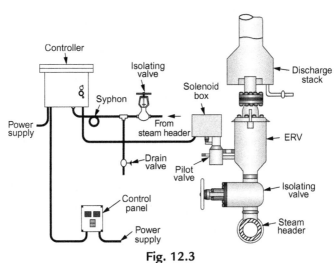

Fig. 12.3

Electromatic relief valve.

Because this valve is a power-operated safety valve, it will fail to operate on power failure; hence, it is not approved by "steam generator codes" as a device to ensure overpressure protection.

Each spring-loaded safety valve is provided with an escape pipe and drain pan. Recommended installation of a safety valve on steam generators is shown in Fig. 12.4. Each safety valve and ERV is provided also with silencers.

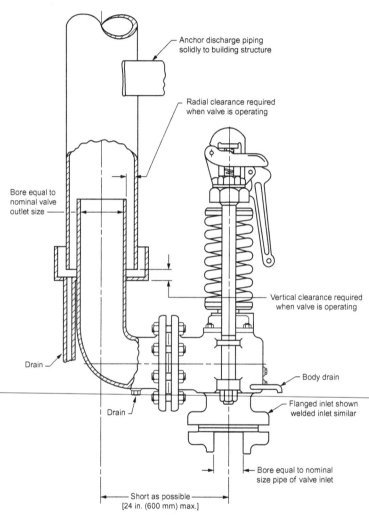

Fig. 12.4

Recommended installation of a safety valve. *Source: From Fig. C4.1–3, P 46, C4.120 Safety Valve Maintenance, ASME B&PVC VII (2010): Boiler and Pressure Vessel Code, Part VII, Recommended Guidelines for the Care of Power Boilers.*

The total relieving capacity of all spring-loaded safety valves used on a steam generator must be equal to or more than the BMCR steam generating capacity of the steam generator, without taking credit for the relieving capacity of ERV.

In order to avoid unnecessary losses and maintenance from frequent popping of the safety valves, the first drum safety valve should be set to relieve at least 5% above the drum maximum operating pressure. The ASME code stipulates that the boiler design pressure must not be less than the lowest safety valve set pressure [1].

Before a steam generator is allowed to be put in normal operation, it is essential to adjust and check floating (setting) of safety valves. Floating of steam generator safety valves is carried out following successful completion of "steam generator initial firing and drying-out of insulation" (Chapter 11), "chemical cleaning of steam generator" (Chapter 6), "steam/air blowing of MS, CRH, HRH pipe lines" (Chapter 7), and "flushing of lube oil piping system" (Chapter 8), and prior to going ahead with turbine rolling. Floating of safety valves must be started with the lowest pressure safety valve/s first, that is, reheater safety valves. Thereafter, floating of the next higher pressure safety valves mounted on superheaters is undertaken followed by floating of boiler drum/steam separator safety valves.

Tables 12.1 and 12.2 show set pressure of safety valves mounted on a typical subcritical steam generator and a supercritical steam generator, respectively.

Table 12.1 Safety valves on typical subcritical steam generator

Location/No. of Valves	Maximum Allowable Pressure (MPa)	BMCR Evaporation: 479.17 kg/s		Maximum Flow Through Reheater: 408.33 kg/s	
		Set Pressure (MPa)			Relieving Capacity as Percentage of Evaporation
		Open	Close	Percentage of Blow-Down	
Boiler drum/6	20.0	20.00	19.20	4	95.10 of BMCR
		20.20	19.20	5	
		20.40	19.38	5	
		20.60	19.57	5	
Super heater outlet/2	18.4	18.40	17.85	3	23.40 of BMCR
ERV on super heater outlet/4	18.4	18.18	17.81	2	30.17 of BMCR
Reheater inlet/4	5.38	5.38	5.22	3	58.58 of reheat flow
		5.43	5.27	3	
		5.48	5.32	3	
		5.54	5.37	3	
Reheater outlet/4	5.38	4.80	4.66	3	44.35 of reheat flow
		4.85	4.71	3	
		4.90	4.76	3	
ERV on reheater outlet/4	5.38	4.71	4.52	4	19.93 of reheat flow
		4.76	4.57	4	

Table 12.2 Safety valves on typical supercritical steam generator

Location/No. of Valves	Maximum Allowable Pressure (MPa)	BMCR Evaporation: 618.05 kg/s		Maximum Flow Through Reheater: 483.83 kg/s	
		Set Pressure (MPa)		Percentage of Blow-Down	Relieving Capacity as Percentage of Evaporation
		Open	Close		
Steam separator outlet/6	28.91	28.91	27.75	4	89.5 of BMCR
		29.67	28.48	4	
Superheater outlet/4	26.75	26.75	25.69	4	17.7 of BMCR
		27.02	25.94	4	
ERV on superheater outlet/4	26.75	26.50	25.70	3	41.5 of BMCR
		26.22	25.70	2	
Reheater inlet/8	5.20	5.20	4.99	4	90.70 of reheat flow
		5.25	5.04	4	
		5.30	5.09	4	
		5.35	5.14	4	
Reheater outlet/2	5.20	4.93	4.74	4	16.50 of reheat flow
ERV on reheater outlet/4	5.20	4.88	4.74	3	61.40 of reheat flow
		4.83	4.74	2	

12.2 Objective

Main objectives of floating of safety valves are to observe and record the following operating characteristics of each safety valve and to carry out necessary adjustment in the spring tension of the valve such that each operating condition matches the design set condition:

i.   opening (popping) pressure;
ii. closing pressure;
iii. the lift of the valve spindle.

A lift indicator should be utilized for personnel safety. If the valve does not operate at its set pressure and does not respond to readjustment, do not attempt to free it by striking the body or other parts of the valve. That valve should be repaired while the boiler is out of service. Any valve that does not perform exactly as the nameplate designates should be reconditioned and retested during the startup of the boiler. Notwithstanding the above, the following variation in set popping pressure is permissible (Table 12.3).

Closing pressure requires valve closure at a specific closing point, depending on the unit design. The valve may perform inconsistently, chatter, or damage itself if the closing pressure is too close to the popping pressure [3].

**Table 12.3 Permissible variation in spring-loaded safety valve popping pressure
(PG 72.2, P68, Section I, BPVC [2])**

Stipulated Pressure (MPa)	Permissible Variation in Pressure (kPa)
≤0.50	±15
>0.50 to ≤2.10	±3%
>2.10 to ≤7.00	±70
>7.00	±1%

Measurement of the spindle travel (lift) is used to determine whether a valve discharges its rated capacity. Travel of the spindle should be equal to or greater than the nameplate identified "lift" value. If the travel is less than the nameplate rating, the capacity must be reduced proportionally in accordance with the reduced travel compared with the full lift [3].

12.3 Precautions

Raising of steam generator pressure and floating of safety valves must be conducted under the direct guidance and supervision of the representative of steam generator manufacturers. Any adjustment of safety valve popping and blow-down pressures must be done gradually by trained personnel only in accordance with the instruction and procedure supplied by the original manufacturer of each safety valve.

Note

Blow-down is the difference between the opening pressure and the closing pressure of a safety valve.

Since the popping pressure of each safety valve is higher than the normal operating pressure of various systems of a steam generator, it is imperative to maintain stringent precautionary measures when this preoperational activity is in progress. Hence, over and above observing "general precautionary measures" laid down under Section 2.1.1: Quality Assurance, the following "specific precautionary measures" also must be observed.

i. Floating of safety valves must be performed when the boiler is under shut-down condition (not in normal operation);

ii. Do not apply gagging devices (Fig. 9.3) on the safety valves until the boiler pressure has been approximately 80% of normal operating pressure for a period of 2 h. This is to prevent damage to the valve spindle due to thermal expansion (Clause no. C4.130, P 47 [3]);

iii. Switch off the power supply to ERVs so as to prevent them from operating;

iv. Minimize simmering of the valve under test by raising and lowering the test pressure as fast as permissible before and after the lift, respectively. This may be achieved by careful manipulation of firing rate and/or opening of boiler startup vent valves;

v. Attempts shall be made to reduce the floating period of safety valves as far as practical to minimize damage to the disk and seat;

vi. While making adjustments on the safety valve when it is mounted on the boiler, the system operating pressure should be lowered to a value at least 10% below the valve set pressure before readjustments are attempted;

vii. Firing rate should be so controlled as to ensure that furnace exit gas temperature does not exceed 813 K (large subcritical steam generator) or 838–883 K (as applicable to supercritical steam generator);

viii. During the execution of floating of safety valves, the general noise level of a plant including surrounding areas would get raised much above the tolerable hearing limit of a human being (typically peak noise levels of 90–100 dBA may be expected during the popping of safety valves). Use earmuffs when floating is in progress;

ix. An audible and a visual alarm must be sounded to clear the area prior to initiating the floating activity.

12.4 Prerequisites

Floating of steam generator safety valves is an extremely risky process. Therefore, before any action is taken to carry out the process, a thorough inspection of systems under purview is essential to ascertain the following (Table 12.4):

Table 12.4 Prerequisites for floating of safety valves

Sl. No.	Areas/Items	Ok ($\sqrt{}$)
1	Protocols jointly signed by the customer, the engineer, the steam generator manufacturer, and the supplier (contractor) are available certifying successful completion of "steam generator initial firing and drying-out of insulation" (Chapter 11), "chemical cleaning of steam generator" (Chapter 6), "steam/air blowing of MS, CRH, HRH steam pipe lines" (Chapter 7) and "flushing of lube oil piping system" (Chapter 8)	
2	Verify that the isolating device, located on the cold reheat pipe lines downstream of HP turbine exhaust NRV, is in place	
3	Verify that the isolating device, located on the hot reheat pipe lines upstream of IP turbine inlet IV, is in place	
4	Verify that each safety valve is erected with spindle axis within a tolerance of ±1 degree from the vertical plane	
5	Ensure that the escape pipe is separately supported and does not touch the drain pan unit and that there is sufficient allowance for expansion in horizontal and vertical directions	
6	Verify that the escape pipe of each safety valve is laid outside the roof of the steam generator	
7	Ensure that the cover plate vent and drain are connected with a pipe of adequate size to safe locations	
8	Check that insulation of inlet and body up to cover plate of each safety valve is complete	
9	Verify that the seals are intact on ring pin, spring adjuster, and overlap collar	

Continued

Table 12.4 Prerequisites for floating of safety valves—Cont'd

Sl. No.	Areas/Items	Ok ($\sqrt{}$)
10	Ensure that the vent valve on the superheater is large enough to provide the minimum flow necessary to prevent overheating	
11	Ensure that gagging devices used during the hydrotest are removed and disks are replaced	
12	Check availability of gagging devices for all safety valves	
13	Check and ensure sufficient approach and maintenance space all around each safety valve as well as overhead	
14	Ensure that pressure gauges to be used for setting safety valve pressures are accurate and calibrated	
15	Verify that all instruments, which were in service during initial firing of the steam generator, are available	
16	Verify that relevant protections, alarms, and interlocks are checked and kept ready	
17	Check that the required quantity of materials (Section 12.4.1) is available	
18	Ensure that safety gadgets (Section 12.4.2) are arranged	
19	Ensure that temporary platforms are provided wherever required	
20	Check that the proper communicating system is established among all concerned personnel stationed at the control room and near the floating area to give the signal to start and stop the floating process	
21	Verify that "no smoking," "danger," "keep off," and other safety tags (in regional and English languages) are displayed at various places around the test area	
22	Verify also that any equipment and systems, which are in service or are energized, are provided with proper safety tags (Appendix C)	

12.4.1 Typical List of Bill of Materials

Materials that are typically required for floating of the steam generator safety valves are listed below. Specification and/or quantity of each of these items would vary from project to project and shall have to be assessed by concerned personnel associated with the floating of safety valve activity.

1. DM water;
2. fuel oil;
3. calibrated pressure gauges;
4. gagging devices;
5. hand-operated hydraulic/pneumatic pump.

Note

Since floating of safety valves follows initial firing of steam generator and steam/air blowing of MS, CRH, and HRH pipe lines, the steam generator is expected to be sufficiently filled with DM water and is under lit-up condition; hence, the required quantity of DM water and fuel oil for executing floating activity will not be much to be critical.

12.4.2 Typical List of Safety Gadgets

Equipment that are generally used for safety of operating personnel and the area covered under safety valve floating are as under:

1. safety shoes of assorted sizes;
2. asbestos gloves;
3. hardhats of assorted sizes;
4. ear muffs;
5. side-covered safety goggles with plain glass;
6. first aid box;
7. sufficient lighting arrangements around the safety valve area;
8. cordoning of the safety valve area;
9. hooters to send warning sound prior to commencing each floating operation;
10. safety sign boards;
11. safety tags.

12.5 Preparatory Arrangements

Prior to carrying out floating of steam generator safety valves, the following preparatory work shall be completed.

a. The steam generator along with its auxiliaries is in operation and pressure can be raised to the required level;
b. The HP bypass system, main steam, cold reheat and hot reheat steam pipe lines, and condensate and feedwater systems are proven to be ready for safe and reliable operation;
c. Turbine gland sealing steam system, lube oil, and jacking oil systems must be ready to be put into operation;
d. Prior to floating of safety valves, the turbine must be put on its turning gear to safeguard the turbine against damage due to unexpected leakage of steam that may cause rotation of the turbine;
e. Arrange the required number of matching gagging devices for each safety valve;
f. Adequate quantity of DM water and fuel oil shall be made available;
g. Quantity and type of safety gadgets, to be required for the process, should be ascertained beforehand and must be kept available as long as the floating process continues (a typical list of safety gadgets is furnished under Section 12.4.2);
h. Tripping of the boiler from "drum high high level" and "drum low low level" shall be bypassed (applicable to drum boiler);
i. Tripping of the boiler circulating pump from level variation in the water separator shall be bypassed (applicable to once-through boiler);
j. Readings of various instruments, used during initial firing of the steam generator, must be recorded in DCS.

12.6 Operating Procedure

Floating of the steam generator safety valves is expected to be carried out in continuation to steam blowing of main steam, cold reheat, and hot reheat steam pipe lines. Hence, all steam pipe lines are expected to be sufficiently warmed up.

As discussed earlier, floating of safety valves must be started with reheater safety valves first. Thereafter, carry out floating of superheater safety valves followed by boiler drum/steam separator outlet safety valves.

Two methods that are adopted in the industry to float safety valves are given below:

i.   The age-old practice followed in the industry is the popping of safety valves at "full set pressure of valve actuation";
ii. Alternatively, the valve may be fitted with a hydraulic or pneumatic lift-assist device and tested on steam at a pressure less than the valve set pressure (PG-73.5.2.2.2, P71, Section I [2])

The following paragraphs describe the procedure of each method separately.

12.6.1 Full Set Pressure Valve Actuation (Popping)

The full set pressure valve actuation (popping) method is the preferred and most reliable technique to ensure that the safety valves are operating properly (Clause No. C4.130, P 47 [3]).

The advantage of the "full set pressure valve actuation" method is that the popping pressure, the blow-down pressure, and the lift of a safety valve can be accurately adjusted and attained (Section 12.2).

This method, however, suffers from the following disadvantages:

i.   because of repeated raising and lowering of steam generator pressure (and also temperature), materials of the steam generator tubes, headers, and so on, are subjected to more combined stresses;
ii. pipe supports experience repeated sudden loading;

iii. spouting of steam during popping of safety valves:
   a.   creates loud noise;
   b.   requires eventual supply of substantial DM water;
   c.   develops sonic velocity through safety valves, resulting in wearing out of valve seats with consequent impairment of valve seat tightness;
iv. time required for completing floating of steam generator safety valves is substantial.

The step-by-step procedure to be followed for the full set pressure valve actuation is described below.

Light up the boiler at a slow rate in accordance with the steam generator manufacturer's recommended cold startup procedure, if the steam generator is not under the lit-up condition. Open SHO stop valve/s and establish flow through HP bypass, cold reheat, and hot reheat pipe lines.

12.6.1.1 *Floating of reheater safety valves (subcritical steam generator: Table 12.1/supercritical steam generator: Table 12.2)*

Increase the firing rate and gradually raise the boiler pressure.

1. When reheater outlet pressure reaches 4.0 MPa or to a value recommended by the steam generator manufacturer, gagging devices may be provided to all safety valves on hot reheat and cold reheat pipe lines barring the lowest pressure safety valve on hot reheat pipe lines;
2. Raise the reheater outlet pressure to the set pressure of lowest pressure safety valve and float the valve;
3. Adjust the popping pressure if so required as explained below (Fig. 12.1):
   a. Remove the lever and the cap;
   b. Loosen the locknut;
   c. Turn the spring adjuster (compression screw nut) clockwise to increase pressure or counter-clockwise to decrease pressure;
   d. Tighten the locknut;
   e. Reassemble the cap and the lever.
4. Repeat step 3 until the popping pressure reaches the design set pressure;
5. If necessary adjust the blow-down (closing pressure) as detailed below (Fig. 12.1):
   a. Remove the upper adjusting ring pin;
   b. Turn the upper adjusting ring to the right to decrease the blow-down or to the left to increase;
   c. Adjust the upper adjusting ring pin;
6. Repeat step 5 until the amount of blow-down becomes equal to the design value;
7. After the proper setting of the popping pressure and the blow-down, reduce reheat pressure to 4.0 MPa or to a value recommended by the steam generator manufacturer;
8. Gag the lowest pressure safety valve on the hot reheat pipe line; remove the gagging device from the next higher pressure safety valve on the hot reheat pipe lines;
9. Raise the reheater outlet pressure;
10. Repeat steps 3–7 for the next higher pressure safety valve on hot reheat pipe lines;
11. Provide gagging device to the lower pressure safety valve on the hot reheat pipe lines; remove the gagging device from the next higher pressure safety valve on the hot reheat pipe line;
12. Repeat steps 9 and 10 for all higher pressure reheat safety valves;
13. Close SHO main steam stop valves along with their equalizing bypass valves.

12.6.1.2 *Floating of boiler drum and superheater safety valves (subcritical steam generator: Table 12.1)*

14. Gradually raise SHO pressure to 15 MPa or to or a value recommended by the steam generator manufacturer. Provide gagging devices on all the safety valves on the boiler drum and on the superheater barring the lowest pressure safety valves on the SHO main steam pipe lines;

15. Raise boiler pressure further to the set pressure of the lowest pressure safety valves on the main steam pipe lines and float them;

16. Repeat steps 3–6;

17. After the proper setting of the popping pressure and the blow-down, reduce superheat pressure to 15.0 MPa or to a value recommended by the steam generator manufacturer;

18. Gag the lowest pressure safety valve on the main steam pipe lines; remove the gagging device from the next higher pressure safety valve on the main steam pipe lines;

19. Raise the SHO pressure;

20. Repeat steps 3–6 and 17 for the next higher pressure safety valve on main steam pipe lines;

21. Continue with gagging, raising pressure and step 20 for all higher pressure safety valves on main steam pipe lines;

22. Once the floating of safety valves on the main steam pipe lines is complete, provide the gagging device on the highest pressure safety valve on the main steam pipe lines; remove the gagging device from the lowest pressure safety valves on the boiler drum;

23. Raise the boiler pressure further to the set pressure of the lowest pressure safety valves on the boiler drum and float them;

24. Repeat steps 16 and 17;

25. Gag the lowest pressure safety valve on the boiler drum; remove the gagging device from the next higher pressure safety valve on the boiler drum;

26. Raise the SHO pressure;

27. Repeat steps 16 and 17 for the next higher pressure safety valve on the boiler drum;

28. Continue with gagging, raising pressure, and step 27 for all higher pressure safety valves on boiler drum;

29. On completion of floating of all the safety valves on reheater, superheater, and the boiler drum, ensure that all gagging devices are removed;

30. Record all observations.

12.6.1.3 *Floating of steam separator outlet and superheater safety valves (supercritical steam generator: Table 12.2)*

Procedure of floating safety valves of the supercritical steam generator is identical to the procedure of floating safety valves of the subcritical steam generator, except the level of floating pressure, which is raised to about 22 MPa or to a value recommended by the steam generator manufacturer.

Floating will start from the lowest pressure safety valve on SHO main steam pipe lines and end up with the highest pressure safety valve located on the steam separator (in lieu of boiler drum) outlet.

To finish the procedure, repeat steps 15–30 above.

12.6.2 Lift-Assisted Valve Actuation

This method may be adopted when all of the following conditions are met (Clause No. PG-73.5.2.2, P71, Section I, BPVC [2]):

i.   Testing of a safety valve at full pressure may cause damage to the valve;
ii. Valve lift is mechanically verified to meet or exceed the required lift;
iii. Blow-down control elements of the valve are set to the specification of the valve original manufacturer;
iv. Valve design is compatible with this alternative test method.

Advantages of adopting the "lift-assisted valve actuation" method are:

i.   The method is very simple; the operator has to operate the hydraulic/pneumatic pump;
ii. Since the steam pressure is maintained at normal operating condition;
    a.   Firing of the steam generator remains undisturbed;
    b.   Materials of the steam generator tubes, headers, and so on, do not get subjected to undue stresses;
    c.   There is less generation of noise;
    d.   There is lower wearing out of valve seats.
iii. Spouting of steam is less with eventual less consumption of DM water.

However, floating of safety valves adopting this method impairs determining the following requirements of C4.120, P43, Section VII [3] and also Section 12.2:

i.   the blow-down pressure;
ii. the lift of the valve spindle.

Steps to be followed for floating of steam generator safety valves adopting this method are described below:

1. Light up the steam generator;
2. Raise steam pressure up to its normal operating pressure (Table 12.5);
3. Remove the lever, cap, and locknut from the selected safety valve (Fig. 12.1);
4. Fix the lift-assist unit (Fig. 12.5) to the yoke of this safety valve;
5. Connect one end of a high pressure hose to the discharge of the hydraulic/pneumatic pump and other end to the lift-assist unit;
6. Manually operate the handle of the hydraulic/pneumatic pump and gradually apply pressure to the lift-assist unit;
7. Keep vigilance on the discharge pressure gauge of the pump;
8. As the pressure is raised to the popping pressure of the selected safety valve, observe whether the valve pops or not;
9. In the event the valve pops at its set pressure, the test of this valve may be construed to be complete;
10. If the valve pops at a pressure lower than the set value, turn the spring adjuster (compression screw nut) clockwise to increase pressure;

11. If the valve fails to pop at its set pressure, turn the spring adjuster counter-clockwise to decrease pressure

12. Repeat step 10 or step 11 until the popping pressure reaches the design set pressure;

13. On successful completion of the popping, carry out the following:

    a. Remove the load-assist unit;

    b. Tighten the locknut;

    c. Reassemble the cap and the lever.

Table 12.5 Normal operating pressure (MPa) of steam

Sl. No.	Location	Subcritical	Supercritical
1	Superheater outlet main steam pipe lines	17.67	23.90
2	Reheater inlet cold reheat pipe lines	4.95	4.58
3	Reheater outlet hot reheat pipe lines	4.67	4.30
4	Boiler drum	19.05	–
5	Steam separator outlet	–	25.86

Note: Refer to Tables 12.1 and 12.2.

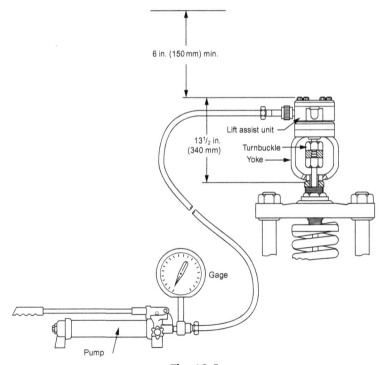

Fig. 12.5

Hydraulic lift assist arrangement of safety valve. *Source: From Fig. C4.1–5, P 49, C4.130 Safety Valve Testing. ASME B&PVC VII (2010): Boiler and Pressure Vessel Code, Part VII, Recommended Guidelines for the Care of Power Boilers.*

12.7 Conclusion

Floating of steam generator safety valves is witnessed by the authorized representative of the boiler inspectorate or regulatory authorities. If the test is observed to be satisfactory in all respects, he approves and certifies the test.

Ensure that gagging devices are removed from each safety valve.

In continuation of Table 12.1 or Table 12.2, a protocol (either Table 12.6 or Table 12.7) jointly signed by the customer, the engineer, the steam generator manufacturer, and the supplier (contractor) may be issued thereafter, which is ratified by the representative of the boiler inspectorate or any other regulatory authority.

Table 12.6 Protocol of floating of safety valves (subcritical steam generator)

Location/No. of Valves	Set Pressure (MPa)		Tested Pressure (MPa)		Percent of Blow-Down (Calculated)		Remark
	Open	Close	Open	Close	Set	Tested	
Boiler drum/6	20.00	19.20			4		
	20.20	19.20			5		
	20.40	19.38			5		
	20.40	19.38			5		
	20.60	19.57			5		
	20.60	19.57			5		
Superheater	18.40	17.85			3		
outlet/2	18.40	17.85			3		
ERV on	18.18	17.81			2		
superheater	18.18	17.81			2		
outlet/4	18.18	17.81			2		
	18.18	17.81			2		
Reheater inlet/4	5.38	5.22			3		
	5.43	5.27			3		
	5.48	5.32			3		
	5.54	5.37			3		
Reheater outlet/4	4.80	4.66			3		
	4.85	4.71			3		
	4.90	4.76			3		
	4.90	4.76			3		
ERV on reheater	4.71	4.52			4		
outlet/4	4.71	4.52			4		
	4.76	4.57			4		
	4.76	4.57			4		
Signed by the customer	Signed by the engineer		Signed by the steam generator manufacturer			Signed by the supplier (contractor)	
			Signed by the representative of regulatory authority				

Table 12.7 Protocol of floating of safety valves (supercritical steam generator)

Location/No. of Valves	Set Pressure (MPa)		Tested Pressure (MPa)		Percent of Blow-Down (Calculated)		Remark
	Open	Close	Open	Close	Set	Tested	
Steam separator outlet/6	28.91	27.75			4		
	28.91	27.75			4		
	28.91	27.75			4		
	29.67	28.48			4		
	29.67	28.48			4		
	29.67	28.48			4		
Superheater outlet/4	26.75	25.69			4		
	26.75	25.69			4		
	27.02	25.94			4		
	27.02	25.94			4		
ERV on superheater outlet/4	26.50	25.70			3		
	26.50	25.70			3		
	26.22	25.70			2		
	26.22	25.70			2		
Reheater inlet/8	5.20	4.99			4		
	5.20	4.99			4		
	5.25	5.04			4		
	5.25	5.04			4		
	5.30	5.09			4		
	5.30	5.09			4		
	5.35	5.14			4		
	5.35	5.14			4		
Reheater outlet/2	4.93	4.74			4		
	4.93	4.74			4		
ERV on reheater outlet/4	4.88	4.74			3		
	4.88	4.74			3		
	4.83	4.74			2		
	4.83	4.74			2		
Signed by the customer	Signed by the engineer		Signed by the steam generator manufacturer				Signed by the supplier (contractor)
Signed by the representative of regulatory authority							

References

[1] S.C. Stultz, J.B. Kitto (Eds.), Steam: Its Generation and Use, 41st ed., The Babcock and Wilcox Company, Barberton, OH, 2005.

[2] ASME BPVC I. Boiler and Pressure Vessel Code, Rules for Construction of Power Boilers, as required by the States of Colorado, Connecticut, Delaware, Indiana, Minnesota, Mississippi, Nevada, New York, North Carolina, Oklahoma, Rhode Island, South Carolina, Tennessee, Texas, et. alia, 2010.

[3] ASME B&PVC VII. Boiler and Pressure Vessel Code, Part VII, Recommended Guidelines for the Care of Power Boilers, as required by the States of Connecticut, Nevada, North Carolina, Oklahoma, Rhode Island, Tennessee, et. alia, 2010.

Clean Air Flow Test of a Pulverizer

13.1 Introduction

The concept of suspension firing of coal dates back to 1824 by *Nicholas Le'onard Sadi Carnot* from the belief that if coal were ground to the fineness of flour, it would flow like oil and would burn as easily and efficiently as gas. Carnot visualized a coal-fired engine, which never came into reality. In 1890 while playing with the idea of a pressurized fuel-injection system, French engineer Rudolf Diesel attempted to use pulverized coal as fuel in his diesel engine, but without any success. It took almost 100 years since Carnot when pulverized coal was tried as fuel in steam generators; it resulted in great success and gained momentum after World War I. By the 1920s pulverized coal firing became so developed that it resulted in more complete coal combustion and became a dependable method of coal firing for commercial production and utilization of steam. Eventually pulverized coal is established as the most efficient way of using coal in steam generators.

As discussed in Chapter 1 the process of pulverization of coal is carried out in two stages: first raw coal is crushed to a size of not more than 15–25 mm, then crushed coal is ground to fine particle sizes in grinding mills or pulverizers. The level of fineness is such that about 70% of fine particles will pass through 200 mesh (0.075 mm) and equal to or more than 98% will pass through 50 mesh (0.300 mm).

Thermal Power Plant. http://dx.doi.org/10.1016/B978-0-08-101112-6.00013-7

Grinding inside a pulverizer is realized by impact, attrition, crushing, or a combination of these. Depending on the pressure existing inside the grinding zone, pulverizers or mills may be categorized as the suction type, where the primary air fan is located downstream of pulverizer, or the pressurized type in which hot or cold primary air fan supplies air to the pulverizer.

A typical coal and primary air flow diagram of a pressurized pulverizer is shown in Fig. 13.1.

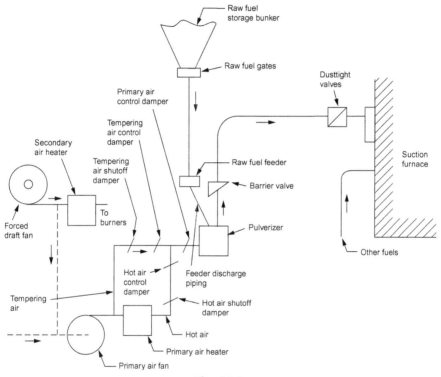

Fig. 13.1

Coal and air flow diagram of a pressurized pulverizer (typical). *Reproduced with permission from Fig. 9.4.5.1.1 (f), P 85–125, Pulverized Fuel Systems, NFPA 85-2015, Boiler and Combustion Systems Hazards Code, Copyright© 2044, National Fire Protection Association. This reprinted material is not the complete and official position of the NFPA on the referenced subject, which is represented only by the standard in its entirety.*

The factors that affect pulverizer performance are:

    i. Raw-coal size
   ii. Fineness
  iii. Volatile matter content in coal
   iv. Moisture content in coal
    v. HHV (GCV) of coal
   vi. Hardgrove Grindability Index (HGI)
  vii. YGP (Abrasive) Index
 viii. Extraneous materials

13.1.1 Classification of Mills [1]

Based on their operating speed, mills are classified as "low," "medium," and "high" speed mills.

13.1.1.1 Low-speed mill

Low-speed mills operate below 75 rpm. Tube mill, also known as ball mill, or drum mill, falls into this category. Its normal operating speed is about 15–25 rpm.

Tube mill (Fig. 13.2) is of drum-type construction with conical ends and heavy-cast wear-resistant liners. This is a very rugged piece of equipment and is less than half-filled with forged alloy-steel balls of mixed size. Grinding in this mill is accomplished partly by impact, as the grinding balls and coal ascend and fall with cylinder rotation, and partly by attrition between coal lumps inside the drum. Primary air is circulated over the charge to carry the pulverized coal to classifiers. In this type of mill, pulverized coal exits from the same side of the mill that solid coal and air enter.

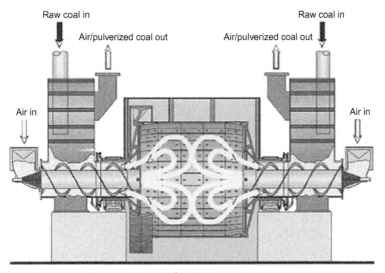

Fig. 13.2
Low-speed mill. *Source: From Foster Wheeler Corporation, 'Hard-coal burn-up increase with adjustable classifier for Ball Mill', POWER-GEN International Conference, Orlando, Florida, USA, 2000. With kind permission from Amec Foster Wheeler.*

Reliability of this type of mill is very high and it requires low maintenance. The disadvantages of tube mills are high power consumption, larger and heavier construction, greater space requirement, and so on.

13.1.1.2 Medium-speed mill

Medium-speed mills are smaller than low-speed units and are generally of the vertical spindle construction. The speed of the grinding section of these mills is usually 75–225 rpm. They operate on the principles of crushing and attrition. Pulverization takes place between two surfaces, one rolling on top of the other. Primary air causes coal feed to circulate between the grinding elements, and when coal becomes fine enough to be airborne, the finished product is conveyed to the burners or the classifier. Medium-speed mills require medium to high maintenance, but their power consumption is low.

This mill is usually of one of two types: ball-and-race mill and roll-and-race mill.

In the ball-and-race mill (Fig. 13.3), balls are held between two races, much like a large ball bearing. The top race or grinding ring remains stationary while the bottom race rotates. As the coal is ground between large diameter balls and the ring, the balls are free to rotate on all axes and therefore remain spherical.

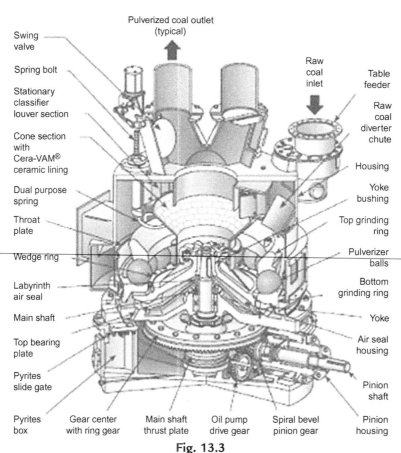

Fig. 13.3

Ball-and-race mill. *Source: From Fig. 5, P 13–3, Chapter 13: Coal Pulverization, Editor: S. C. Stultz and J. B. Kitto, STEAM Its Generation and Use (41st Edition). Courtesy of The Babcock & Wilcox Company.*

The grinding elements of a roll-and-race mill consist of three equally spaced, spring-loaded heavy conical (Fig. 13.4) or toroidal rolls (Fig. 13.5) suitably suspended inside near the periphery. These rolls travel in a concave grinding ring or bowl (with heavy armoring). The main drive shaft turns the table supporting the grinding ring, which in turn transmits the motion to the rolls. There is no metal-to-metal contact between grinding elements, since each roller rests on a thick layer of coal. Thus the maintenance is minimized.

Mills with conical rolls are known as bowl mills. As the coal is ground between large diameter rolls and the bowl, rolls revolve about their own axes, and the grinding bowl revolves about the axis of the mill.

In the toroidal rolls type, grinding occurs only under the rollers in the grinding ring.

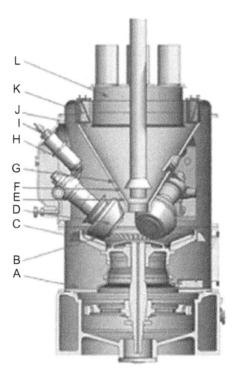

A. Pyrite sweep conditions/clearances

B. Grinding element condition/clearances

C. Throat dimensions/opening

D. Roll/journal condition

E. Feed pipe clearances

F. Inverted cone/conical baffle clearances

G. Classifier cone condition

H. Button clearance/spring height

I. Preload of spring canisters

J. Outlet cylinder height in relation to classifier blades

K. Classifier blade condition/ length/stroke synchronized angles

L. Outlet smooth, free of any obstructions or spin arresting protrusions into the spinning two-phase mixture of coal and air

Fig. 13.4

Conical roll-and-race mill. *Source: From P4, Pulverizers 101: Part I, by Dick Storm, PE, Storm Technologies Inc, POWER Magazine, 08/01/2011. Reproduced with permission from Storm Technologies.*

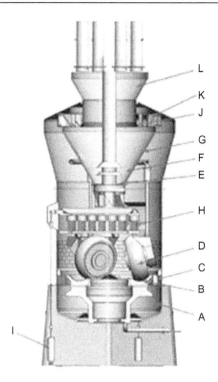

A. Pyrite sweep conditions/clearances   B. Grinding element condition/clearances
C. Throat dimensions/opening   D. Roll/journal condition
E. Feed pipe clearances   F. Inverted cone/conical baffle clearances
G. Classifier cone condition   H. Button clearance/spring height
I. Preload of spring canisters   J. Outlet cylinder height in relation to classifier blades
K. Classifier blade condition/   L. Outlet smooth, free of any obstructions or spin arresting
length/stroke synchronized angles   protrusions into the spinning two-phase mixture of coal and air

Fig. 13.5

Toroidal roll-and-race mill. *Source: From P4, Pulverizers 101: Part I, by Dick Storm, PE, Storm Technologies Inc, POWER Magazine, 08/01/2011. Reproduced with permission from Storm Technologies.*

13.1.1.3 High-speed mill

These mills use central horizontal shafts having number of arms; hammer beater of different designs are attached to these arms to beat the coal to be pulverized (Fig. 13.6). The beaters revolve at speeds above 225 rpm in a chamber equipped with high-wear-resistant liners. Both impact and attrition is combined in this mill to pulverize coal, which is pulverized by the rubbing of coal on coal, by the impact of coal on impeller clips, and also by the rubbing of peg with peg. A classifier returns the coarse coal particles for further classification.

Its capital cost per unit output is low, it requires minimum space, and its parts are lightweight to facilitate maintenance. However, its maintenance cost is so high that this type of mill has long been discontinued from service.

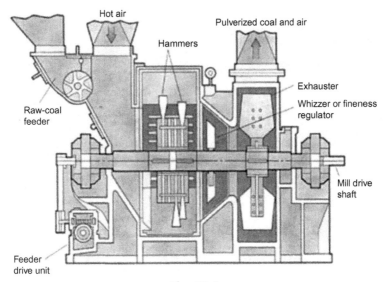

Fig. 13.6

High speed mill. *Source: From Fig. 6–18, P 6–20, Chapter 6: Auxiliary Equipment, Editor: Carl Bozzuto, Clean Combustion Technologies (5th Edition), Alstom.*

13.2 Objective

Each pulverizer may be provided with 4–10 coal-air discharge pipes from its top. Configuration of each coal-air discharge pipe layout is quite complex with vertical run, horizontal run, bends, changes in direction, and so on, and similarity between two coal-air pipes, in particular length and flow resistance, is seldom attainable. The purpose of the "clean air flow test of a pulverizer" is to achieve the design rate of primary air flow through each pulverizer at any time and to ensure equal flow through each coal-air discharge pipe of each of the pulverizers.

It is widely recognized within the industry that balanced coal-air flow to the burners of a pulverized coal-fired steam generator results in more efficient combustion, reduced emissions of air pollutants, and fewer problems of waterwall tube wastage and burner damage due to locally reducing zones in the furnace [2].

Hence, clean air flow test is performed to:

i. Establish similar system resistance for each coal-air discharge pipe and balancing of coal-air flow through each coal pipe from pulverizer discharge to individual coal burner;

ii. Provide a correlation between velocities of coal-air and clean air through each coal-air discharge pipe;

iii. Balance coal-air discharge pipe-to-pipe "air-to-fuel ratio" and "fuel flow" with minimum relative variation;

iv. Optimize clean air flow through each coal-air discharge pipe to determine the minimum velocity of coal-air through each pipe that would establish improved flame stability at lower loads and reduce fuel-line stoppages;

v. Determine variation in air flow is within $\pm 5\%$ and the coal flow within $\pm 10\%$ through each coal-air pipe [3];

vi. Achieve maximum efficiency of steam generators with minimum pollution emissions.

13.3 Precautions

Consequent benefits of conducting the clean air flow test are ensuring stable flame in each burner and efficient combustion in the furnace. To reap these benefits to their full potential, this test must be conducted in accordance with the instruction and procedure supplied by the original manufacturer of pulverizers and under the direct guidance and supervision of the representative of the original manufacturer. Any adjustment of velocity to establishing smooth flow through each coal-air pipe, equal load sharing, and so on, must be done by trained personnel only.

In order to ensure fine particles of coal remain in suspension in coal-air mixture, it is essential to maintain a velocity of about 17 m/s in each coal-air pipe. In the event this velocity falls below 15 m/s, fire or explosion in a coal-air pipe or pulverizer may be confronted.

Another aspect that is crucial is, irrespective of load on the pulverizer, a minimum primary air flow (typically as high as 70% of full-load primary air flow) shall always be maintained. Furthermore, it is observed in the industry that with a 25% drop in load from a 100% pulverizer load, primary air flow gets dropped by only 10% from 100% primary air flow.

In addition to the above, the following precautionary measures must be observed while conducting the clean air flow test:

i.   Variation in pulverizer air flow must be kept as minimum as attainable;

ii. Temperature of air must remain reasonably constant;

iii. Impulse lines, pitot tubes, and so on, must be free from any leakage;

iv. There must not be any plugging in impulse lines, pitot tubes, and so on.

Notwithstanding the above precautionary measures, it is also essential to observe "general precautionary measures" discussed under Section 2.1.1: Quality Assurance.

13.4 Prerequisites

Prior to carrying out the clean air flow test of pulverizers, a thorough inspection of systems under purview is essential to ascertain the following (Table 13.1):

Table 13.1 Areas/items to be checked and satisfied

Sl. No.	Areas/Items	Ok (√)
1	Protocols jointly signed by the customer, the engineer, the steam generator manufacturer, and the supplier (contractor) certifying erection completion of following equipment are available (Fig. 13.1): i. raw coal bunkers; ii. raw coal feeders; iii. pulverizers; iv. interconnecting pipe between raw coal bunkers & raw coal feeders with intermediate raw coal gates and also between raw coal feeders and pulverizers; v. coal-air pipes; vi. dampers/gates/valves on tempering and hot primary air ducts and coal-air pipes; vii. coal burners.	
2	Trial run of the following equipment is successful, and respective protocols jointly signed by the customer, the engineer, the steam generator manufacturer, and the supplier (contractor) are available: i. ID fans; ii. FD fans; iii. air preheaters (regenerative type); iv. PA fans; v. seal air fans.	
3	Prior to commencing the clean air flow test, ensure that all draft/impulse lines are pressure tested and related protocol is available.	
4	Verify calibration certificates of required instruments.	
5	Ensure that associated instrumentation and control are installed and operable.	
6	Verify that straight run of pipe of at least $10D$ upstream of and $5D$ downstream from the location of the desired air flow measurement is available (where D is the internal diameter (ID) of each coal-air pipe).	
7	To facilitate isokinetic sampling of coal-air, the pitot tube is marked for the various traverse positions as given below: Location 1: $0.044D+X$ Location 2: $0.146D+X$ Location 3: $0.296D+X$ Location 4: $0.704D+X$ Location 5: $0.854D+X$ Location 6: $0.956D+X$ where D is ID of each coal-air pipe, and X is the stub height on the pipe for traversing the pitot tube.	

Table 13.1 Areas/items to be checked and satisfied—cont'd

Sl. No.	Areas/Items	Ok ($\checkmark$)
8	A representative reading of air flow through each coal-air pipe can be measured by traversing the pitot tube across the cross-section of the pipe. Check that the number of the traverse point complies with either of the following requirements: i. If 10*D* and 5*D* distances of item 6 above are satisfied, traverse of pitot tube should be made along at least two diameters at right angles to each other, ie, the required number of measuring points are 24; ii. If requirement of item 4 cannot be met, traverse of pitot tube should be made along at least four diameters at 45 degrees to one another, ie, the required number of measuring points are 48 (Fig. 13.7).	

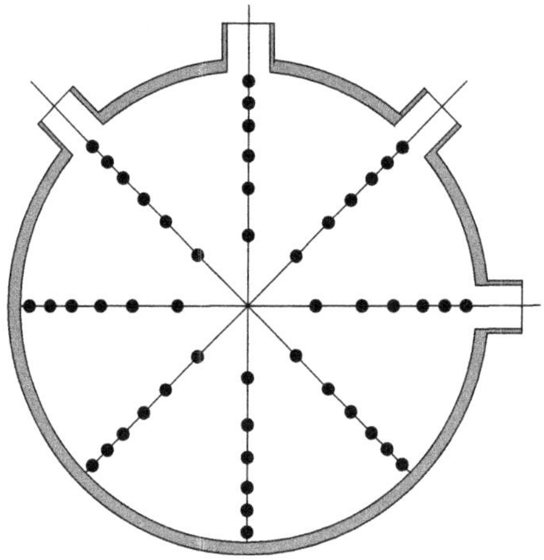

Fig. 13.7
Traverse points for isokinetic sampling.

9	Temporary platforms are provided wherever required.	
10	Apparatus/instruments generally required for isokinetic sampling (Section 13.4.1) are arranged.	
11	Safety gadgets (Section 13.4.2) are arranged.	

13.4.1 Typical List of Apparatus/Instruments

Apparatus/instruments that are typically required for isokinetic sampling during the clean air flow test of a pulverizer are listed below. Specification and/or quantity of each of these items would vary from project to project and shall have to be assessed by concerned personnel associated with the test.

    i. pitot tubes;

    ii. barometer;

   iii. manometers;

   iv. thermometer;

    v. orifices to be used in certain coal-air pipes to balance the loss and attain the same flow resistance in all coal-air pipes.

13.4.2 Typical List of Safety Gadgets

Equipment generally used for safety of operating personnel during the clean air flow test of a pulverizer are given below:

1. safety shoes of assorted sizes;
2. hand gloves;
3. hardhats of assorted sizes;
4. side-covered safety goggles with plain glass;
5. first aid box;
6. sufficient lighting arrangements;
7. safety sign boards;
8. safety tags.

13.5 Preparatory Arrangements

Prior to carrying out the clean air flow test of a pulverizer, the following preparatory work shall be completed:

1. Close raw coal gates between raw coal pipes from raw coal bunkers and raw coal feeders. Fill the raw coal pipe with coal up to the above gate. Such arrangement will ensure that all air from the PA fan will pass through coal-air pipes without any leakage through raw coal pipes;
2. Open the pulverizer classifier vanes to the full open position;
3. Close the following dampers/gates:
       i. regenerative air preheater primary air inlet and outlet damper;
      ii. hot air shut-off dampers/gates to pulverizers;
     iii. hot air control dampers to pulverizers.
4. Open the following dampers/gates:
       i. pulverizer discharge gates;
      ii. barrier valves and dust tight valves on coal-air pipes;
     iii. tempering air shut-off dampers/gates to pulverizers;
     iv. tempering air control dampers to pulverizers;
      v. primary air control dampers to pulverizers;
     vi. seal air supply dampers to pulverizers.

13.6 Operating Procedure

   i. Start one ID fan;
  ii. Start one FD fan;
 iii. Start one PA fan;
 iv. Maintain about (−) 40 Pa of furnace draft throughout the clean air flow test;
  v. Start the seal air fan of the associated pulverizer under test;
 vi. Start the pulverizer under test;
 vii. Maintain constant primary air differential pressure at the pulverizer inlet corresponding to 25% pulverizer load by adjusting the cold air damper and the PA fan control vane/blade pitch;
viii. Insert the pitot tube in one of the pulverizer coal-air discharge pipes and take six readings for each traverse;
 ix. Repeat step (viii) in line with the requirement of Section 13.4 (8);
  x. Repeat steps (vii), (viii), and (ix) for the remaining coal-air discharge pipes of this pulverizer;
 xi. Calculate the average differential pressure for each discharge pipe;
 xii. Calculate the average differential pressure for all the discharge pipes;
xiii. Check that the variation between the calculated values of steps (xi) and (xii) is within ±5%. If the variation is more than ±5%, adjust the orifice in the discharge pipes and repeat the test;
xiv. Continue with step (xiii) until the variation is brought within ±5%;
 xv. Repeat steps (vii) through (xiv) for constant primary air differential pressure at the pulverizer inlet corresponding to 50%, 75%, and 100% pulverizer load;
xvi. Record barometric pressure, ambient temperature, air pressure, and temperature at the pitot tube traverse, and air pressure and temperature at the venturi (permanent measuring device, located upstream of each pulverizer) on the primary air line.

13.7 Conclusion

1. Once the average differential pressure is obtained, the velocity and flow of clean air through each coal-air pipe is calculated as shown below (assuming pitot tube coefficient as 1):

$$V = \sqrt{2g\Delta P} = \sqrt{2 \times 9.81 \times \frac{\Delta P}{9.81 \times 10^3}} = 0.0447\Delta P$$

$$W = \frac{\pi}{4} \times D^2 \rho V = 7.857 \times 10^{-7} D^2 \rho V$$

where V: velocity of clean air through each coal-air pipe (m s^{-1});
W: flow of clean air through each coal-air pipe (kg s^{-1});
ΔP: measured differential pressure in the pitot tube (Pa);
ρ: specific weight of air at the pitot tube traverse (kg m^{-3});
D: ID of each coal-air pipe (mm).

Note

9.81×10^3 Pa $= 1$ mwc.

2. Total the clean air flow through all coal-air pipes of a pulverizer. With an air tight system, the total air flow must be equal to the design primary air flow through the venturi;

3. If there is any discrepancy between the total clean air flow and the design primary air flow through the pulverizer, then to make these two flows equal, the primary air differential pressure for the design primary air flow shall be corrected as

$$\text{Design PA } \Delta P\,(\text{corrected}) = \text{Measured PA } \Delta P \times \left(\frac{\text{Design PA Flow}}{\text{Measured PA Flow}} \right)^2$$

4. On completing steps 1 through 3, prepare a log sheet per Table 13.2, typically for a pulverizer having "4 coal-air discharge pipes."

Table 13.2 Log sheet

Sl. No.	Time	Burner No.		ΔP Across Pitot Tube (Pa) 1	2	3	4	5	6
1		1	Plane 1						
			Plane 2						
		2	Plane 1						
			Plane 2						
		3	Plane 1						
			Plane 2						
		4	Plane 1						
			Plane 2						

Average ΔP Through Each Pipe (Pa) I	Average ΔP Through Four Pipes (Pa) II	Variation Between I and II (%)	Average Velocity Through Each Pipe (m s^{-1})	Average Flow Through Each Pipe (kg s^{-1})

Total Flow Through Each Pipe (kg s^{-1}) III	Design Primary Air Flow Through Pulverizer (kg s^{-1}) IV	Variation Between III and IV (kg s^{-1})	Measured PA ΔP (Pa)	Design (Corrected) PA ΔP (Pa)	Remarks
Signed by the customer	Signed by the engineer	Signed by the steam generator manufacturer		Signed by the supplier (contractor)	

References

[1] D.K. Sarkar, Thermal Power Plant—Design and Operation, Elsevier, Amsterdam, Netherlands, 2015.
[2] H. Bilirgen, H. Caram, J. DuPont, E. Levy, J. Farren, A. Stockdale, New coal-flow balancing technology ready for field trial, Lehigh Energy Update 27 (2) (2009).
[3] R.G. Mudry, Coal Pipe Balancing Case Study, Airflow Sciences Corporation, Livonia, MI, 2006.

CHAPTER 14

Condenser Flood and Vacuum Tightness Tests

Chapter Outline

14.1 Introduction

Thomas Newcomen was the innovator in developing the commercially successful atmospheric steam engine way back in 1712. This engine did not have separate condensation chamber; thus, efficiency of this engine was very poor due to the loss of heat resulting from the condensation of steam within the working cylinder. It took more than 50 years since Newcomen that another innovator, James Watt, while repairing a model Newcomen engine, found how inefficient this engine was. Consequent to this finding, in the year 1765 Watt conceived the idea of equipping this engine with a separate condensation chamber, which he designated as a "condenser." As a result while the condensation chamber continued to be cold, the cylinder remained hot, thereby eliminating loss of heat from the cylinder of a Newcomen engine during condensation of steam and enhancing the efficiency of the Watt engine. Pressure inside the condensation chamber of the Watt engine always remained subatmospheric.

Thermal Power Plant. http://dx.doi.org/10.1016/B978-0-08-101112-6.00014-9

The function of a condenser in a conventional steam power plant (Chapter 1) is to create the lowest possible turbine back pressure while condensing steam. The condenser pressure is lowered to subatmospheric condition by evacuating air from the condenser shell. A liquid ring vacuum pump and/or noncondensing type single-stage starting (hogging) air ejector sucks air from the condenser shell including LP turbine casing and discharges air to the atmosphere creating subatmospheric condition or vacuum in the condenser. During normal operation of a plant, this pump/steam-jet (holding) air ejector must remain in service to continuously remove air ingress to a condenser to maintain condenser vacuum.

In a conventional steam power plant, by the condensation of exhaust steam from steam turbines at subatmospheric pressure, "steam pressure drop" across steam turbines from the inlet to the exhaust is enhanced; as a result, the amount of heat available also enhances for conversion to mechanical power. Heat liberated from the exhaust steam is removed by a cooling medium: either water (Fig. 1.25) or air, or a combination of water and air.

14.1.1 Circulating Water (CW) Cooled Condenser [1]

The CW system could be either the "recirculating" type or "once through" type. In the "recirculating" type (Fig. 14.1), the clarified water is supplied to the CW sump. From this sump, CW pumps supply cooling water to the condenser. Hot water from the condenser outlets is then carried to the hot water section of the cooling tower, where the downward flow of water is cooled by the upward flow of air, thereby rejecting heat to the atmosphere. The cooling tower may be of the "induced draft" type or of the "natural draft" type.

Cooled water is collected at the cold-water basin of the cooling tower. From the cold-water basin, water travels to the CW sump. Since the cooling water travels through a closed cycle, only the makeup water is replenished from the clarified water reservoir to the CW sump.

The "once through" condenser cooling water system (Fig. 14.2) can only be adopted in coastal areas. In this type, sea is the source of water. CW pumps are installed near or on the coast, from which water is pumped through CW piping. The hot water from the condenser is returned to the sea downstream of the CW pump house, such that the maximum temperature difference between the supply water and the return water is restricted to 7 K to protect marine life.

In both "recirculating" and "once through" types, cooling water enters the water box of the condenser at the bottom and subsequently flows through the tubes of the condenser.

Steam enters the condenser shell through the steam inlet connection from the LP turbine exhaust and is spread out lengthwise over the outside of the tubes. The cooling water absorbs the latent heat of the exhaust steam, causing steam to condense with consequent rapid change from a vapor phase to a liquid phase. This phase change results in a great reduction in specific volume of the fluid. Consequent to this reduction in volume, a vacuum is created in the condenser by condensation of the exhaust steam. The vacuum produced by condensation of steam will sustain as long as air ingress to the condenser could be prevented.

The condensed steam or the condensate is then directed to the condenser hotwell and warm CW exits through the condenser water box outlet at the top. With this configuration, the CW flow through each shell of the condenser and the pressure (vacuum) in each shell would be identical, since the CW inlet temperature in each shell remains unchanged. This condenser is called a "single-pressure" condenser (Fig. 14.3).

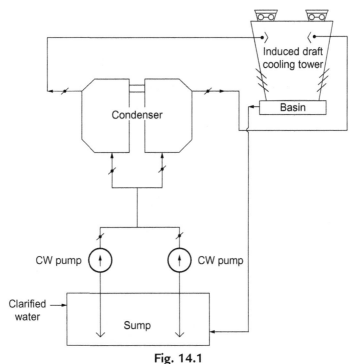

Fig. 14.1
Recirculating type CW system.

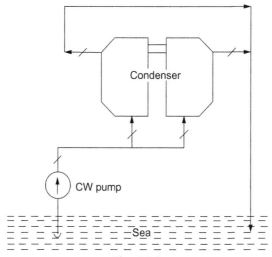

Fig. 14.2

Once through type CW system. *Source: From Fig. 9.10, P 326, Chapter 9: Steam Power Plant Systems. D.K. Sarkar, Thermal Power Plant–Design and Operation, 2015, Elsevier; Amsterdam, Netherlands.*

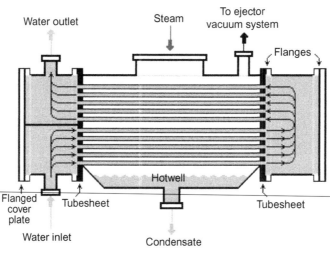

Fig. 14.3

Single-pressure condenser. *Source: Surface condenser, https://en.wikipedia.org/wiki/Surface_condenser.*

In a different configuration of condenser, cooling water enters into one shell only, say shell 1 of the condenser, passes through the tubes of this shell, and then exits through its outlet water box. The outlet cooling water from shell 1, at a higher temperature than its inlet temperature, enters shell 2 as "cold water," passes through the tubes of this shell and comes out through condenser outlets (Fig. 14.4). With the new configuration of CW flow, the pressure in shell 1 would be less than the pressure in shell 2 because of dissimilar "cold water" inlet temperatures. Thus, the nomenclature given to this condenser is "multipressure" condenser.

The net result of the new configuration of CW flow is always a lower average back pressure for the multipressure condenser compared to that obtained with the same water flow and surface in a conventional single-pressure condenser. A multipressure condenser would hence yield the following benefits:

   i. With lower average back pressure, the turbine heat rate will improve;

  ii. If the average back pressure of multipressure condenser is selected same as a single-pressure condenser, the turbine heat rate will remain unchanged, but the surface area and cooling water flow will be less. Thus, capital expenditure and running cost will be reduced.

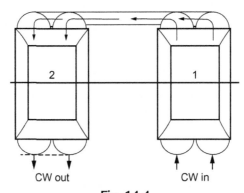

Fig. 14.4
Multipressure condenser.

14.1.2 Air-Cooled Condenser (ACC) *(Fig. 14.5)*

Air is used as cooling medium in condensers located either in deserts or in those areas where substantial quantity of CW is rarely available. Exhaust steam from the turbine is condensed on the inside of the ACC by air flowing past the outside of the tubes. This system requires additional auxiliary power to run the cooling fans and suffers from reduced efficiency due to elevated temperature prevailing within the condenser shell. These condensers are installed in an open area free from tall obstructions around it.

Advantages of ACC are:

    i. complete elimination of CW pumps and CW piping;

   ii. clarified water makeup is not required;

  iii. cooling tower vapor plume is eliminated;

  iv. no problems resulting from cooling tower blow-down problems.

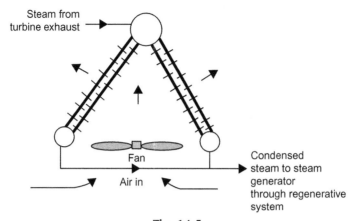

Fig. 14.5
Air-cooled condenser.

Disadvantages of ACC are:

  i.   lower cycle efficiency arising out of lower vacuum (higher pressure) in condenser;
 ii.   consequent to the above, more fossil fuels need to be burned for maintaining the same generation of power with additional rise in greenhouse gas (GHG), sulfur oxides (SOx), nitrogen oxides (NOx), and suspended particulate matter (SPM) emissions;
iii.   requires a vast area for installation of ACC;
 iv.   higher noise levels;
  v.   initial investment cost is more than the cost of the water cooling type;
 vi.   higher operating cost.

14.1.3 Hybrid or Combination Water and Air Cooling (Fig. 14.6)

Combination cooling is attractive in those areas where there is a shortfall of adequate cooling water for working exclusively with water cooling. This system is a compromise between foregoing cooling systems and shares part benefit of both. Adopting this system reduces the cooling tower size with consequent reduction in cost. Its performance is better than the performance that could be achieved with air cooling only.

14.2 Objective

A common cause of poor condenser vacuum is excessive air ingress or leakage. Any fall in condenser vacuum would have consequent rise in turbine heat rate vis-à-vis plant heat rate. As a result, cost of power generation would increase with eventual loss in revenue earning.

The condenser flood test and vacuum tightness test are conducted to identify locations or areas, eg, all connections to condenser, gland seals of valves located on vacuum services, relief

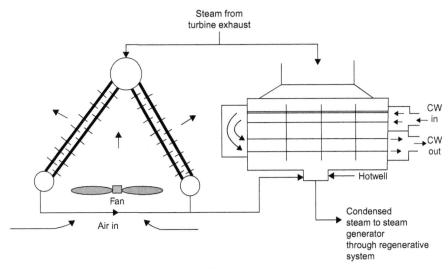

Fig. 14.6
Hybrid cooling system.

valves, gauge glasses, flow control valves, flanges, LP turbine gland seals, and so on, through which air ingress to the condenser shell may take place. The objective is to eliminate these defective locations and areas, achieving a vacuum-tight system to maintain the designed rated vacuum. As a result, the designed turbine heat rate is maintained.

14.3 Precautions

A condenser flood test and vacuum tightness test must be conducted in line with the guidelines, instructions, and procedures supplied by the original manufacturer of the condenser and under the direct guidance and supervision of the representative of the original manufacturer. All adjustments in the potential source of air leakages must be undertaken by trained personnel only.

In addition to the above, prior to performing the condenser flood and vacuum tightness tests, it is imperative to observe "general precautionary measures" discussed under Section 2.1.1: Quality Assurance.

14.4 Prerequisites

The following areas/items must be thoroughly inspected before carrying out the condenser tests (Table 14.1):

Table 14.1 Areas/items to be checked and satisfied

Sl. No.	Areas/Items	Ok (√)

1	Verify the completeness of installation as follows:
	  i. Steam dome units are in place;
	  ii. LP feedwater heater/s are inserted into the steam dome cavity (optional);
	  iii. Shell internals, eg, steam baffle plates, part of the air extraction line, various bracing assemblies, and so on, are inserted, housed properly, and welded in place;
	  iv. Welding between the bottom plate and the hotwell is complete;
	  v. Welding between two tube plate halves is complete;
	  vi. Side walls are properly assembled;
	  vii. Manhole covers are installed;
	  viii. If the tubesheet is flanged to the shell, the waterboxes on either end of the shell are bolted to the tube sheets and shell flanges;
	  or
	  If the tubesheet is welded to the shell, and the tubesheet outside diameter is larger than the shell, the waterbox is bolted to the tubesheet with "through bolts";
	  or
	  If the tubesheet is welded to the shell and to the waterbox, then the waterbox covers are bolted with through bolts';
	  ix. Tube support plates are fixed;
	  x. Condenser tubes are inserted properly and tube ends are expanded to roll the tubes onto the tube plates.	

2	Protocols jointly signed by the customer, the engineer, the turbine manufacturer, the condenser manufacturer, and the supplier (contractor) certifying erection completion of the following equipment are available:
	  i. condenser;
	  ii. waterbox units are aligned in the horizontal and vertical planes and alignment data are recorded;
	  iii. complete turbine with associated casings and gland sealing arrangement;
	  iv. liquid ring vacuum pump and/or hogging ejector and steam-jet (holding) air ejector;
	  v. condenser inlet and outlet CW piping;
	  vi. all other interconnecting piping with condenser, eg, feedwater heater vents and drains, steam drains, makeup water connections, condensate spray, LP bypass dump, air evacuation, and so on, along with hangers, supports, and associated valves;
	  vii. condensate piping.	

3	Trial run of the following equipment is successful and respective protocols jointly signed by the customer, the engineer, the steam turbine manufacturer, and the supplier (contractor) are available:
	  i. condensate extraction pumps (CEPs);
	  ii. liquid ring vacuum pumps (if provided);
	  iii. gland seal steam condenser steam exhaust fans;
	  iv. condensate transfer pump/s.	

Table 14.1 Areas/items to be checked and satisfied—cont'd

Sl. No.	Areas/Items	Ok (√)
4	Construction materials—temporary scaffolding, wooden planks, welding rod ends, cotton wastes, and so on, are removed from the condenser shell, tube bundles, and hotwell.	
5	Condenser shell, tube banks, and hotwell are made free of accumulated dirt by water washing prior to boxing up the condenser.	
6	Ensure that no one is working inside the condenser.	
7	The vacuum-breaking valve (vacuum breaker) is operable.	
8	Installation of gland sealing water lines to valves coming under vacuum services are complete.	
9	Verify that the atmospheric relief valve is properly sealed with water and is operable.	
10	Check all valves are free to operate.	
11	The turbine barring gear operation is verified.	
12	Level gauge glasses are installed and working.	
13	Ensure that associated instrumentation and controls are installed and operable.	
14	Temporary platforms are provided wherever required.	
15	Material and apparatus/instruments generally required for testing (Section 14.4.1) are arranged.	
16	Safety gadgets (Section 14.4.2) are arranged.	

14.4.1 Typical List of Bill of Material and Apparatus/Instruments

Material and apparatus/instruments that are typically required for the condenser flood and vacuum tightness tests are listed below. Specification and/or quantity of each of these items would vary from project to project and shall have to be assessed by concerned personnel associated with the test.

i. DM water;
ii. fluorescein solution;
iii. quartz lamp;
iv. wax crayon;
v. tube expander rollers;
vi. level gauge glasses on hotwell, HP feedwater heaters, and LP feedwater heaters;
vii. level switches and level transmitters on hotwell HP feedwater heaters and LP feedwater heaters;
viii. flexible polythene tube for monitoring the water level in the condenser steam space;
ix. calibrated vacuum gauge;
x. mercury manometer.

14.4.2 Typical List of Safety Gadgets

Equipment generally used to ensure safety of operating personnel during condenser flood and vacuum tightness tests are given below:

1. safety shoes of assorted sizes;
2. hand gloves;
3. hardhats of assorted sizes;
4. side-covered safety goggles with plain glass;
5. first aid box;
6. sufficient lighting arrangements;
7. safety sign boards;
8. safety tags.

14.5 Preparatory Arrangements

a. Close the following valves:
- i. drain before HP turbine control valves;
- ii. drain after HP turbine control valves;
- iii. drain before IP turbine control valves;
- iv. drain after IP turbine control valves;
- v. drain before HP turbine exhaust NRVs;
- vi. HP turbine casing drain;
- vii. drain before extraction line NRVs;
- viii. drain after extraction line NRVs;
- ix. gland sealing steam header drain;
- x. main steam pipe line drains;
- xi. main steam pipe line strainer drains;
- xii. HP turbine ESV before seat drains;
- xiii. cold reheat pipe line drains;
- xiv. cold reheat steam supply line to deaerator drain;
- xv. cold reheat steam supply line to BFP drive turbine drain (if provided);
- xvi. hot reheat pipe line drains;
- xvii. hot reheat pipe line strainer drains;
- xviii. IP turbine IV before seat drains;
- xix. turbine gland sealing steam pressure control station drain;
- xx. LP bypass drains;
- xxi. condensate supply temperature control valves to steam drain flash tank manifolds;
- xxii. condensate extraction pump (CEP) suction valves;
- xxiii. CEP recirculation control valve/s;
- xxiv. hotwell drains;
- xxv. condensate normal makeup line valves;
- xxvi. condensate emergency makeup line valves;

xxvii. gland steam condenser minimum recirculation flow line valves;

xxviii. drain and vent valves to HP flash tanks;

xxix. HP flash tank drains to condenser;

xxx. emergency drains from HP feedwater heaters;

xxxi. HP feedwater heater vent valves;

xxxii. drain and vent valves to LP flash tanks;

xxxiii. LP flash tank drains to condenser;

xxxiv. emergency drains from LP feedwater heaters;

xxxv. LP feedwater heater vent valves;

xxxvi. drain cooler normal drain valve;

xxxvii. vacuum breaker;

xxxviii. air evacuation valves;

xxxix. all pressure/vacuum measuring instrument root valves.

b. Prior to filling water in the condenser steam space, restrain the movement of condenser spring supports by changing the position of "locking screws" to tighten them. This arrangement will take the load off condenser springs by locking the spring assembly;

c. All the hangers of drain flash tanks and connected pipe lines in a vacuum system must be locked.

14.6 Operating Procedure

Condenser flood and vacuum tightness tests of a new unit are described separately below.

14.6.1 Condenser Flood Test

a. Open all manually operated valves on makeup lines from condensate storage tank to hotwell;

b. Start condensate transfer pump and open its discharge valve;

c. Gradually open hotwell fill line valve and start taking water to the hotwell;

d. Open vent valves of LP feedwater heaters, HP feedwater heaters, drain cooler, and gland steam condenser to evacuate air from the system;

e. Check for leakages in the system after every 1 m rise in water level;

f. In the event any leakage is detected, stop the condensate transfer pump and drain the system down to below the location of the leakage to attending it;

g. Start the condensate transfer pump for filling water afresh;

h. Once the water level reaches about 1 m above the top row of the condenser tube banks, stop taking water to the condenser;

i. Stop the condensate transfer pump and close its discharge valve;

j. Add 20 ppm of a fluorescein solution to the water;

k. Leave the water in the condenser for at least 24 h so that even the minutest leakage could be detected;

l. Inspect the face of tubesheets of the condenser thoroughly with a quartz lamp to detect any leaks;

m. Mark the location of any leaks with wax crayon;

n. Carefully expand the defective tube ends just enough to stop the leak;

Note

Before expanding the tube ends, moisten the location with at least two drops of glycerin only, so as to ensure that the expander rollers rotate smoothly and with minimum wear.

---

o. On completion of successful testing, neutralize the water-fluorescein mixture;

p. Drain the test water and dry the condenser;

q. Loosen the spring-locking screws (Section 14.5, step b) before the condenser is put in service.

Note

In a running unit, condenser tube leak may be suspected if there is a rising tendency of both specific and cation conductivity of condensate at the hotwell and/or at the CEP discharge pipe line. The following steps may be adhered to, to attend the leakage:

i. Shut down the unit;

ii. Stop CW pumps;

iii. Leave the waterboxes bolted on to the shell to reinforce the tubesheets;

iv. Open manholes;

v. Fill the condenser shell side with a fresh supply of water 1 m above the tube banks;

vi. Examine the face of tubesheets for any leaks;

vii. Water coming out of a tube end indicates that the tube is ruptured inside the condenser;

viii. As a temporary measure, plug the defective tube from both ends, taking care to prevent any damage to the tubesheet;

ix. Replace defective tube/s during the next available shut down;

x. In the event a crack is detected between the tube and the tubesheet, the leakage may be arrested by expanding the tube as described earlier.

14.6.2 Condenser Vacuum Test

i. Start the turbine auxiliary/lube oil pump;

ii. Start the jacking oil pump (if provided);

iii. Put the turbine on turning gear;

iv. Start one CEP;

v. Start flow through the CEP minimum flow recirculation line;

vi. Line up turbine gland sealing steam supply and dump system;

vii. Start gland steam condenser with steam exhauster;

viii. Start both vacuum pumps or noncondensing type single-stage starting (hogging) air ejector;

ix. On achieving about a 20 kPa vacuum, supply steam to turbine gland seals;

x. Maintain pressure and temperature on the turbine gland steam supply header at 102.3 kPa and >433 K, respectively;

xi. Verify whether the desired vacuum, as recommended by the turbine manufacturer, is maintained;

xii. In case the desired vacuum cannot be achieved, check all air ingress points, eg, glands of valves located on interconnecting piping with the condenser, vacuum-breaking valve, all atmospheric vent valves on the shell side (steam side) of the condenser, feedwater heater vents and drains connections, steam drains connections, makeup water connections, condensate spray connections, LP bypass dump line connections, air evacuation connections, and so on;

xiii. Identify the defective point/s and rectify the defect/s;

xiv. Repeat steps xii and xiii until the desired vacuum is achieved;

xv. Stop one vacuum pump or change over from the hogging air ejector to the steam-jet (holding) air ejector;

xvi. Verify whether the desired vacuum can be sustained;

xvii. If step xvi fails, repeat steps xii and xiii until desired vacuum is sustained.

Note

In a running unit, air leakages to the condenser may be detected by adopting the "Helium leak testing" technique, which is the fastest and most-effective method for identifying condenserleaks. This method, however, requires a highly skilled operator for effective results.

An alternative technique to the above is the "ultrasound detection" technique. Although this technique offers favorable results with less-skilled technicians, noise emanating from steam leakages may interfere with the end result.

14.7 Conclusion

The condenser flood and vacuum tightness tests are carried out under the guidance and supervision of representatives of the turbine and condenser manufacturers.

Condenser flood test is certified to be completed after satisfactorily attending to all the identified leakages in all respects.

Condenser vacuum tightness test may be construed to be satisfactory once the following criterion is fulfilled:

Start both vacuum pumps/hogging air ejector to arrive at the recommended vacuum. Then stop both vacuum pumps/hogging air ejector and record the fall of vacuum in 10 min. If the rate of fall of vacuum remains within 267–400 Pa of Hg per minute, the test is successful.

After attending to all the leakages, it is desirable to evacuate the LP turbine and condenser rapidly by means of either the hogging air ejector or a vacuum pump to fulfill the following recommendation of the Heat Exchange Institute (HEI).

For evacuation of air from atmospheric pressure to 33.86 kPa absolute pressure Hg in about 1800 s, the capacity of the evacuating equipment shall be as laid down in Table 14.2 [2].

Note

As per HEI, standard condition corresponds to pressure 101.3 kPa (14.7 psia) and to temperature 294 K (70°F).

A protocol (Table 14.3) jointly signed by the customer, the engineer, the steam turbine manufacturer, the condenser manufacturer, and the supplier (contractor) may be issued thereafter.

Table 14.2 Rapid evacuation equipment capacity

Sl. No.	Exhaust Steam From LP Turbine to Be Condensed (kg s^{-1})	Required Capacity of the Evacuating Equipment (N m^3 s^{-1})
1	63–126	0.153
2	126–252	0.307
3	252–378	0.460
4	378–504	0.613
5	504–630	0.767

Table 14.3 Protocol of condenser flood test and vacuum tightness test

Test		Remarks		
Condenser flood test Condenser vacuum tightness test		Condenser flood test is completed after satisfactorily attending to all of the identified leakages in all respects. On completion of a condenser vacuum tightness test, vacuum is raised to the recommended value; evacuation is stopped and the fall of vacuum is recorded for 10 min. The rate of fall of vacuum is noted to be within 267–400 Pa of Hg per minute.		
Signed by the customer	Signed by the engineer	Signed by the turbine manufacturer	Signed by the condenser manufacturer	Signed by the supplier (contractor)

References

[1] D.K. Sarkar, Thermal Power Plant—Design and Operation, Elsevier, Amsterdam, Netherlands, 2015.
[2] Heat Exchange Institute, Inc., Table 8, Section 6.6, Standards for steam surface condensers, 1995.

Generator Drying Out and Air-Tightness Tests

15.1 Introduction

A generator, also known as an alternator, is the essential entity of every power-generating plant. It is a machine that converts mechanical energy into electrical energy by electromagnetic induction. All generators, used in various power plants are connected to the grid (or infinite bus-bar) in parallel. These generators may be of different capacities, but their voltage and frequency essentially have to be identical and constant. The main generator system transmits generator output power to a high-voltage (132/220/400/700 kV) grid through generator transformers, isolators, and generator circuit breakers (Fig. 1.12) [1].

A generator comprises of a rotating DC field, or "rotor," and stationary armature windings, or "stator." In its commonest form, a large number of conductors mounted on an armature (rotor) are rotated in a magnetic field produced by either permanent magnets or field coils

Thermal Power Plant. http://dx.doi.org/10.1016/B978-0-08-101112-6.00015-0

through which excitation current passes to generate the field. The magnetic field induces AC voltage in the stator windings. Armature windings of the generator generate the electric current.

Electrical insulation is provided on generator winding to prevent winding faults as well as to prevent current flow between the copper conductors and the core of the stator or rotor. Hence, service life of a generator depends on the condition of electrical insulation between the copper and the core. Conditions of insulation may deteriorate on the environment of operation. Insulation may become brittle due to excessive temperature rise. Buildup of dirt or moisture on windings jeopardizes also the insulating effect.

In view of the above, it is essential to test the conditions of insulation periodically. Two parameters that categorically predict the conditions of insulation are insulation resistance (IR) and polarization index (PI). While IR is a measure of the conductivity of the insulation system that evaluates the condition of insulation between conductors and ground, PI assesses the health of the insulation system and indicates any drop in IR value over time.

Ideally, the value of IR should be infinite, which is difficult to achieve in practice. So getting a large value of IR to the tune of Mohm or higher, it may be inferred that any leakage of current will not be of much significance. Any fall in IR value over a time period signifies that conditions of insulation have started deteriorating.

PI indicates if there is any variation in the IR value and is expressed as the ratio of IR measured after 10 min (IR10) to the IR measured after 1 min (IR1). A low value of PI indicates that the windings may have been contaminated (with oil, dirt, and so on), absorbed moistures, became brittle, or gotten damaged. PI testing must be performed above the dew point and corrected for temperature to 313 K [2].

The above practice notwithstanding, IEEE 43-2000 states, "When the IR1 is higher than 5000 MΩ, the PI may or may not be an indication of the insulation condition and is therefore not recommended as an assessment tool." Under this condition, an additional parameter named the insulation resistance profile (IRP) may serve as a helpful criterion. IRP is obtained by plotting IR values over a period of typically 10 min. Fig. 15.1 depicts the IRP of a healthy insulation system [3].

The recommended procedure for measuring IR and typical IR characteristics is described in the latest edition of IEEE 43. This document also describes how these characteristics indicate winding condition. It recommends minimum acceptable values of IR for AC and DC rotating machine windings [4].

With the above background and prior to narrating the procedure for generator drying out and the air-tightness test, it is imperative to describe the following associated systems of a large generator.

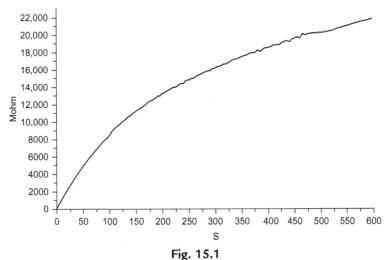

Fig. 15.1

IRP of a healthy insulation system.

15.1.1 Generator Air/Hydrogen Cooling System [1]

Each side of the generator enclosure is provided with stator end shields to keep the enclosure gas tight. When the machine is running, heat is generated within the generator enclosure. In small-sized generators, air is circulated through the generator stator and rotor to dissipate the heat. This air is circulated by axial fans installed on both sides of the rotor shaft ends. The cold air cools the stator core, and the hot air is cooled by circulating cooling water (Fig. 15.2).

As the unit size increases, hydrogen gas (Chapter 16), instead of air, is circulated to dissipate the heat. The pressure of hydrogen is normally maintained between 0.2 and 0.4 MPa, but could be raised to 0.6 MPa. Hydrogen is cooled by circulating cooling water through the hydrogen coolers (Fig. 15.3).

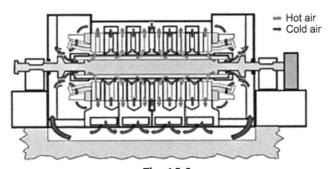

Fig. 15.2

Air-cooled generator. *Source: From Fig. 9.23, P 343, Chapter 9: Steam Power Plant Systems. D.K. Sarkar, Thermal Power Plant—Design and Operation, 2015, Elsevier.*

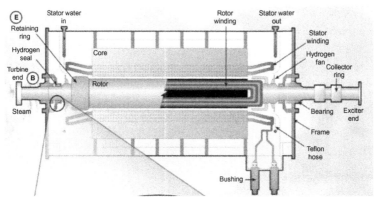

Fig. 15.3

Hydrogen-cooled generator. *Source: Hydrogen Purity and Moisture Monitoring in Hydrogen Cooled Generators. Michael Hicks, Utility Product Specialist, Environment One Corporation, April 16–17, 2013. https://www.aegislink.com/presentations/2013_eums/pdfs/11_hydrogen_dew_point_monitoring.pdf.*

15.1.2 Generator Stator Cooling Water System

Heat generated in the stator winding is dissipated by circulating DM water through stator conductors. The dissolved oxygen content of DM water is maintained at ≤ 80 µg L^{-1} and the pH is between 8.0 and 8.5. DM water is supplied to the stator cooling water tank, from which a pump drawing water passes it through the cooler and filter and then supplies the water to stator conductors. The return from the conductors is drained to the stator cooling water tank (Fig. 15.4).

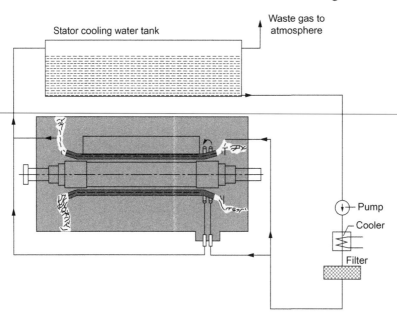

Fig. 15.4

Generator stator cooling water system. *Source: From Fig. 9.25, P 344, Chapter 9: Steam Power Plant Systems. D.K. Sarkar, Thermal Power Plant–Design and Operation, 2015, Elsevier.*

15.1.3 Generator Seal Oil System [1]

To prevent leakage of hydrogen through shaft seals, a continuous film of oil is maintained between the rotor and stationary seal rings. One seal oil pump draws seal oil from the seal oil tank and supplies the oil to the hydrogen-side shaft seals through a cooler and filter. Drainage from the hydrogen-side seal ring is returned to the seal oil tank. Another seal oil pump receives suction from the seal oil storage tank and supplies oil to the air-side shaft seals through the cooler and filter. Drainage from the air-side seal ring is returned to the seal oil storage tank. The seal oil tank is connected with the seal oil storage tank through an overflow float valve. The seal oil on the hydrogen side, being at a higher pressure than hydrogen, prevents the continuous loss of hydrogen. Seal oil also prevents entry of air into the hydrogen-cooled enclosure (Fig. 15.5).

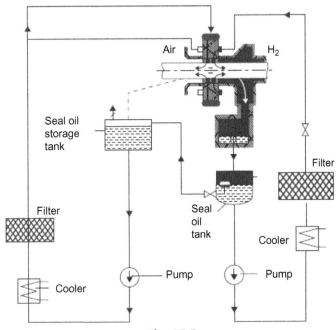

Fig. 15.5

Generator seal oil system. *Source: From Fig. 9.27, P 345, Chapter 9: Steam Power Plant Systems. D.K. Sarkar, Thermal Power Plant–Design and Operation, 2015, Elsevier.*

15.2 Objective

The purpose of conducting generator drying out is to eliminate moisture film which may develop on the surface of the insulation of generator stator windings during transport, storage, and erection of the generator. The presence of any moisture on the surface of the insulation will result in reduction of IR.

To ensure safe operation of the generator, it is essential that hydrogen does not escape through any leakage in the generator casing and the gas circuit. To achieve this condition, the air-tightness test is performed on the generator and the gas system with oil-free dry instrument air prior to introducing hydrogen in to the generator.

15.3 Precautions

In order to harness the benefits described under Section 15.2, generator drying out and the air-tightness test must be conducted by trained personnel in accordance with the instructions and procedures supplied by the original manufacturer of the generator and under the direct guidance and supervision of the representative of the original manufacturer.

Before conducting generator drying out and the air-tightness test, observe the following specific precautionary measures:

i.   Verify that necessary statutory clearances (these are project and country specific, hence need to be assessed by the personnel associated with the test) are obtained;
ii. Ensure that coupling bolts between turbine and generator are not fitted;
iii. Ensure that the generator rotor is not in contact with the turbine rotor.

Along with the above specific precautionary measures, "general precautionary measures," as described under Section 2.1.1: Quality Assurance, must be followed before executing the current activities.

15.4 Prerequisites

Undertake a thorough inspection of systems under purview before carrying out generator drying out and the air-tightness test to ascertain the following (Table 15.1):

15.4.1 Typical List of Apparatus/Instruments

Apparatus/instruments that are typically required for drying out and the air-tightness test of large generators are listed below. Specification and/or quantity of each of these items would vary from project to project and shall have to be assessed by concerned personnel associated with the test.

i.   resistance temperature detectors (RTDs)
ii. 100 V megger
iii. 250 V megger
iv. 5 kV megger
v.   forced air heater of 2 kW rating
vi. continuity tester

vii. winding temperature recorder

viii. U-tube manometer

ix. precision pressure gauge

x. soap solution

Table 15.1 Areas/items to be checked and satisfied

Sl. No.	Areas/Items	Ok (√)

1	Verify availability of protocols jointly signed by the customer, the engineer, the generator manufacturer, and the supplier (contractor) certifying erection completion of the following equipment and system:
	
	  i. Generator checks include the following:
	    a. generator placed on foundation
	    b. bearings fitted
	    c. rotor threaded in
	    d. end-shield assembled
	    e. alignment complete and foundation bolts tightened
	    f. generator field connection complete
	  ii. Exciter is erected and aligned, and coupling bolts are tightened
	  iii. Excitation system including AVR erection is complete
	  iv. Generator lube oil system erection is complete and flushed (Chapter 6) and is in operation
	  v. Generator seal system erection is complete, seal oil system is flushed, and initial circulation is complete and in operation
	  vi. Generator stator cooling water system erection is complete, flushed, and drained completely. The water circuit is completely dried with circulation of hot dry air
	  vii. Hydrogen and carbon dioxide gas systems, including piping, valves, and valve rack erection, are complete
	  viii. Hydrogen coolers, seal oil coolers, and stator water coolers erection are complete. Piping and valve connections in the cooling water side are complete	

2	A trial run of the following equipment are successful, and respective protocols jointly signed by the customer, the engineer, the generator manufacturer, and the supplier (contractor) are available:
	
	  i. seal oil pumps
	  ii. stator cooling water pumps
	  iii. exhausters	

3	All thermometers, pressure gauges, H_2 gas analyzer, and so on, are calibrated and installed. Calibration certificates are verified	
4	Check that both "line" and "neutral" side connections of the generator to the bus duct are open	
5	Ensure that the generator body earthing is done as recommended	
6	Ensure that access to the generator and its auxiliaries are free from any obstacle	

Continued

Table 15.1   Areas/items to be checked and satisfied—cont'd

Sl. No.	Areas/Items	Ok (√)
7	Verify that the required statutory clearances have been obtained	
8	The generator bearing pedestal insulation value is measured and found to be >1 MΩ	

9	Instrument air of the following quality is available:
	  i. Air must be free of corrosive contaminants and hazardous (flammable or toxic) gases
	  ii. Maximum total oil or hydrocarbon content, exclusive of noncondensable, must be ≤ 1 mg kg^{-1}
	  iii. Air must be practically free of dust. Maximum particle size in air stream must not exceed 5 μm
	  iv. Oxygen content of the expended air shall be between 20% and 21% V V^{-1}
	  v. The dew point at line pressure is at least 15 K below the minimum possible generator casing temperature. In no case should the dew point at line pressure exceed 283 K
	
	Note
	
	If the generator casing temperature exceeds 313 K, the dew point at line pressure may be raised up to 293 K
	
	  vi. Air pressure must be between 0.6 and 0.9 MPa gauge	

10	Temporary platforms are provided wherever required	
11	Apparatus/instruments generally required for drying out and air-tightness testing (Section 15.4.1) are arranged	
12	Safety gadgets (Section 15.4.2) are arranged	

15.4.2 Typical List of Safety Gadgets

Equipment generally used for safety of operating personnel during generator drying out and the air-tightness test are given below:

1. safety shoes (provided with insulation material at the bottom) of assorted sizes;
2. hand gloves;
3. hardhats of assorted sizes;
4. side-covered safety goggles with plain glass;
5. first aid box;
6. sufficient lighting arrangements;
7. safety sign boards;
8. safety tags.

15.5 Preparatory Arrangements

Preparatory work for generator drying out and the air-tightness test are described below separately.

15.5.1 Generator Drying Out

15.5.1.1 Checking of RTDs

1. Check continuity of each RTD with a continuity tester;
2. Measure resistance of each RTD with a Wheatstone Bridge (Fig. 15.6) and the corresponding ambient temperature;
3. Measure IR value with 100 V megger;

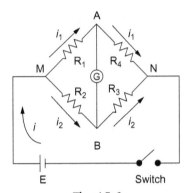

Fig. 15.6
Wheatstone Bridge.

Record the values of steps 1, 2, and 3 in Table 15.2.

Table 15.2 RTD checking

RTD No.	Location	Ambient Temperature (K)	Continuity	Resistance (Ω)	IR Value (MΩ)
⋮					

15.5.1.2 Checking of bearing and seal insulation

Enclose IR values of the following areas from erection log sheets as "attachments":

1. generator bearings before rotor threading; this value should be above 10 MΩ;
2. generator bearings immediately after rotor threading; this value is expected to be less than 10 MΩ;

3. inner and outer labyrinth rings;
4. pedestal bearing of exciter before fixing of exciter rotor; this value should be above 10 MΩ;
5. after fixing of each oil pipes;
6. final value between rotor shaft and earth.

Measure IR value between generator rotor and earth with a 100 V megger, and record the value (Table 15.3). This value should be above 1 MΩ. In the event a lower value is obtained, dry out all bearings, seals, oil pipes, and so on, until the IR value exceeds 1 MΩ.

Table 15.3 Bearing insulation

Measured Between	Ambient Temperature (K)	IR Value (MΩ)
Generator stator and earth		

15.5.1.3 Measurement of IR of stator winding

While meggering one phase of stator winding, the other two phases not under testing must be shorted and earthed. Using a 5 kV power megger, measure the IR of each phase of stator winding. Record 1-min (IR1) and 10-min (IR10) megger readings in Table 15.4, and calculate the PI (PI = IR10/IR1). On completion of meggering, discharge the tested winding for at least 15 min by connecting it to earth.

The measured IR value must be above 1 MΩ kV^{-1} of generator-rated voltage at 313 K and calculated PI should be greater than 2.0. If the results fail to meet these requirements, it is recommended to dry out the generator.

Table 15.4 Insulation resistance of stator winding

Measured Between	Ambient Temperature (K)	Voltage Applied (kV)	IR, 1 min (MΩ)	IR, 10 min (MΩ)	$PI = \dfrac{IR10}{IR1}$
R–E					
Y–E					
B–E					

Note: R—red phase, Y—yellow phase, B—blue phase, E—earth.

15.5.1.4 Measurement of IR of rotor circuit

There are two sliprings on the generator shaft, one slipring connected to the rotor winding and the other in contact with the shaft. Measure the IR between these two sliprings with a 250 V megger, and record the value in Table 15.5. The measured IR value must be above 1 MΩ at 313 K.

Table 15.5 Measurement of insulation resistance of rotor circuit

Measured Between	Ambient Temperature (K)	Voltage Applied (V)	IR Value (MΩ)
Two sliprings			

15.5.2 Generator Air-Tightness Test

1. Verify that turbine and generator lube oil pump/s are running and the bearing lubrication system is in service;
2. Start the seal oil pumps and put the seal oil system in operation;
3. Fill the stator winding with stator cooling water and stator cooling water tank up to the normal level;
4. Close the following valves:
   a. valve connecting the waste gas pipe with the generator;
   b. vent valve in the waste gas pipe;
   c. stator cooling water supply line valves;
   d. all valves on the H_2 gas supply to the generator;
5. Connect a U-tube manometer on the waste gas pipe of the stator cooling water tank;
6. Verify that the H_2 supply line inlet valve to the generator is closed. Connect compressed air upstream of this valve;
7. Connect a U-tube manometer or a precision pressure gauge in the H_2 supply line inlet to the generator.

15.6 Operating Procedure

The procedure for drying out the generator is completely different from the generator air-tightness test procedure; hence; they are dealt with separately in following paragraphs.

15.6.1 Generator Drying Out

Present-day generators are usually provided with epoxy resin insulation on the stator windings. The advantage of epoxy resin is that it seldom absorbs any moisture; hence, under normal circumstances it may not be required to resorting to assisted drying out of insulation. Notwithstanding this attractive characteristic, moisture does form a film on the surface of the insulation of the stator windings during transportation, storage, and erection. This film of moisture can be removed easily by heating the generator enclosure at a temperature just above the ambient temperature.

Remove manhole covers on both sides of generator end shields. Insert one forced air heater of 2 kW rating through each manhole, and keep the air heaters running. Note down the temperature inside the generator enclosure. On reaching a steady temperature state, measure and record IR value of all three phases connected together. Continue with this process and record IR values in Table 15.6 until the measured IR value exceeds $1\ M\Omega\ kV^{-1}$ of generator-rated voltage at 313 K and PI value becomes greater than 2.0.

Do not switch off the forced air heaters.

Put the stator cooling water system into operation, but do not put the cooler of this system (Fig. 15.4) into service. Allow the system to run for about 6 h.

Switch off the forced air heaters and take them out. Put back the manhole covers.

Table 15.6 Generator drying out

Reading No.	Date/Time	Ambient Temperature (K)	Generator Enclosure Temperature (K)	IR, 1 min (MΩ)	IR, 10 min (MΩ)	$PI = \dfrac{IR10}{IR1}$
1						
2						
⋮						

15.6.1.1 Resistance measurement of stator winding

Measure resistance of each phase of the stator winding with a Kelvin Double Bridge (Fig. 15.7), and record the results in Table 15.7.

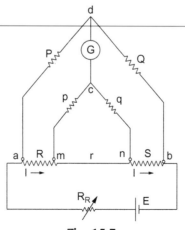

Fig. 15.7
Kelvin Double Bridge.

Table 15.7 Measurement of resistance of stator winding

Measured Between	Ambient Temperature (K)	Resistance (Ω)	Remarks
R1–R2 Y1–Y2 B1–B2			

15.6.1.2 Capacitance and tanδ measurement

Measure capacitance and tanδ of the stator winding with a Schering Bridge (Fig. 15.8), and record the results in Table 15.8.

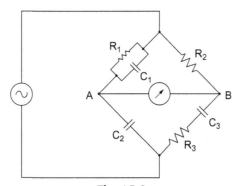

Fig. 15.8
Schering Bridge.

Table 15.8 Capacitance and tanδ measurement

Sl. No.	Temperature (K)	Measured Between Terminals	Test Voltage	Capacitance (Farad)		Measured tanδ
				Standard	Measured	
1						
2						
⋮						

15.6.2 Generator Air-Tightness Test

1. Open the H_2 supply line inlet valve to the generator (Step 6, Section 15.5.2). Allow the instrument air to enter the generator casing until the casing air pressure rises to about 0.4 MPa. Close the valve;
2. Verify that the seal oil pressure is higher than the generator casing air pressure;
3. Keep the generator casing pressurized for 48 h;
4. Take hourly readings of pressure and temperature (Table 15.9);

5. During the pressurizing period, keep vigilance on the U-tube manometer on the waste gas pipe of the stator cooling water tank (Step 5, Section 15.5.2). Even a small increase in pressure recorded in this manometer indicates a leak in the stator cooling water system;
6. Identify the leaked area/s with soap solution and rectify the defect;

Note

Suspected areas of leakages are flange joints, welding joints, welds, bushings, valve glands, relief valves, manifold connections, drain connections, instrument connections, gas dryer, and so on.

7. Loss of air from the generator casing may be calculated using the following formula:

$$V_{LA} = \frac{K \times 24}{Z} \times V_G \times \left[\frac{P_1 + P_{B1}}{T_1} - \frac{P_2 + P_{B2}}{T_2} \right]$$

where V_{LA}: loss of air per 24-h period (m^3) (s.t.p.),
K: a constant (to be supplied by the generator manufacturer),
Z: duration of leakage test (h),
V_G: generator volume (m^3),
P_1: pressure inside the generator casing at the start of the leakage test (Pa g),
P_2: pressure inside the generator casing at the end of the leakage test (Pa g),
P_{B1}: barometric pressure at the start of the leakage test (Pa),
P_{B2}: barometric pressure at the end of the leakage test (Pa),
T_1: temperature of air inside the generator casing at the start of the leakage test (K), and
T_2: temperature of air inside the generator casing at the end of the leakage test (K).
If V_{LA} is found to be less than V_{LP}, which is the permissible amount of leakage allowed by the generator manufacturer, then the H_2 gas system is considered sufficiently tight.
If, however, V_{LA} is higher than V_{LP}, identify the leakage with soap solution, rectify the defect, and repeat the leakage test and subsequent calculation until acceptable results are achieved.
8. Record all readings per Table 15.9.

Table 15.9 Readings of generator air-tightness test

Reading No.	Time	Stator Cooling Water Tank Pressure (Pa)	Casing Air Pressure (Pa g)		Barometric Pressure (Pa g)		Casing Air Temperature (K)	
			Start (P_1)	Stop (P_2)	Start (P_{B1})	Stop (P_{B2})	Start (T_1)	Stop (T_2)
1								
2								
⋮								
47								
48								

15.7 Conclusion

On successful completion of generator drying out and the air-tightness test, a protocol (Table 15.10) may be signed jointly by the customer, the engineer, the generator manufacturer, and the supplier (contractor).

Table 15.10 Protocol for generator drying out and the air-tightness test

Type of Test	Findings		Remarks
Generator drying out	IR value of insulation of stator windings is greater than 1 MΩ kV^{-1} of generator-rated voltage at 313 K, and PI value is greater than 2.0		Generator is cleared for H$_2$ filling (Chapter 16)
Generator air-tightness test	$V_{LA} < V_{LP}$		
Signed by the customer	Signed by the engineer	Signed by the generator manufacturer	Signed by the supplier (contractor)

References

[1] D.K. Sarkar, Thermal Power Plant—Design and Operation, Elsevier, Amsterdam, Netherlands, 2015.
[2] IEEE Standard No. 43-2000: Recommended Practice for Testing Insulation Resistance of Rotating Machines.
[3] D.L. McKinnon, PdMA Corporation Tampa, FL, United States. IRP of a Healthy Insulation System.
[4] IEEE Standard No. 43–2013: Recommended Practice for Testing Insulation Resistance of Rotating Machines.

CHAPTER 16

Filling of Generator With Hydrogen and Protection Stability Test of Generator

16.1 Introduction

From Chapter 15, we know that when the machine is running, heat is generated within the generator enclosure. In small-sized generators, air is circulated through the generator stator and rotor by axial fans mounted on both sides of the rotor shaft ends to dissipate the heat. The cold air cools the stator core, and the hot air is cooled by circulating cooling water (Fig. 15.2). As the unit size increases, hydrogen gas instead of air is circulated through the generator through different paths by shaft-mounted fans for cooling of the stator core laminations and the rotor. Hydrogen in turn is cooled by circulating cooling water through the hydrogen coolers (Fig. 15.3).

Thermal Power Plant. http://dx.doi.org/10.1016/B978-0-08-101112-6.00016-2

Advantages of hydrogen over air at a given pressure and temperature include the following (Table 16.1):

i. The specific weight of hydrogen is 1/14th the specific weight of air.
ii. The thermal conductivity of hydrogen is almost seven times that of air.
iii. The absolute viscosity of hydrogen is about half the absolute viscosity of air.

Hence hydrogen reduces the windage and friction losses on the spinning generator shaft more expeditiously than air, thereby increasing the generator's efficiency.

On the flammability scale of NFPA 704, hydrogen possesses the highest rating of 4 [1]. Hydrogen gas risks autoignition when mixed even with as low as 4% air. At 101.33 kPa pressure, the flammability limits of hydrogen in air range from 4% to 75% and in oxygen from 4% to 95%. The explosive limits of hydrogen in air range from about 18% to 60%.

On large generators an increase in hydrogen purity from 95% to 98% can reduce windage and friction losses by typically about 32%. Hence, as far as possible, the purity of hydrogen for circulation has to be maintained more than 97%.

Table 16.1 Properties of air and hydrogen

At 101.33 kPa Pressure and 273 K Temperature			
Type of Gas	Specific Weight (kg m^{-3})	Thermal Conductivity (W m^{-1} K^{-1})	Absolute Viscosity (Pa s)
Air	1.2933	24.20×10^{-3}	1.73×10^{-5}
Hydrogen	0.0902	172.58×10^{-3}	0.84×10^{-5}

Note

1 Pa s = 10 Poise.

In a thermal power plant, the generator plays the most important role; it is also an expensive equipment in the complete power plant. Besides, when in service, the generator gets subjected to various types of stresses. Hence, for the reliability of uninterrupted power generation, it is extremely important that the generator is protected in all respects for its safety and/or for minimizing its damage, minor or severe, by providing various types of fault prevention relays so that it can serve safely. These relays are categorized as electromagnetic relay, static relay, and mechanical relay. In large thermal power plants, static numerical relays are universally adopted.

Protection relays which are generally provided for generator safety are given in Table 16.2. The device number of each relay is included in this table [2]. The physical location of some of these relays in the plant is depicted in Fig. 16.1.

Table 16.2 Generator protection relay

Sl. No.	Name of Relay	Device No.
1	Generator over voltage	59G
2	Generator under voltage	27, 27N, 27-3N
3	Generator over current	51G, 51N, 51V
4	Generator over frequency	81O
5	Generator under frequency	81U
6	Generator volts/Hertz—over flux	24
7	Generator differential protection	87G
8	Bus differential protection	87B
9	Overall differential protection	87U
10	Generator loss of excitation	40G
11	Generator over excitation	24G
12	Generator stator earth fault	64G1, 64G2
13	Generator restricted earth fault	64R
14	Generator rotor earth fault	64F1, 64F2
15	Generator inter turn fault	95G
16	Generator pole-slip or out-of-step protection	78G
17	Generator backup impedance protection	21G1, 21G2
18	Generator negative phase sequence protection	46G, 47
19	Generator stator high temperature protection	49
20	Reverse power relay	32G1, 32G2
21	Local breaker backup (LBB) relay	50Z

Actuation of any of these protective relays trips the generator automatically with or without time delay, depending on the nature of emergency. Tripping of the generator, initiated through generator lockout relay (industrially known as 86G relay), essentially means disconnecting it from the grid by opening the generator circuit breaker.

In a running plant the protection relays remain inoperative for a long time, sometimes ages, before a fault occurs; but whenever a fault occurs, the relays must respond instantly and correctly. As such, these protective relays must remain vigilant whenever the generator is on. It is of paramount importance that these relays remain highly dependable. It is the dynamic testing of protection relays that ensures the reliability and dependability of these relays. The test is conducted by simulating fault conditions and observing relay operation. This method of testing protection relays is termed the protection stability test of the generator.

16.2 Objective

i. The purpose of filling hydrogen (H_2) in the generator, as already explained, is to dissipate the heat generated within the generator enclosure and at the same time ensure minimum loss due to windage and friction. The procedure for filling the generator with hydrogen constitutes a systematic method of evacuating air from the generator

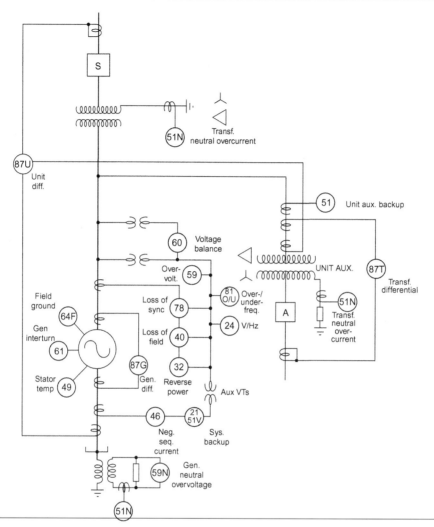

Fig. 16.1
Location of protective relays.

enclosure through a vent connection by supplying carbon dioxide (CO_2) to the generator enclosure from a bottom distribution piping, then purging CO_2 by supplying H_2 to the generator enclosure from a top distribution piping. As described in Chapter 15, H_2 inside the generator enclosure must be kept under a sealed condition, maintaining a constant pressure of about 0.4 MPa and a purity of more than 97%.

ii. The healthiness and operation stability of generator protection relays are extremely important for ensuring the safety of the generator to ensure continuous generation of electric power. Hence, all applicable tests must be conducted to establish the integrity of the protection relays prior to synchronizing the machine with the grid. The procedure to carry out the protection stability test of relays is narrated in detail in this chapter.

16.3 Generator Hydrogen Filling

The sequence of hydrogen filling is described in the following paragraphs.

16.3.1 Prerequisites

Prior to filling the generator with hydrogen, carry out a thorough inspection of the gas filling systems to ascertain the following (Table 16.3):

Table 16.3 Areas/items to be checked and satisfied

Sl. No.	Areas/Items	Ok ($\sqrt{}$)
1	Verify the availability of protocol jointly signed by the customer, the engineer, the generator manufacturer, and the supplier (contractor) certifying successful completion of the generator drying out and air-tightness test	
2	Ensure that coupling bolts between turbine and generator are fitted	
3	Ensure that the hydrogen and carbon dioxide gas system, including piping, gas racks, and valve racks erection, is complete	
4	The casing pressure gauge and purity meters to monitor air in hydrogen, air in carbon dioxide, carbon dioxide in hydrogen, and so on are calibrated and installed. Ensure that calibration certificates are verified	
5	An adequate number of H_2 cylinders for one filling of the generator and for raising H_2 pressure is available and that H_2 cylinders are connected to the H_2 cylinder rack	
6	An adequate number of CO_2 cylinders for one filling and one purging of the generator is available and that CO_2 cylinders are connected to the CO_2 cylinder rack	
7	Ensure that temporary platforms are provided wherever required	
8	The materials/instruments generally required for filling the generator with hydrogen and testing protection relays (Section 16.3.2) are arranged	
9	The safety gadgets (Section 16.3.3) are arranged	

16.3.2 Typical List of Bill of Materials/Instruments

The specification and/or quantity of each of the following vary from project to project and have to be assessed by concerned personnel associated with the test beforehand.

1. The number of CO_2 cylinders needed to replace air from the generator enclosure and purge H_2
2. The number of H_2 cylinders needed to replace CO_2 from the generator enclosure and to raise H_2 gas pressure and purity inside the generator enclosure to rated value as recommended
3. H_2 gas leakage detector must be used to confirm no leakage from the feeding system and generator casing

16.3.3 Typical List of Safety Gadgets

The equipment generally used for the safety of operating personnel is given here:

1. Safety shoes of assorted sizes
2. Hand gloves
3. Hard hats of assorted sizes
4. Gas masks
5. First aid box
6. Sufficient lighting arrangements
7. Safety sign boards
8. Safety tags

16.3.4 H_2-CO_2 System (Fig. 16.2)

The gas system is comprised of the following arrangement:

16.3.4.1 H_2 cylinder rack

High-pressure H_2 cylinders are connected to the manifold on the cylinder rack. Valves on cylinders and the manifold allow for the replacement of cylinders during the operation. High-pressure hydrogen is first reduced to an intermediate pressure through two parallelly connected pressure reducers and then passed through another pressure reducer on the valve rack for expansion to the pressure required for generator operation, which is monitored by a pressure gauge mounted on the valve rack. The hydrogen then passes through a filter and enters into the upper distribution piping inside the generator enclosure. Relief valves on the intermediate side of pressure reducers are connected to the outlet pipe system, through which any excess hydrogen is vented to the atmosphere.

16.3.4.2 CO_2 cylinder rack

The arrangement of CO_2 cylinders in the rack is similar to the arrangement of H_2 cylinders in the rack. At the gas valve rack, liquid CO_2 is expanded and evaporated. The CO_2 pressure required for generator operation is monitored by a pressure gauge mounted on the valve rack. Gaseous CO_2 is then passed through a filter before entering into the lower distribution piping inside the generator enclosure.

16.3.4.3 Gas valve rack and gas monitoring equipment

The gas valve rack is provided with a casing flow meter, detector, gas purity meter, and sampling points for chemical analysis of the gas.

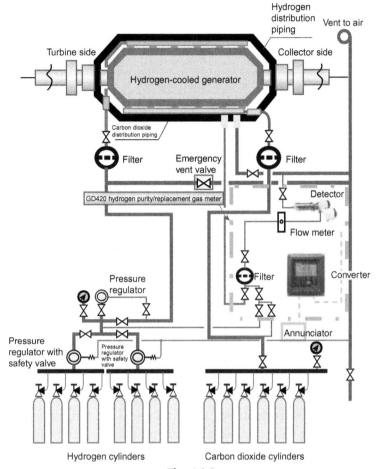

Fig. 16.2

Gas manifolds with supply connections to generator enclosure. *Reproduced with kind permission from Yokogawa Electric Corporation.*

16.3.4.4 Gas dryer

A small amount of hydrogen circulating in the generator for cooling is typically passed through a gas dryer. Inlet and outlet points of the gas dryer are so tapped that the gas is circulated due to differential pressure developed by the fan. The dryer is filled with absorbent material which can be regenerated anytime the generator is running (Fig. 16.3).

16.3.5 Precautions

Beyond observing the general precautionary measures discussed in Section 2.1.1: Quality Assurance, following specific precautionary measures must be observed prior to filling the generator enclosure with H_2.

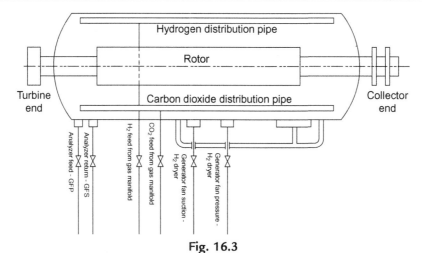

Fig. 16.3

H_2 and CO_2 distribution pipes in generator and other connections. *Source: From Hydrogen Purity and Moisture Monitoring in Hydrogen Cooled Generators. Michael Hicks. Utility Product Specialist. Environment One Corporation. April 16–17, 2013. https://www.aegislink.com/presentations/2013_eums/pdfs/11_ hydrogen_dew_point_monitoring.pdf.*

16.3.5.1 H_2 and CO_2 cylinders

i. H_2 & CO_2 cylinders should always be stored upright and provided with appropriate chains or straps to secure cylinders in place connected to a wall.

ii. Cylinders must be located in a dry, well-ventilated area away from flames, sparks, or any source of heat or ignition, where they will not get subjected to mechanical or physical damage, heat, or electrical circuits, to prevent possible explosion or fire.

iii. Empty gas cylinders should be separated from full or partially full cylinders.

iv. Cylinders should not be exposed to open flame or to any temperature above 325 K.

v. ~~Never drop, bang, or strike cylinders against each other nor against solid objects.~~

vi. Cylinders should not be placed on the ground or on surfaces where water can accumulate.

vii. Cylinders should be provided with physical protection from vehicle damage.

viii. Cylinders must be covered with canopies of noncombustible construction.

ix. Use gloves when handling cylinders and valves.

x. Use the cylinder valve in the fully open position.

xi. Before opening the valve, keep the valve outlet away from you.

xii. Before disconnecting the cylinder, close the cylinder valve and depressurize the system.

xiii. Do not open valve fully until satisfied that all connections are properly made.

xiv. Do not handle cylinders and valves with wet hands.

xv. Do not use the cylinder valve as a flow controller.

xvi. Do not interfere with pressure regulator or relief valve settings.

xvii.   Do not use the valve cover to lift a cylinder because they may get unattached, causing the cylinder to drop on a hard surface, resulting in a probable explosion.

xviii.   Do not lift a cylinder by its cap, which may come loose and cause harm.

16.3.5.2 Hydrogen and hydrogen cylinders

i.   It is evident from Section 16.1 that H_2 cannot be directly introduced inside the generator replacing air, as this will lead to the potential risk of autoignition and to the formation of an explosive mixture. Hence an inert gas like CO_2 is used as an intermediate medium. During the filling of the generator with hydrogen replacing air, air is first replaced with CO_2, and then CO_2 is replaced with H_2. In contrast, while replacing hydrogen with air, H_2 is first scavenged with CO_2, and then CO_2 is replaced with air.

ii.   In order to prevent the explosion of probable leakage of hydrogen gas, all welding work or any other hot work (eg, gas cutting) around the generator and cylinder storage area shall be strictly prohibited.

iii.   Display "No Smoking" signs within 8 m of the area.

iv.   Hydrogen is a colorless, odorless, tasteless, highly flammable, nontoxic gas. Due to the high flammability and explosive nature of hydrogen, storing and the storage area of H_2 cylinders must comply with the requirements of local fire and building codes. In addition, the following special safety precautions must also be met:

1. H_2 cylinders should not be installed within 3 m of windows, doors, or other building openings, or within 15 m of ventilation intakes.

2.   Storage areas should have a minimum of 25% of the perimeter open to the atmosphere, incorporating a chain-link fence with lattice construction for the full height of a cylinder, covering the width of all cylinders.

3.   Cylinders must be located at least 8 m from open flames, ordinary electrical equipment, and/or any ignition source.

4.   Cylinders should never be stored near oxygen, chlorine, or other oxidizing gases. These materials must be separated by at least 6 m, or an appropriate gas cylinder cabinet must be used.

5. H_2 cylinders must never be stored underground of the building.

6.   When attaching a regulator or control valve, do not partially open, or "crack," the hydrogen cylinder valve. This may lead to self-ignition of the hydrogen.

16.3.5.3 Carbon dioxide and carbon dioxide cylinders

i.   Carbon dioxide is stored in cylinders as a liquid and an increase in temperature will cause an increase in pressure. Check that cylinders are fitted with bursting discs to protect against overpressure.

ii.   A number of cylinders must be connected to a manifold to discharge CO_2 together because higher flows from a single cylinder may cause the cylinder to become very cold and collapse the pressure inside it.

iii. Because CO_2 gas is heavier than air, it will tend to sink to the lowest level when released to the atmosphere; if inhaled that may become fatal. Even as little as 15% concentration of CO_2 gas can cause unconsciousness in less than 15 min.

iv. Empty CO_2 cylinders must be stored with valves tightly closed.

16.3.6 Preparatory Arrangements

Preparatory work for filling the generator with hydrogen is described here:

i. Close the following valves:
   1. Isolating valves for H_2 filter line
   2. Isolating valves for CO_2 filter line
   3. Valve on vent gas line
   4. Sampling valves

ii. Connect one full cylinder of H_2 to the H_2 cylinder rack

iii. Set the following pressure reducers to their minimum set values:
   1. H_2 pressure reducer at the H_2 cylinder rack
   2. H_2 pressure reducer at the H_2 line

iv. Open the following valves to set the flow path of measuring gas to the purity monitor:
   1. Valve at H_2 cylinder
   2. Valve at H_2 system
   3. Valve before H_2 pressure reducer at H_2 cylinder rack
   4. Valve after H_2 pressure reducer at H_2 cylinder rack
   5. Valve before H_2 pressure reducer in H_2 system
   6. Valve after H_2 pressure reducer in H_2 system
   7. Isolation valve of H_2 high-pressure switch
   8. Isolation valve of H_2 purity meter system

v. Set H_2 pressure reducers 1 and 2 at H_2 cylinder rack to approximately 0.8 MPa (or as recommended by the generator manufacturer) and tighten the locks firmly.

vi. Set H_2 pressure reducers 1 and 2 in H_2 system to rated gas pressure as recommended by the generator manufacturer and tighten the locks firmly.

vii. Open the H_2 supply valve to gas purity meter.

viii. The gas side of purity meter system is scavenged by raising the gas flow to the maximum range of the flow meter for a period of 30–60 min, as recommended by the generator manufacturer.

ix. Reduce gas flow to normal operating condition.

x. Adjust electrical zero of the gas purity meter using both measuring gas and comparison gas per recommendation of the generator manufacturer until the purity of 100% H_2 is displayed on the purity meter.

xi. Close all valves opened previously (steps iv and vii). The purity meter is now ready for the filling or removal of CO_2 and H_2 and generator operation with H_2.

16.3.7 Operating Procedure

16.3.7.1 Replacement of air from generator enclosure with CO_2

For the filling of CO_2, the generator may remain at standstill condition or be running on turning gear per recommendation of the generator manufacturer. In the event that turning gear is put in operation, gas consumption during filling will be more.

i. Open vent valve of generator enclosure to atmosphere for evacuation of air with CO_2.
ii. Switch on heater of CO_2 flash evaporator and raise temperature to typically about 403 K or as recommended.
iii. Open CO_2 filling line valve to generator casing to replace air per recommendation of the generator manufacturer.
iv. Select online purity meter to monitor air-in-carbon dioxide and switch on the meter.
v. Once the purity of carbon dioxide exceeds 95% and the pressure of carbon dioxide inside the generator casing reaches a little above atmospheric, stop CO_2 filling and close the vent valve to atmosphere of step i.

16.3.7.2 Replacement of CO_2 from generator enclosure with H_2

Open the following valves:

1. Valve after H_2 pressure reducer in H_2 system
2. Valve before H_2 pressure reducer in H_2 system
3. Isolation valve of H_2 high-pressure switch
4. Valve after H_2 pressure reducer at H_2 cylinder rack
5. Valve before H_2 pressure reducer at H_2 cylinder rack
6. Isolation valve of H_2 purity meter system
7. Valve at H_2 cylinder rack manifold
8. Drain valve of generator enclosure to atmosphere for evacuation of CO_2

16.3.7.3 H_2 filling into generator enclosure

i. Open valves on H_2 cylinders one by one.
ii. Open H_2 filling line valve to generator casing to replace CO_2 per recommendation of the generator manufacturer.
iii. Select online purity meter to monitor carbon dioxide-in-hydrogen and switch on the meter.
iv. Once the purity of hydrogen exceeds 97% and the pressure of hydrogen inside the generator casing reaches a little above atmospheric, close the drain valve to atmosphere of Section 16.3.7.2 step (8).
v. Continue filling the generator casing with hydrogen until the pressure inside the casing reaches a rated pressure of typically 0.4 MPa.

vi.   On completion of filling the generator casing with hydrogen, record the values of H_2 gas pressure, H_2 gas purity, seal-oil pressure, and so on.

vii. In the event that filling the generator casing with gases was undertaken when the machine was at standstill condition, place the machine on turning gear per recommendation of the generator manufacturer.

16.4 Protection Stability Test of Generator

The generator is protected from faults, which may cause severe damage to the generator that requires immediate isolation of the unit from the grid. These faults generally include faults inside and outside the protective zone of the generator, generator transformer, and unit auxiliary transformer, as well as faults pertaining to the associated prime-mover. This section, however, is restricted to faults inside and outside the protective zone of the generator, on actuation of which the generator is tripped automatically.

During dynamic testing of the generator, the speed of the generator is raised to its rated value by actually running the associated prime-mover, and then performing tests by open circuiting and shorting at the transmission side isolators in order to achieve generator open circuit characteristics (OCCs) and short circuit characteristics (SCCs), respectively. Healthiness of all current transformer (CT), potential transformer (PT), and associated circuits is also confirmed. The characteristics achieved by dynamic testing are then compared with the actual design characteristics of the generator.

16.4.1 Prerequisites

Before the stability test of protection relays is carried out, ascertain the following (Table 16.4):

Table 16.4 Areas/items to be checked and satisfied

Sl. No.	Areas/Items	Ok ($\sqrt{}$)
1	Check that all protection relays are installed in their respective locations and that the corresponding protocol certifying completeness of erection of these relays is available	
2	Ensure that all drawings related to the generator protection system supplied by the generator manufacturer are available	
3	Verify that CTs and PTs are installed at their respective locations as per the protection diagram	
4	Ensure that breakers, isolators, cables, connecting buses, and so on are properly commissioned	
5	Ensure that factory test certificates of all protection relays are available	
6	Temporary platforms are provided wherever required	
7	Ensure that instruments generally required for the testing of protection relays (Section 16.4.1) are calibrated and installed. Verify calibration certificates	
8	Safety gadgets (Section 16.3.3) are arranged	

16.4.2 Typical List of Instruments

  i.   Voltmeters of required range as recommended
 ii.   Ammeters of required range as recommended
iii.   Portable and online bearing vibration measuring instruments

16.4.3 Precautions

While carrying out the protection relay stability test, observe the general precautionary measures discussed in Section 2.1.1: Quality Assurance.

16.4.4 Preparatory Arrangements

The preparatory work for stability testing of generator protection relays is described below:

  i.   Bypass generator trip command to trip associated prime-mover.
 ii.   Check open-close operation of generator circuit breaker.
iii.   Remove local breaker backup (LBB) isolation link.
 iv.   Set all relays to minimum settings.
  v.   Keep open isolators and breakers.
 vi.   Check continuity and phase sequence of the cable and bus duct.
vii.   Check insulation resistance of the cable.
viii. Check that the polarity and connection of all CTs are OK.
 ix.   Test relay characteristic.

16.4.5 Test Procedure

The following paragraphs discuss the procedure to be adopted during the test. Although the procedure is developed under four headings, the test under each heading takes care of multiple relays.

16.4.5.1 Generator differential relay

This relay is used for protection against generator differential. A stability test and operation of a generator differential relay is checked by simulating a fault, as described below (Fig. 16.4).

  i.   Open generator cable links after generator terminal CT.
 ii.   Create three-phase short circuit at position A after generator terminal CT, with shorting links suitable to carry a current as recommended by generator manufacturer.
iii.   Open terminals of trip circuit signal and connect a neon lamp across these terminals so as to obviate any disturbance.
 iv.   Place excitation on manual mode.
  v.   Verify that the relay is set at minimum value.

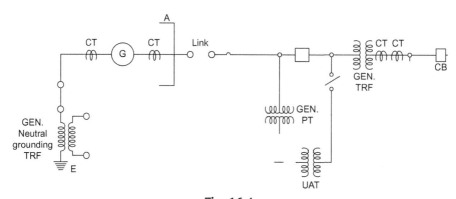

Fig. 16.4

Test setup of generator differential protection.

    vi.   Start the prime-mover and raise its speed to rated value.

   vii.   Do not excite.

  viii.   Measure current in different relays in all three phases of both sides and operating coils.

    ix.   If no abnormalities are observed, close field breaker and excite slowly.

     x.   Record excitation current, stator current, and current in different relays at 25%, 50%, 75%, and 100% of full load current.

    xi.   Verify operation of overall differential relay (87U).

   xii.   Verify operation of generator differential relay (87G).

  xiii.   Increase stator current until overcurrent relay (51G) operates. Record the value.

  xiv.   Record all readings per Tables 16.5 and 16.6.

   xv.   Decrease stator current to minimum and trip field breaker.

  xvi.   Trip the prime-mover.

 xvii.   Remove shorting at position A.

Table 16.5 Stability test reading of generator differential relay

| | | | | Bearing Vibration | | | | | | |
| | | | | DE | | | NDE | | | |
Sl. No.	Excitation Voltage (V)	Excitation Current (A)	Stator Current (A)	V	H	A	V	H	A	Remarks
1										
2										

Table 16.6 Stability test reading of generator differential relay

Type of Relay/ Relay No.	Phase		Terminal No.	Without Excitation	With Excitation				Relay Operation
					25%	50%	75%	100%	
Overall differential/87U		R							
		Y							
		B							
		N							
Backup impedance/21G1		R							
		Y							
		B							
		N							
Negative phase sequence/46G		R							
		Y							
		B							
Indicator/46A Field failure/40G		N							
		R							
		B							
Pole slipping/78G		Y							
Generator differential/87G	R	Spill Volt.							
	Y	Spill Volt.							
	B	Spill Volt.							
		N							
Inter turn fault/ 61N	R	Spill Volt.							
	Y	Spill Volt.							
	B	Spill Volt.							
		N							
Reverse power/ 32G2		R							
		Y							
		B							
		N							
Reverse power/ 32G1		R							
		Y							
		B							
		N							
Backup impedance/21G2	R	Spill Volt.							
	Y	Spill Volt.							
	B	Spill Volt.							
		N							

16.4.5.2 Overall differential relay

A stability test and operation of an overall differential relay is checked by simulating a fault, as described below (Fig. 16.5).

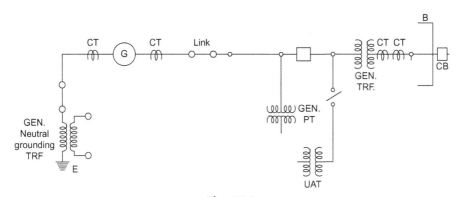

Fig. 16.5

Test setup of overall differential protection.

   i. Remove shorting at position A (Fig. 16.4) and connect generator bus duct link after terminal CT.
  ii. Create three-phase short circuit at position B with conductor suitable to withstand short circuit current.
 iii. Close generator transformer side isolator of generator breaker.
  iv. Open terminals of trip circuit signal to generator circuit breaker and connect a neon lamp across these terminals so as to obviate any disturbance.
   v. Place excitation on manual mode.
  vi. Verify that all relays are set at the minimum value.
 vii. Start the prime-mover and raise its speed to rated value.
viii. Do not excite.
  ix. Measure current in all relays.
   x. If no abnormalities are observed, close field breaker and excite slowly.
  xi. Raise stator current up to 25% of full load current.
 xii. Record current in different relays.
xiii. Verify operation of 87U.
 xiv. Record this operating value.
  xv. Record all readings in Table 16.7.
 xvi. Trip field breaker.
xvii. Trip the prime-mover.
xviii. Remove shorting at position B.

Table 16.7 Stability test reading of overall differential relay

Type of Relay/Relay No.	Phase	Terminal No.	CT Secondary Current at Stator (A)			Relay Operation
			Without Excitation	With Excitation		
Overall differential/87U	R					
	Y					
	B					
Overcurrent/51G	R					
	Y					
	B					
LBB/50Z	R					
	Y					
	B					
Stator earth fault/64G2 & 64G1						
Earth fault/51N	N					
Restricted earth fault/64R	Spill					
Bus differential/87B	R					
	Y					
	B					
Backup impedance/21G1	R					
	Y					
	B					
	N					
Backup impedance/21G2	R					
	Y					
	B					
	N					
Pole slipping/78G	Y					
Field failure/40G	R					
	B					
P.T. secondary voltage	R-Y					
	Y-B					
	B-R					
Mvar						
Negative sequence (%)						

16.4.5.3 Generator earth fault relay

A stability test and operation of a generator earth fault relay is checked by simulating a fault, as described below (Fig. 16.6).

i. Create short circuit from R phase to earth at position A.
ii. Open terminals of trip circuit signal to generator circuit breaker and connect a neon lamp across these terminals so as to obviate any disturbance.
iii. Place excitation on manual mode.
iv. Verify that all relays are set at the minimum value.

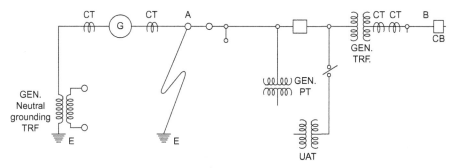

Fig. 16.6
Test setup of generator earth fault protection.

v.   Start the prime-mover and raise its speed to rated value.
vi.   Close field breaker and excite slowly.
vii.   Measure voltage in 64G.
viii.   Increase stator voltage until 64G operates. Record the trip value.
ix.   Verify operation of 87G.
x.   Trip the prime-mover.
xi.   Remove shorting from R phase.
xii.   Create short circuit to earth from Y phase and B phase one after another and repeat the test.
xiii.   Record all readings in Table 16.8.

Table 16.8 Stability test reading of generator earth fault relay

Single Phase to Earth Created in	Stator		Voltage in 64G	Voltage in 87G			Remarks
	Current	Voltage		R	Y	B	
R							
Y							
B							

16.4.5.4 Open circuit characteristic test

The test of generator OCC is narrated below (Fig. 16.7).

i.   Remove all earthing and shorting of Sections 16.4.5.1, 16.4.5.2, and 16.4.5.3.
ii.   Normalize gen bus duct link to make circuit up to generator transformer.
iii.   Open generator circuit breaker, along with its isolators.
iv.   Place excitation on manual mode.
v.   Verify that all relays are set at the minimum value.
vi.   Start the prime-mover and raise its speed to rated value.
vii.   Do not excite.

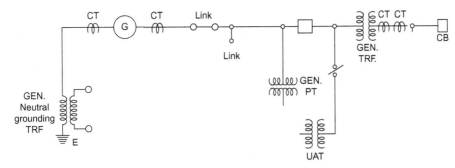

Fig. 16.7
Test setup of generator open circuit characteristic.

viii.   Measure generator terminal voltage.

  ix.   Close field breaker and excite slowly.

   x.   Record excitation current and generator voltage at 25%, 50%, 75%, and 100% of rated voltage. Compare these values with the OCC curve of the test certificate.

  xi.   Record current in different relays.

 xii.   When rated voltage is attained, measure voltage across the pressure coil of 64G.

xiii.   Increase generator voltage and reduce machine speed until overfluxing relay operates.

xiv.   Record overfluxing relay operates at ... voltage and ... Hz.

All readings are to be recorded per Tables 16.9 and 16.10.

Table 16.9 Open circuit characteristic of generator

Sl. No.	Field Voltage	Field Current	Generator Voltage	Remarks

Table 16.10 Open circuit characteristic of generator

Sl. No.	Relay No.	Terminal No.	Relay Operated at Generator Voltage					Remarks
			0%	25%	50%	75%	100%	
1	32G1/G2							
2	37G							
3	40G							
	40/27G							
4	59G							
5	99GT							
6	98G							
	95G							
7	21GR							
	21GY							
	21GB							
8	64G2							
9	46G							
10	50G1							

16.5 Conclusion

On completion of the filling of the generator with hydrogen and stability testing of generator protection relays, two separate protocols (Tables 16.11 and 16.12) may be signed jointly by the customer, the engineer, the generator manufacturer, and the supplier (contractor), certifying that the "generator is ready for synchronizing, loading, and commencement of commercial operation."

Table 16.11 Protocol for filling of generator with hydrogen

Type of Test	Findings	Remarks
Evacuating air from generator enclosure with carbon dioxide Displacing carbon dioxide from generator enclosure with hydrogen	Purity of carbon dioxide in air > 95% Purity of hydrogen in carbon dioxide > 97% Pressure of hydrogen inside the generator enclosure is about 0.4 MPa	Generator enclosure may be filled with hydrogen Raise pressure of hydrogen inside the generator enclosure Record the following parameters: 1. H_2 gas pressure 2. H_2 gas purity 3. Seal-oil pressure, etc.
Signed by the customer · Signed by the engineer	Signed by the generator manufacturer	Signed by the supplier (contractor)

Table 16.12 Protocol for stability test of generator protection relays

Type of Relay	Findings	Remarks
Generator differential relay	All salient data are properly recorded and all associated relay operations are in order	Characteristics achieved by dynamic testing are comparable to design characteristics of protection relays
Overall differential relay	All salient data are properly recorded and all associated relay operations are in order	Characteristics achieved by dynamic testing are comparable to design characteristics of protection relays
Generator earth fault relay	All salient data are properly recorded and all associated relay operations are in order	Characteristics achieved by dynamic testing are comparable to design characteristics of protection relays
Relays related to OCC of generator	All salient data are properly recorded and all associated relay operations are in order	Characteristics achieved by dynamic testing are comparable to design characteristics of protection relays

(Continued)

Table 16.12 Protocol for stability test of generator protection relays—Cont'd

Type of Relay	Findings	Remarks	
Generator is now ready for synchronizing, loading, and commencement of commercial operation			
Signed by the customer	Signed by the engineer	Signed by the generator manufacturer	Signed by the supplier (contractor)

References

[1] NFPA 704: Standard System for the Identification of the Hazards of Materials for Emergency Response, 2012.
[2] C37.102–2006, IEEE Guide for AC Generator Protection.

CHAPTER 17

Completion Test of a Thermal Power Plant

Chapter Outline

17.1 Introduction

Completion test of a power plant is undertaken upon successful execution of all preoperational activities discussed in Chapter 3 through Chapter 16. Once preoperational activities are satisfactorily executed the complete unit is placed on trial run. During this period all equipment, along with their subsystems and supporting equipment, are operated as an integral unit and operational parameters pertaining to each system and equipment are recorded. All adjustments and repairs to any equipment or system are carried out by the supplier (contractor) and recorded accordingly during this period, if necessary, by shutting down the plant. On completion of a satisfactory trial run of the plant, certified jointly by the customer, the engineer, equipment manufacturers, and the supplier (contractor), the plant is placed under initial operation. Further adjustments, if required, and final testing of all controls and protections as well as miscellaneous tests, as discussed in Section 17.5, are carried out covering the full load range of the plant during this period.

Thermal Power Plant. http://dx.doi.org/10.1016/B978-0-08-101112-6.00017-4

17.2 Objective

During the trial run and initial operation of the plant, all steady-state operating parameters and generator output, as specified and/or contractually agreed with the customer, are to be demonstrated by the representatives of equipment manufacturers and the supplier (contractor) to the customer and the engineer. Once the demonstration is construed to be satisfactory the plant is placed under a reliability run test, followed by performance/acceptance tests.

> **Note**
>
> Because the equipment, systems, and the plant as a whole of coal-fired steam power plants are more complex and complicated than the equipment, systems, and so on of gas turbine power plants or diesel power plants, the contents of this chapter are more relevant to a coal-fired steam power plant.

17.3 Precautions

General precautionary measures, as discussed in Section 2.1.1: Quality Assurance, and specific precautionary measures, as discussed in previous chapters, must be observed strictly before going ahead with conducting a completion test.

17.4 Prerequisites

Prior to carrying out the completion test of the plant, a thorough inspection must be fulfilled as part of the quality assurance (QA) process (Table 17.1).

Table 17.1 Areas/items to be checked and satisfied

Sl. No.	Areas/Items	OK (√)
1	Verify availability of protocols jointly signed by the customer, the engineer, equipment manufacturers, and the supplier (contractor) certifying successful completion of preoperational activities discussed in Chapter 3 through Chapter 16	
2	Verify that an adequate number of skilled and unskilled laborers, fitters, welders, as well as competent personnel who are conversant with all attributes of various testing activities are deputed to execute trial run, initial operation, and completion test of the plant	
3	Ensure that performance guarantee (PG)/performance acceptance (PA) test procedures of various equipment are approved by the customer/the engineer and are available	
4	Ensure that the firefighting system fulfills all regulatory requirements and is functional in all respects	
5	Check availability of all consumable materials; tools and tackles; matching flanges; impulse piping and valves; thermowells (both screwed and welded); calibrating instruments/devices, including flow devices, special test instruments, tong testers, Megger, and so on; and any special equipment as required for trial run, initial operation, and completion test of the plant	

Table 17.1 Areas/items to be checked and satisfied—cont'd

Sl. No.	Areas/Items	OK (√)
6	Ensure that all tests are conducted under steady-state operating conditions	
	Note Some tests may need to be conducted under transient operating conditions also.	
7	Ensure that auto-control loops pertaining to the steam generator, steam turbine, and regenerative systems, including coordinated control of the steam generator and steam turbine, are functional and operational	
8	Ensure that flame scanners are functioning properly as designed	
9	Verify that protections and interlocks of all equipment and systems are in place	
10	Ensure that all tests that are to be demonstrated to the client are performed successfully and applicable protocols are jointly signed	
11	Verify that analysis of as-fired coal and ash is available	
12	Ensure that the quality of water (Tables 3.1 and 3.2) and steam (Tables 3.3 and 6.1) fulfills the specified requirements of steam generator and steam turbine manufacturers	

17.5 Miscellaneous Tests

During the trial run of the plant, "start permissive" and "interlocks and protections" of each individual equipment and system, along with "unit protections" pertaining to the steam generator-steam turbine generator, are checked. Reports and protocols of trial runs and checking of interlocks and protections are maintained.

Certain tests that are typically followed in the industry during initial operations of the plant, as described in Table 17.2, are to be completed either separately or in conjunction with other tests to the full satisfaction of the customer.

Table 17.2 Tests to be completed during initial operation

Sl. No.	Type of Test	Parameters to be Demonstrated
1	Capacity, head developed, power consumption, and margins envisaged on capacity and head of FD, ID, and PA fans	Comply with predicted performance
2	Load sharing of ID, FD, and PA fans	Equal load sharing while running in parallel
3	Maximum permissible rotor vibration of pumps, fans, and blowers	Within good range as per VDI 2056
4	Operation of turbine governor	Droop of steam turbine governor is functioning between 3% and 5% of rated speed
5	Voltage control by the automatic voltage regulator (AVR)	Generator voltage is controlled within ±5% of rated voltage
6	Capability of generator	Operate at rated voltage and frequency at power factors and reactive conditions per the generator capability curve furnished by the generator manufacturer and also to match the grid requirement

Continued

Table 17.2 Tests to be completed during initial operation—cont'd

Sl. No.	Type of Test	Parameters to be Demonstrated
7	Minimum sustainable load	Capability of the plant to operate at specified minimum generator net electrical output at specified operating conditions while all plant auxiliaries operate under stable conditions without any disturbance
8	Load throw-off	From full load to house load without disturbing stable operating conditions of all plant auxiliaries
9	Load rejection test	Ability of the plant to operate under steady conditions during part load rejection of the plant. Observe the following while this test is in progress: 1. Transient overspeed of the turbine does not lead to overspeed trip of the turbine 2. All running auxiliaries associated with the power plant remain in service without any disturbance to themselves or to the plant 3. The steam generator remains in service along with associated controls 4. Deviation of operating parameters remains within respective trip limit
10	Automatic load runback capability of the unit (steam generator-steam turbine generator) ensuring smooth and stable runback operation	Unit must run back to about 60% TMCR load safely without tripping in the event of tripping of any one of the 2x60% major auxiliaries, such as ID fan, FD fan, P.A. fan, air preheater, BFP, CEP, and CW pump
11	Maximum permissible time from start-up to synchronization of a large reheat steam power plant not exceeding predicted time period	1. Cold start-up: typically 8 h 2. Warm start-up: typically 150 min 3. Hot start-up: typically 45 min
12	Maximum permissible time from synchronization to full load during hot start-up of a large reheat steam power plant not exceeding predicted time period	Typically 45 min
13	Response to load ramp rate: minimum ± 3% per minute (between 30% and 50% load); minimum ± 5% per minute (between 50% and 100% load)	Steam generator-steam turbine generator capable of increasing or decreasing unit output without compromising design life of pressure parts
14	Above the minimum sustainable operating load response to step load changes: minimum ± 10% per minute	Steam generator-steam turbine generator capable of increasing or decreasing unit output steadily without compromising design life of pressure parts
15	On-load closure test of turbine valves (ESVs, IVs, and CVs)	Operation successful
16	Accuracy of AVR	Within 0.5%
17	AVR reference voltage adjustable range for all loads	−15% to +10% of nominal voltage

17.6 Reliability Run Test

The objective of the reliability run test is to observe the behavior of the plant along with its auxiliaries as an integrated unit to demonstrate continuous operation of the plant without any interruption. It is expected that successful completion of this test would ensure smooth, trouble-free continuous operation of the plant in the future.

On completion of all miscellaneous tests (Section 17.5) the reliability run of the unit is commenced for a specified period, typically from 14 to 30 days of continuous operation with all auto-control loops in service. Out of these 14–30 days, the unit must run on full load for a period of at least 72 h without any interruption. In case any major equipment trips during the reliability run, it must not lead to a tripping of the whole unit.

During the reliability run test period, repair of any equipment or system will not be permitted. However, minor adjustment of any equipment, cleaning of filter elements, changeover of coolers/dryers, replacement of printed-circuit cards, or any other actions necessary during normal operation of the plant are permissible with prior knowledge of the customer. It is to be ensured that such adjustments do not in any way interfere with or prevent the commercial use of the plant by the customer, or result in reducing the output, decreasing the efficiency, or exceeding the environmental limits.

In the event of any interruption occurring anywhere within the plant, due to or arising from faulty design, material, or workmanship, a new reliability test for a period of 14–30 days shall commence after the supplier (contractor) has rectified the defect. Any or a combination of the following conditions may be interpreted as an interruption:

i. Any equipment/system failure that results in a reduction in generation capacity from the required power output
ii. Deviation greater than $\pm 2.5\%$ of the required power output
iii. Switching over to standby equipment without any valid reason
iv. Any auto-control loop that runs on manual mode instead of auto mode
v. Failure of an auxiliary/system to start

Notwithstanding the previous conditions, if the reliability run test gets interrupted due to any of the following reasons, the test shall be allowed to resume as if the interruption had not occurred:

1. In the event of interruption in load dispatch
2. Analysis of as-fired coal grossly deviates from the analysis of design range of coals, resulting in improper functioning of the control loops on auto
3. If for reasons beyond the control of the supplier (contractor), which are attributable to force majeure, such as:
   i. natural phenomena, including but not limited to floods, droughts, earthquakes, epidemics, cyclones, lightning, storms, and plagues

   ii. legal strikes, legal lockouts, and other generalized labor action, excluding such events which are site-specific and attributable to the supplier (contractor)

  iii. act of terrorism or sabotage; act of any government, including but not limited to war (declared or undeclared), priorities, quarantines, and embargoes

  iv. radioactive contamination or ionizing radiation or chemical contamination

   v. fire or explosion, except as may be attributable to the supplier (contractor)

  vi. air crash or shipwreck

 vii. an act of God, provided either party shall within 7 days of occurrence notify the other in writing of such a cause

During the reliability run test it is imperative to demonstrate certain operating parameters/conditions, which vary from project to project. For a specific project, operating parameters/conditions of concern must be specified beforehand by the customer. Some typical operating parameters/conditions are described in Table 17.3.

Table 17.3 Parameters to be demonstrated during reliability run test

Sl. No.	Type of Test	Conditions/Parameters to be Demonstrated
1	Superheater outlet steam temperature at 60–100% TMCR unit load	Rated main steam temperature
2	Reheater outlet steam temperature at 60–100% TMCR unit load	Rated hot reheat steam temperature
3	Steam temperature imbalance between superheater and reheater outlets (in case of more than one outlet) under all loads, including transients and rated steam conditions and condenser vacuum with 0% make-up	Not exceeding 10 K
4	Superheater/reheater outlet steam temperature deviation between 60% and 100% TMCR unit load	Within +5 and −10 K of rated steam temperature
5	Superheater/reheater outlet steam temperature deviation during loading/unloading at the rate of 5% of TMCR load per minute	Within +5 and −10 K of rated steam temperature
6	Steam generation with light fuel oil only	From 0% to 10% BMCR
7	Steam generation with heavy fuel oil only	From 0% to 30% BMCR
8	Minimum achievable steam generation when firing coals only, from the range of coals specified, with any combination of mills/adjacent mills in service, without the support of oil firing for stable and efficient operation	Not exceeding 40% BMCR
9	Operation of steam generator at 80%, 60%, 50%, 40% of BMCR load	Part load stable operations without any disturbance
10	Pressure drop at superheater at 100% TMCR unit load	Not exceeding predicted value

Table 17.3 Parameters to be demonstrated during reliability run test—cont'd

Sl. No.	Type of Test	Conditions/Parameters to be Demonstrated
11	Pressure drop at reheater at 100% TMCR unit load	Not exceeding predicted value
12	Spray water flow to superheater attemperation system at all loads up to and including BMCR	Not exceeding the value considered for design while maintaining the rated superheater outlet steam temperature
13	Maximum furnace exit gas temperature (FEGT)	Minimum 60 K below the minimum initial deformation temperature of ash
14	The flue gas temperature at the entry and exit of various steam generator heating surfaces and also the variation across the cross section perpendicular to gas flow	Not exceeding the values considered for the heating surfaces design
15	The air-heater air-in-leakage and maximum drift in air leakage after 3000 h of operation from taking over date	Not exceeding the predicted value
16	Life of pulverizer wear parts for the entire range of coal characteristics specified	Not less than predicted hours of operation
17	Capacity output (kg s^{-1}) of each pulverizer with coal fineness of not less than 70% through 200 mesh screen and not less than 98% through 50 mesh screen, applying corrections for the variation in coal characteristics (ie, hardgrove grindability index, total moisture, etc.)	Not less than predicted output
18	Electrostatic precipitator (ESP) air-in-leakage at the guarantee point condition	Not more than 1% of total gas flow at ESP inlet
19	Maximum pressure drop across ESP at the guarantee point flow condition	Not exceeding 200 Pa
20	Gas distribution in the various stream and fields of ESP	Must be uniform
21	Maximum continuous output at generator terminals corresponding to VWO flow, at rated steam conditions and condenser vacuum with 3% makeup flow	Comply with the predicted value
22	Maximum continuous output at generator terminals corresponding to all HP feedwater heaters out of operation at rated steam conditions and condenser vacuum with 3% makeup flow	Comply with the predicted value
23	Maximum continuous output at generator terminals corresponding to one HP heater train out of operation under rated steam conditions and condenser pressure with 3% makeup flow	Comply with the predicted value
24	Continuous output at generator terminals with 0.85 power factor under rated steam conditions at worst condenser pressure with 3% makeup flow	100% TMCR unit load

Continued

Table 17.3 Parameters to be demonstrated during reliability run test—cont'd

Sl. No.	Type of Test	Conditions/Parameters to be Demonstrated
25	Constant pressure operation of the unit in conjunction with the steam generator, HP-LP bypass system, and instrumentation and control system	From no-load condition to VWO condition
26	Modified sliding pressure operation of the unit in conjunction with the steam generator, HP-LP bypass system, and instrumentation and control system	Constant pressure operation from no-load to 35%/40% TMCR unit load; Sliding pressure operation from 35%/40% TMCR to 85%/90%TMCR unit load; Constant pressure operation from 85%/90% TMCR unit load to VWO condition
27	Changeover from constant pressure mode to modified sliding pressure mode and vice versa	Must be bumpless
28	Start-up, loading, unloading, and shutdown characteristics for the steam turbine generator for cold start conditions, warm start conditions, and hot start conditions under constant pressure and sliding pressure mode	Turbine operational parameters like vibration, absolute and differential expansion, eccentricity, steam-metal temperature mismatch, and so on must be within design limits
29	Sudden total loss of all external load conditions	The steam turbine generator unit shall not trip on overspeed but shall continue in operation under the control of its speed governor to supply power for the unit auxiliary load with HP-LP bypass in operation while staying within the prescribed permissible limits of steam-metal temperature mismatch, exhaust hood temperature, absolute and differential expansion, vibration, and eccentricity
30	Turbine overspeed during turbine trip from any load	Less than overspeed trip setting (109–111%)
31	Maximum permissible satisfactory rotor vibration of steam turbine	Typically 76 μm
32	Generator capability	Maximum permissible temperature of different parts of the generator for continuous operation
33	Frequency range	Within 47.5–51.5 Hz
34	Excitation system performance	a. The excitation system shall be capable of supplying field forcing for 30 s in case of all types of faults, including close in faults of the generator b. The ceiling voltage shall not be less than 150% of the machine excitation voltage c. Nominal exciter response ratio shall not be less than 2.0 per second.
35	Operation at reduced generator hydrogen pressure	Generator shall be capable of operating at reduced capacity
36	Automatic online turbine testing system	On-load testing of turbine protective equipment without disturbing normal operation and keeping all protective functions operative during the test

Table 17.3 Parameters to be demonstrated during reliability run test—cont'd

Sl. No.	Type of Test	Conditions/Parameters to be Demonstrated
37	HP-LP bypass system should come into operation automatically under the adjoining conditions	a. Generator circuit breaker opening b. HP-IP turbine stop valves closing due to turbine tripping c. Sudden reduction in demand to house load
38	Lube oil purification system	Capacity and the purity of purified oil at the outlet of the centrifuge and the outlet of the polishing filter
39	Performance of the condenser	a. The back pressure achieved at design CW flow and inlet temperature and cleanliness factors at VWO condition and 3% makeup b. Temperature of condensate at outlet of condenser shall not be less than the saturation temperature corresponding to the condenser pressure at all loads c. Oxygen content in the condensate at hotwell outlet shall not exceed 0.015 mL kL^{-1} over the entire load range d. When one half of the condenser is isolated, condenser capability shall be demonstrated to take at least 60% TMCR load e. Air leakage into the condenser under full load condition shall not exceed more than 50% of design value taken for sizing of the vacuum pumps f. The capacity of each vacuum pump in free dry air under standard conditions at a specified condenser pressure and specified subcooled temperature g. The air and vapor mixture from the air cooling zone of the condenser shall be equal to the specified subcooled temperature corresponding to the specified condenser pressure h. Combined pressure drop in condenser tube, waterbox, and inlet and outlet piping, measured between intake and discharge point of the CW system with cleanliness factor of 0.9 and "condenser on-load tube cleaning system" in service, shall be within the design value

Continued

Table 17.3 Parameters to be demonstrated during reliability run test—cont'd

Sl. No.	Type of Test	Conditions/Parameters to be Demonstrated
40	Feedwater heaters and deaerator performance	a. TTDs and DCAs of feedwater heaters in line with 100% TMCR heat rate guarantee heat balance b. Difference between saturation temperature of steam entering the deaerator and temperature of feedwater leaving the deaerator c. Dissolved oxygen content in feedwater measured at deaerator outlet without any chemical dosing shall not exceed 0.005 mL kL^{-1} at all loads from no-load to VWO condition with 3% cycle makeup with normal pressure and overpressure with incoming condensate presumed to be saturated with oxygen d. Free carbon dioxide in deaerator effluent shall be untraceable at all loads from zero to VWO condition with 3% cycle makeup with normal pressure e. Continuous and efficient operation and performance of feed heating plant without undue noise and vibrations at all loads and duty conditions
41	Condensate extraction pumps (CEPs)	a. Each CEP shall be capable of delivering flow and total dynamic head corresponding to run-out point b. Running of pumps within the specified permissible limits of vibration, noise level, and parallel operation of the pumps
42	Boiler feed pump (BFP) performance	a. Each BFP shall be capable of delivering flow and total dynamic head corresponding to run-out point b. Running of pumps within the specified permissible limits of vibration, noise level, and parallel operation of pumps c. Cold start-up/hot start-up of the unit using steam turbine driven BFP (if provided) with motive steam supply from auxiliary steam header
43	Condensate polishing unit performance	a. Effluent quality at outlet of each vessel at its rated design flow and design service length between two regenerations b. Pressure drop across the polisher service vessel, as specified, in clean and dirty condition of resin at rated design flow

Table 17.3 Parameters to be demonstrated during reliability run test—cont'd

Sl. No.	Type of Test	Conditions/Parameters to be Demonstrated
44	Closed cycle cooling water system performance	a. Running of pumps within the prescribed permissible limits of vibration, noise level, and parallel operation of the pumps b. Design heat load of heat exchangers and inlet and outlet temperatures on the primary and secondary sides, pressure drop across heat exchangers on the primary and secondary water circuit
45	Noise (Sound pressure shall be measured all around the equipment at a distance of 1.0 m horizontally from the nearest surface of any equipment/machine and at a height of 1.5 m above the floor level in elevation)	a. Steam generator, auxiliaries and systems, and other continuously operating equipment shall perform without the noise level (individually or collectively) exceeding 85 dBA over the entire range of output and operating frequencies b. Noise level for turbine generators shall not exceed 90 dBA over the entire range of output and operating frequencies c. For short-term exposure, noise levels shall not exceed the limits as stipulated in the following Occupational Safety & Health Administration (OSHA) standard [1]

Duration per Day (h)	Sound Level (dB(A)) Slow Response
8.00	90
6.00	92
4.00	95
3.00	97
2.00	100
1.50	102
1.00	105
0.50	110
0.25 or less	115

Sl. No.	Type of Test	Conditions/Parameters to be Demonstrated
46	Utility (eg, DM water, clarified water, compressed air, etc.) consumption	Must be within design limits
47	Maximum pressure dew point	233 K
48	Maximum permissible SOx in flue gas	Not exceeding 200 mg Nm^{-3} (for solid fuels ≥600 MWth—world bank norm)
49	Maximum permissible NOx in flue gas	Not exceeding 200 mg Nm^{-3} (for solid fuels >50 MWth—world bank norm)
50	Maximum permissible CO in flue gas	Not exceeding 100 mL kL^{-1}
51	Free chlorine in effluent	Not exceeding 1.0 mg L^{-1}
52	Suspended solids in effluents	Not exceeding 100 mg L^{-1}
53	Surface temperature of thermal insulation	Not exceeding 333 K
54	Turbine hall EOT crane performance	Overload test, travel and hoist speed checks, and so on
55	Passenger lifts performance	Overload test, travel and hoist speed checks, and so on

In addition to these, the representatives of equipment manufacturers and the supplier (contractor) must demonstrate the following modes of operation to the satisfaction of the customer before conducting a performance guarantee/acceptance (PG/PA) test of various equipment, systems, and the unit as a whole:

i.   Operation of each system by remote manual control
ii. Operation of all systems in integrated manner on auto control
iii. Operation of the entire unit with auto-control loops fully implemented, including different modes of load control with the help of control system to verify the integrated performance of the total unit "control and instrumentation" and to verify whether all important parameters remain within stipulated permissible limits under all operating conditions. In case during tests or otherwise it is observed that the behavior/response of drives/actuators/valves, and so on is not satisfactory/acts as a limitation/restriction in achieving the permissible limits, the supplier (contractor) shall carry out all required modifications, rectification, and so on in his system so that the permissible limits can be achieved.

17.7 Performance Guarantee (PG)/Performance Acceptance (PA) Tests

17.7.1 General Requirements

The PG/PA tests for various equipment/systems/plants are carried out to meet the guaranteed ratings and performance requirements as specified by the customer for various equipment and also as claimed by the equipment supplier (contractor) while signing the contract.

The guaranteed performance parameters, which are subject to a penalty, shall be without any tolerance and all margins required for instrument inaccuracies and other uncertainties shall be included in the guaranteed figures. All the guarantees shall be tested together as far as practicable. During these tests, the plant shall be fully on automatic control under steady load condition.

Following the construction of the plant, a series of preoperational and commissioning tests are conducted, some of which are described in Chapter 3 through Chapter 16. The PG/PA tests are the final tests to prove accurately that the integrated unit will deliver the guaranteed parameters (Table 17.4) efficiently and reliably throughout its economic life.

The PG/PA tests are conducted by the equipment supplier (contractor) with full involvement of the customer. The customer provides the necessary operating inputs and associates his or her supporting staff with the equipment supplier (contractor) to carry out the various activities related to the tests. The supplier (contractor) also provides the necessary labor/supporting staff, and so on.

Per general practice in the industry, the equipment supplier (contractor) submits the PG/PA test procedures within 12 months from the date of the letter of award of the contract for the approval of the customer/the engineer. These procedures include the detailed methodology to conduct various tests so as to verify the guarantees offered by the equipment supplier (contractor).

For the convenience of all concerned it is advisable to conduct PG/PA tests within a period of 3 months after the successful completion of initial operation.

The tests are conducted at the specified load points, and as near the specified cycle conditions as practicable. Proper corrections in calculations to take into account the conditions which deviate from the guaranteed conditions may be applied in the test report (Detailed treatment on "correction factors" is presented in Appendix A). Prior to conducting PG/PA tests all applicable correction curves must be submitted by the equipment supplier (contractor) to the customer/the engineer for their information.

The PG/PA tests are carried out generally with the plant operating at the guaranteed point conditions corresponding to 100%, 80%, 60%, and 50% TMCR load or to loads as specified by the customer.

All instruments required for PG/PA tests must be of the type and accuracy required by the relevant test codes to be followed. All instrument inaccuracies must be computed as per these codes and values corrected to the advantage of the client. No negative tolerances are allowed.

Prior to conducting the tests all test instruments shall be calibrated in an independent test laboratory. The protecting tubes, pressure connections, and other test connections required for conducting guarantee tests shall conform to the relevant codes.

All test grade instruments, equipment, tools, and tackles required for the successful completion of the PG/PA tests are arranged by the equipment supplier (contractor) free of cost.

On completion of each PG/PA test the equipment supplier (contractor) submits a detailed test report within preferably 1 month's time of completion of the test for the approval of the customer/engineer. Should the assessment of these test reports by the customer/engineer show any deterioration from the guaranteed values, the equipment supplier (contractor) shall modify the equipment as required to enable it to meet the guarantees. In such case, PG/PA tests shall be repeated after 1 month from the date the equipment is ready for retest, and all costs for modifications, including labor, materials, and the cost of additional testing to prove that the equipment meets the guarantees, shall be borne by the equipment supplier (contractor).

The PG/PA tests are categorized as given here:

(a) Category "A" (Table 17.4), which attract liquidated damages (LDs)
(b) Category "B" (Tables 17.2 and 17.3), which are mandatory yet do not attract LDs

Table 17.4 Parameters included under category "A"

Sl. No.	Guaranteed Parameter	Operating Condition
1	Efficiency of steam generator in percentage	With rated steam pressure and temperature at superheater outlet and at 100% TMCR unit load
2	Capacity of steam generator in kg s^{-1} of steam	With rated steam pressure and temperature at superheater outlet and with any combination of pulverizers working as decided by the customer, the coal being fired from within the specified range
3	Turbine cycle HR in kcal kWh^{-1}	Under rated steam conditions and maximum condenser vacuum with 0% makeup at 100% TMCR unit load with all feedwater heaters in service
4	Continuous TG output at 100% TMCR unit load	Under rated steam conditions and condenser vacuum, design CW temperature with 0% makeup and all feedwater heaters in service
5	Suspended particulate matter downstream of ESP and/or bag filter	Not exceeding 50 mg N m^{-3} (solid fuels—world bank norm) at 100% TMCR unit load
6	Collection efficiency of ESP and/or bag filter in percentage	At 100% TMCR unit load
7	Power[a] consumed in kilowatt by all the auxiliaries and equipments for continuous plant operation at 100% TMCR unit load (To be measured during the PG/PA test)	Under rated steam conditions and condenser vacuum with 0% makeup

[a]Auxiliary power consumption of plant-related continuously operating auxiliaries shall be measured along with the plant PG/PA test. For intermittently operating auxiliaries, power consumption is computed based on actual hours of operation during the plant PG/PA test. All auxiliaries must be taken into consideration while guaranteeing the auxiliary power consumption. Plant auxiliary power consumption (APC) may be calculated using either of the following methods.

Method 1: $APC = P_G - P_{GT} + P_{SA}$

where

APC = guaranteed auxiliary power consumption,

P_G = power measured at generator terminals,

P_{GT} = power measured downstream of generator transformer,

$P_G - P_{GT}$ = power consumed by unit auxiliaries along with transformer losses,

P_{SA} = power consumed by station auxiliaries (contribution to the unit), may be measured downstream of station transformers.

Method 2: $APC = P_U + T_L$ [2]

where

APC = guaranteed auxiliary power consumption,

P_U = power consumed by unit auxiliaries plus power consumed by station auxiliaries (contribution to the unit),

T_L = transformer losses (may be accounted from works test reports).

Auxiliaries that are typically considered in the industry are listed in Table 17.5 but are not limited to these.

17.7.2 Liquidated Damages (LDs) for Shortfall in Performance

In case during performance guarantee test(s) it is found that the equipment/system has failed to meet the guarantees, all necessary modifications and/or replacements shall be carried out to make the equipment/system comply with the guaranteed requirements and the same shall be demonstrated by the equipment supplier (contractor) by conducting another performance guarantee test. However, if the specified performance guarantee(s) are still not met, but are achieved within the "acceptable shortfall limit" specified, the equipment may be accepted by the customer after levying LDs. If, however, the demonstrated guarantee(s) continue to be more than the

Table 17.5 List of auxiliaries considered for computing guaranteed auxiliary power consumption

Sl. No.	Equipment	Sl. No.	Equipment
1	ID fans	29	Chemical dosing pumps
2	FD fans	30	Condenser on-load tube cleaning system
3	PA fans	31	Oil pumps for HP and LP bypass system
4	Pulverizers	32	Ventilation of power house building, ESP
5	Pulverizer rejects handling system		building, and so on
6	Lube oil pumps for fans/air heaters,	33	Chlorination
	pulverizer system, and so on	34	Circulating water pumps
7	ESPs (with TR sets and hopper heaters of all	35	Cooling tower fans
	ESP fields in service, insulator heater of all	36	Auxiliary cooling water pumps
	ESP fields/penthouse fan (as applicable),	37	Closed circuit cooling water pumps
	and rapping system under normal	38	Service air and instrument air compressors
	operation)		
8	Air preheaters	39	Air drying plant for compressors
9	Seal air fans	40	Air conditioning plant compressors
10	Scanner air fans	41	Air washers
11	Igniter air fans	42	Air handling units
12	Coal feeders	43	Cooling tower and pumps of air
13	Fuel oil pressurizing pumps		conditioning plant
14	Fly ash exhausters	44	Raw water pumps
15	Ash conveying blowers	45	Cooling tower makeup pumps
16	Ash water pumps	46	Pretreatment plant
17	Ash slurry pumps	47	DM plant
18	Ash water recovery pumps	48	DM water transfer pumps
19	Boiler water circulation pumps	49	Clariflocculators
20	Boiler feed pumps	50	DM plant supply pumps
21	Lube oil pumps of BFPs	51	Battery chargers
22	Condensate extraction pumps	52	Lighting
23	GSC exhauster	53	Switchgear
24	Condenser air evacuation pumps	54	UPS
25	Main oil tank vapor extractor	55	Air conditioning units of miscellaneous
26	Turbine lube oil pumps and oil purifiers		rooms
27	Seal oil pumps	56	Power consumption of any other
28	Stator cooling water pumps		continuously operating auxiliary and
			equipment for unit operation

stipulated acceptable shortfall limit, even after the aforementioned modifications/replacements have been completed within a specified period after the tests, typically 90 days of notification by the customer, the customer will have the right to take either of the following actions:

i. Reject the equipment/system/plant and recover the payment already made or accept the equipment/system/plant only after levying LDs to those guarantees covered under category "A."

ii. Reject the equipment/system/plant and recover the payment already made or accept the equipment/system/plant only after assessing and deducting from the contract price an amount equivalent to the deficiency of the equipment/system as assessed by the customer, for those guarantees covered under category "B."

The LDs shall be calculated prorate for the fractional parts of the unit unless stated otherwise by the customer.

Example 17.1

The guaranteed efficiency (E_{SG}) of a coal-fired steam generator (SG) of a 500 MW (TMCR load) unit is 87% and the guaranteed average turbine heat rate (HR) is 2053 kcal kWh^{-1}. Other salient parameters of this unit are furnished in Table 17.6.

Calculate the LDs to be levied for the following deviations from guarantee conditions, considering 50% rise in capitalized cost:

  i. Shortfall in the efficiency of the steam generator: 0.10%
 ii. Shortfall in steam generation capacity: 0.1 kg s^{-1}
iii. Increase in turbine HR: 1 kcal kWh^{-1}
 iv. Shortfall in generator output: 1 kW
  v. Excess auxiliary power consumption: 1 kW

Table 17.6 Salient parameters of a 500 MW unit

Sl. No.	Item	Symbol	Unit	Value
1	Plant economic life	n	Years	30
2	Plant load factor	PLF	%	85
3	Rate of interest	i	%	10
4	Coal cost	C_{COST}	US\$ kg^{-1}	55×10^{-3}
5	Design coal GCV	C_{GCV}	kcal kg^{-1}	5500
6	Coal handling plant loss	L_{CHP}	%	0.5
7	Main steam flow at 500 MW	Q_{MS}	kg s^{-1}	461.39
8	Unit running hours at 100% TMCR	H_{100}	% of total annual running hours	80
9	Cost of generation	G_{COST}	US\$ kWh^{-1}	0.063

Solution

Unit running hours per year, $H = 365 \times 24 \times PLF = 7446.0$.

$$\text{Present worth factor (PWF)} = \frac{(1+i)^n - 1}{i(1+i)^n} = 9.43$$

i. LDs to be levied for shortfall in steam generator efficiency by 0.10%

1.	Specific fuel consumption at 87.0%, $SFC_{87.0}$ (kg kWh^{-1})	$\dfrac{HR \times (1 + L_{CHP})}{C_{GCV} \times E_{SG}} = \dfrac{2053 \times (1 + 0.005)}{5500 \times 0.87}$	0.4311943574
2.	Specific fuel consumption at 86.9%, $SFC_{86.9}$ (kg kWh^{-1})	$\dfrac{2053 \times (1 + 0.005)}{5500 \times 0.869}$	0.4316905534
3.	Increase in "specific fuel consumption" due to 0.1% decrease in boiler efficiency, SFC_{EX} (kg kWh^{-1})	$SFC_{86.9} - SFC_{87.0}$	0.000496196
4.	Annual cost of extra fuel due to item 3, ExC_{COST} (US\$ kW^{-1})	$H \times C_{COST} \times SFC_{EX} = 7446 \times 55 \times 10^{-3}$ $\times 0.000496196$	0.203
5.	Capitalized cost for extra fuel, $ExCC_{COST}$ (US\$ kW^{-1})	$PWF \times ExC_{COST} = 9.43 \times 0.203$	1.91
6.	LDs to be levied for 0.1% shortfall in SG efficiency (US\$)	$1.5 \times 1.91 \times 500 \times 10^3$	1,432,500.00 say 1.45×10^6

Continued

Example 17.1—cont'd

ii. LDs to be levied for shortfall in steam generation capacity by 0.1 kg s^{-1}

Loss due to shortfall in SG output is conceivable only when the unit is running at 100% TMCR.

1.	Plant running hours per year at 100% TMCR, H_{100}	0.8×7446	5956.8
2.	Generator output at 461.39 kg s^{-1} steam flow, G_{FL} (kW)		500,000.00
3.	Generator output at 461.29 kg s^{-1} steam flow, G_{SL} (kW)	$\dfrac{500,000.00 \times 461.29}{461.39}$	499,891.63
4.	Drop in generator output, G_{Drop} (kW)	$G_{FL} - G_{SL}$	108.37
5.	Specific fuel consumption at 87.0%, $SFC_{87.0}$ (kg kWh^{-1})	(from i, 1)	0.4311943574
6.	Savings in coal consumption due to less generation of power, FC_S (kg h^{-1})	$G_{Drop} \times SFC_{87.0} = 108.37$ $\times 0.4311943574$	46.73
7.	Annual savings in cost of coal due to less generation of power, Sav_{COST} (US$)	$H_{100} \times C_{COST} \times FC_S = 5956.8 \times$ $55 \times 10^{-3} \times 46.73$	15,309.87
8.	Annual loss in earning due to less generation of power, $Earn_{Loss}$ (US$)	$G_{Drop} \times H_{100} \times G_{COST} = 108.37$ $\times 5956.8 \times 0.063$	40,668.92
9.	Total loss in revenue due to less generation of power, Rev_{Loss} (US$)	$Earn_{Loss} - Sav_{COST}$	25,359.05
10.	Capitalized cost for less generation of steam by 0.1 kg s^{-1}, $RevCC_{COST}$ (US$)	$PWF \times Rev_{Loss} = 9.43 \times 25,359.05$	239,135.84
11.	*LDs to be levied for 0.1 kg s^{-1} shortfall in steam generation (US$)*	$1.5 \times 239,135.84$	358,703.76 say 0.36×10^6

iii. LDs to be levied for increase in turbine HR: 1 kcal kWh^{-1}

1.	Excess coal consumption for 1 kcal kWh^{-1} increase in HR, considering 0.5% coal handling plant loss, FC_{EX} (kg kWh^{-1})	$\dfrac{1 \times (1 + L_{CHP})}{C_{GCV} \times E_{SG}} = \dfrac{1 \times (1 + 0.005)}{5500 \times 0.87}$	0.00021
2.	Annual cost of excess coal due to item 1, ExC_{COST} (US$ kW^{-1})	$H \times C_{COST} \times FC_{EX} = 7446 \times 55$ $\times 10^{-3} \times 0.00021$	0.0860013
3.	Capitalization cost of excess coal, $ExCC_{COST}$ (US$ kW^{-1})	$PWF \times ExC_{COST} = 9.43 \times 0.0860013$	0.811
4.	*LDs to be levied for 1 kcal kWh^{-1} increase in turbine HR (US$)*	$1.5 \times 0.811 \times 500 \times 10^3$	608,250.00 say 0.61×10^6

Continued

Example 17.1—cont'd

iv. Shortfall in generator output: 1 kW

1.	Plant running hours per year at 100% TMCR, H_{100}	(from ii, 1)	5956.8
2.	Capitalization cost for 1 kW shortfall in generator output, GenCC$_{COST}$ (US$)	PWF $\times H_{100} \times G_{COST} = 9.43$ $\times 5956.8 \times 0.063$	3538.88
3.	LDs to be levied for 1 kW shortfall in generator output (US$)	$1.5 \times$ GenCC$_{COST}$	5308.31 say 0.00531×10^6

v. Excess auxiliary power consumption: 1 kW

1.	Capitalization cost for 1 kW increase in auxiliary power consumption, AuxCC$_{COST}$ (US$)	PWF $\times H \times G_{COST} = 9.43$ $\times 7446.0 \times 0.063$	4423.59
2.	LDs to be levied for 1 kW shortfall in generator output (US$)	$1.5 \times$ AuxCC$_{COST}$	6635.39 say 0.00664×10^6

17.7.3 Performance Guarantee/Acceptance Test Codes

In the following paragraphs, various testing codes that are used during PG/PA tests of major equipment of thermal power plants are brought up. Inclusion of test procedures followed in the industry to conduct each test is beyond the scope of this chapter. Nevertheless, for the convenience of readers, procedures to conduct PG/PA tests of a steam generator by the energy-balance method and of a steam turbine, ESP, cooling tower are briefly described in Appendix A.

17.7.3.1 Steam generator

There are two generally accepted methods for determining the efficiency of a steam generator, namely the input-output method and the energy-balance method, also known as the heat-loss method [3].

Efficiency calculation by the input-output method follows the equation given below:

$$\text{Efficiency} = \frac{\text{output}}{\text{input}} \times 100 \qquad (17.1)$$

Parameters which need to be measured for evaluating energy inputs to and energy outputs from the steam generator are presented in Table 17.7.

Table 17.7 Salient parameters to be measured for input-output method

Sl. No.	Name of Flowing Medium	Parameters	Remark
1	Feedwater at economizer inlet	Flow, pressure, and temperature	Input
2	Spray water inlet to superheater and reheater attemperator	Flow, pressure, and temperature	Input
3	Main steam at superheater outlet	Flow, pressure, and temperature	Output
4	Cold reheat steam at reheater inlet	Flow[a], pressure, and temperature	Input
5	Hot reheat steam at reheater outlet	Flow[b], pressure, and temperature	Output
6	Steam generator blow down	Flow, pressure, and temperature	Output
7	Auxiliary steam supply from steam generator	Flow, pressure, and temperature	Output
8	Energy input from any other source	Flow, pressure, and temperature	Input
9	Fuel	Flow and HHV	Input

[a]Typically cold reheat steam flow to reheater is calculated, not measured, as given here:

$$M_{CR} = M_{MS} - M_{GL} - M_{CV} - M_{EX1} - M_{AUX} \qquad (17.2)$$

where

M_{CR}: cold reheat steam flow to reheater (kg s^{-1}),

M_{MS}: main steam flow to HP turbine (kg s^{-1}),

M_{GL}: cumulative leakage steam flow from HP turbine gland seals (calculated) (kg s^{-1}),

M_{CV}: cumulative leakage steam flow from HP control valves,

M_{EX1}: extraction steam flow from cold reheat line to HPH 1 (kg s^{-1}),

AUX: auxiliary steam flow from cold reheat line (kg s^{-1}).

[b]Hot reheat steam flow from reheater also is calculated as under:

$$M_{HR} = M_{CR} + M_{RSP} \qquad (17.3)$$

where

M_{HR}: hot reheat steam flow from reheater (kg s^{-1}),

M_{RSP}: reheat attemperator spray water flow (kg s^{-1}).

The input-output method is very appropriate while firing natural gas or fuel oil in the furnace because flow of these fuels can be measured accurately. On the contrary it is practically not possible to measure accurately flow of solid fuels. ASME PTC 4: 2008 stipulates that "It is not recommended that coal-fired units be tested using the input-output Method because of the large uncertainties of measuring coal flow." Hence, for firing solid fuels in a furnace, the energy-balance method is more appropriate.

The steam generator efficiency by energy-balance method is determined as per the requirements of the latest version of ASME PTC4 or BS EN 12952-15, using higher heating value (HHV) or gross calorific value (GCV) of the fuel. The latest version of German code DIN EN 12952-15, which uses lower heating value (LHV) or net calorific value (NCV) of the fuel, can also be used to determine the steam generator efficiency. Efficiency calculated on the basis of LHV is higher than the efficiency based on HHV because LHV ignores the heat loss due to moisture (cumulative effect of moisture in fuel, moisture in air, and moisture resulting from combustion of H_2 in fuel) in flue gas.

Usually, the efficiency test of the steam generator is done simultaneously with the PG/PA test of the steam turbine generator set.

In the energy-balance method, the efficiency of a steam generator, following ASME PTC 4, is calculated using the following relationship:

$$E_{SG} = 100 - QpL + QpB \tag{17.4}$$

where

E_{SG}: efficiency of steam generator (%),
100: energy input to the furnace by the combustion of fuel (%),
QpL: heat loss from the steam generator (%),
QpB: heat credits to steam generator (%).

Heat loss from the steam generator (QpL) constitutes the following:

$$QpL = \begin{aligned} &QpLDFg + QpLH2F + QpLWF + QpLWA + QpLUb \\ &+ QpLCO + QpLPr + QpLRs + QrLSrc \end{aligned} \tag{17.5}$$

where

QpLDFg: heat loss in dry flue gas,
QpLH2F: heat loss due to moisture formed from the combustion of hydrogen in fuel,
QpLWF: heat loss due to moisture content in the fuel,
QpLWA: heat loss due to moisture in air,
QpLUb: heat loss due to sensible heat of unburned carbon in residue,
QpLCO: heat loss due to carbon monoxide in flue gas,
QpLPr: heat loss due to pulverizer rejects,
QpLRs: heat loss due to sensible heat of residue (bottom ash and fly ash),
QrLSrc: heat loss due to surface radiation and convection.

Heat credits to the steam generator (QpB) are calculated as:

$$QpB = QpBDA + QpBWA + QpBF + QrBX \tag{17.6}$$

where

QpBDA: heat credit due to dry air entering steam generator,
QpBWA: heat credit due to moisture in entering air,
QpBF: heat credit due to sensible heat in fuel,
QrBX: heat credit due to steam/electric driven auxiliary equipment.

A detailed procedure to evaluate steam generator efficiency by the energy-balance or heat-loss method is discussed in Appendix A.

17.7.3.2 Steam turbine

The PG/PA test of a steam turbine is carried out to prove that the turbine is capable of meeting the guaranteed turbine heat rate (HR) as claimed by the turbine manufacturer/supplier while signing the contract. This test may be conducted by adopting guidelines presented in the latest version of test codes ASME PTC 6/BS 752/DIN 1943/IEC 953-1 [IS 14198 (Part 1)], and so on. It is preferable to conduct the PG/PA test of steam turbines within 8 weeks after the turbine is first loaded (Clause No. 3-3.1, p. 8 [4]).

For turbines operating in a reheat-regenerative cycle, the performance is expressed as HR or gross HR, defined as:

$$\mathrm{HR} = \frac{\text{Net heat input to the cycle}}{\text{Power output}} \times 3600 \, (\mathrm{kJ\,kWh^{-1}}) \tag{17.7}$$

where

$$\text{Net heat input to the cycle} = (M_{MS} \times H_{MS} - M_{FW} \times H_{FW}) + (M_{HR} \times H_{HR} - M_{CR} \times H_{CR})$$
$$- (M_{SSP} \times H_{SSP} + M_{RSP} \times H_{RSP}) \, (\mathrm{kJ\,h^{-1}}) \tag{17.8}$$

Power Output $= P_G$, measured at generator terminals (kW).

M_{MS}: flow of main steam at HP turbine inlet (kg s^{-1}).

$$M_{MS} = M_{FW} + M_{SSP} - M_{UNL} \tag{17.9}$$

H_{MS}: enthalpy of main steam at HP turbine inlet (kJ kg^{-1}),
M_{FW}: flow of feedwater at economizer inlet (kg s^{-1}),
H_{FW}: enthalpy of feedwater at economizer inlet (kJ kg^{-1}),
M_{SSP}: flow of superheater attemperator spray water (kg s^{-1}),
H_{SSP}: enthalpy of superheater attemperator spray water (kJ kg^{-1}),
M_{UNL}: unaccounted for leakage (kg s^{-1}),
M_{HR}: flow of hot reheat steam at IP turbine inlet (kg s^{-1}) (Ref. equation 17.3),
H_{HR}: enthalpy of hot reheat steam at IP turbine inlet (kJ kg^{-1}),
M_{CR}: flow of cold reheat steam at reheater inlet (Ref. equation 17.2) (kg s^{-1}),
H_{CR}: enthalpy of cold reheat steam at HP turbine exhaust (kJ kg^{-1}),
M_{RSP}: flow of reheater attemperator spray water (kg s^{-1}),
H_{RSP}: enthalpy of reheater attemperator spray water (kJ kg^{-1}).

A detailed procedure of this test is discussed in Appendix A.

In addition to the steam generator and steam turbine, some typical test codes that are generally followed in the industry for evaluation of PG/PA tests of major equipment of thermal power plants are summarized in Table 17.8. It is recommended to follow the latest version of all codes.

Table 17.8 Codes followed for PG/PA tests of some major equipment

Sl. No.	Name of Equipment	Details of Test Code
1	Steam generator	ASME PTC 4/BS EN 12952-15/DIN 1942
2	Coal pulverizers	ASME PTC 4.2
3	Air heaters	ASME PTC 4.3
4	Gas turbine HRSG	ASME PTC 4.4
5	Steam turbine	ASME PTC 6/BS 752/DIN 1943/IEC 953 _ 1 [IS 14198 (Part 1)]
6	Industrial type steam turbines	BS 5968
7	Steam turbines in combined cycles	ASME PTC 6.2
8	Combined cycle module	ASME PTC-46
9	Compressors and exhausters	ASME PTC-10
10	Fans	ASME PTC-11
11	Electrostatic precipitator (ESP)	EPA 17 (method-17 of EPA)/ASME PTC 38
12	Feedwater heater	ASME PTC-12.1
13	Steam condenser	ASME PTC-12.2
14	Deaerator	ASME PTC-12.3
15	Water cooling towers	CTI Code ATC 105/ASME PTC 23/BS 4485-PART 2
16	Air cooled heat exchangers	ASME PTC 30
17	Gas turbines	ASME PTC 22

17.7.4 Performance Test Procedures

It is general practice in the industry that PG/PA test procedures pertaining to each test are made available by the equipment supplier (contractor) within 12 months from the date of the letter of award of the contract for the approval of the customer/the engineer. PG/PA test procedures usually comprise the following [2]:

a. Object of the test
b. Various guaranteed parameters and tests as per contract
c. Method of conducting tests and relevant test codes
d. Duration of test, frequency of readings, and number of test runs
e. Method of calculation
f. Correction curves
g. Instrument list delineating range, accuracy, least count, and location of instruments
h. Scheme showing measurement points
i. Sample calculation
j. Acceptance criteria
k. Any other information necessary for conducting the tests

17.8 Conclusion

Upon successful demonstration of miscellaneous parameters/conditions during the trial run and reliability run tests, PG/PA tests of major equipment are carried out in accordance with the contract agreement. In the event that PG/PA test results fulfill the contract agreement conditions, a protocol (Table 17.9) may be jointly signed by the customer, the engineer, equipment manufacturers, and the supplier (contractor) certifying compliance of test results to contract. The plant may now be declared ready for commercial operation.

Table 17.9 Protocol certifying completeness of PG/PA tests

Sl. No.	Guaranteed Parameter	Remarks
1	Efficiency of steam generator in percentage with rated steam pressure and temperature at superheater outlet and at 100% TMCR unit load	Complies with contract agreement condition
2	Capacity of steam generator in kg s^{-1} of steam with rated steam pressure and temperature at superheater outlet and with any combination of pulverizers working as decided by the customer, the coal being fired from within the specified range	Complies with contract agreement condition
3	Turbine cycle HR in kcal kWh^{-1} under rated steam conditions and maximum condenser vacuum with 0% makeup at 100% TMCR unit load with all feedwater heaters in service	Complies with contract agreement condition
4	Continuous TG output at 100% TMCR unit load under rated steam conditions and condenser vacuum, design CW temperature with 0% makeup and all feedwater heaters in service	Complies with contract agreement condition
5	Suspended particulate matter downstream of ESP and/or bag filter at 100% TMCR unit load	Not exceeding 50 mg Nm^{-3}
6	Collection efficiency of ESP and/or bag filter in percentage at 100% TMCR unit load	Complies with contract agreement condition
7	APC at 100% TMCR unit load for continuous unit operation under rated steam conditions and condenser vacuum with 0% makeup	Complies with contract agreement condition

Continued

Table 17.9 Protocol certifying completeness of PG/PA tests—cont'd

Sl. No.	Guaranteed Parameter	Remarks	
The plant is declared ready for commercial operation			
Signed by The customer	Signed by The engineer	Signed by The steam generator manufacturer The steam turbine manufacturer The generator manufacturer The ESP manufacturer	Signed by The supplier (contractor)

References

[1] OSHA's Noise Standard Defines Hazard, Protection. Education 2000 Resource Guide, http://multimedia.3m.com/mws/media/918620/oshas-noise-standard-defines-hazard-protection.pdf.
[2] Central Electricity Authority (CEA), India. Standard Technical Specification for Sub-critical Thermal Power Project—2 × (500 MW or above) Main Plant Package.
[3] ASME PTC 4-2008: Fired Steam Generators—Performance Test Codes.
[4] ASME PTC 6-2004: Steam Turbines—Performance Test Codes.

Brief Description on Performance Guarantee/Performance Acceptance Tests

A.1 Introduction

From Chapter 17 it is known that performance guarantee/performance acceptance (PG/PA) tests of a plant are expected to be conducted as soon as the reliability run test of the plant is completed satisfactorily. PG/PA tests are carried out to meet the guaranteed ratings and performance requirements. The plant shall be fully on automatic control under steady-state specified load points and as near the specified cycle conditions as practicable during the test. In the event the test fails to achieve guarantee conditions, they will be subjected to penalty.

Prior to conducting PG/PA tests, all measuring instruments must be calibrated in a laboratory accepted by all concerned.

This appendix deals with the PG/PA tests of a steam generator (by the energy-balance method), steam turbine, electrostatic prtecipitator and cooling tower.

A.2 PG/PA Test of Steam Generator Efficiency

The efficiency of a steam generator may be evaluated following either the input-output method or the energy-balance/heat-loss method. However, for a pulverized coal-fired steam generator, evaluation of efficiency by the energy-balance/heat-loss method provides the following advantages:

  i. It is not essential to measure steam flow from the steam generator, the measurement of which leads to more uncertainty in the end result.
 ii. Analysis of flue gas and measurement of flue gas temperature can be carried out very accurately.
iii. Major contributors to various heat losses can be accurately assessed.
 iv. Because the measured quantity of various heat losses is a small fraction of the total heat input, uncertainty in the end result is also very small.

It is a general practice in the industry to adopt the following sequence of activities for conducting the efficiency test.

   I. Inspection and tuning of the steam generator as narrated here:
      a. Check availability of all instruments, including the distributed control system (DCS), required for evaluation of the test.
      b. Check the calibration of all primary and secondary instruments.
      c. Check and correct steam leakage.
      d. Check and, if required, correct the pressure setting of all safety valves.
      e. Check and, if required, correct the working condition of wall blowers and soot blowers.
      f. Check the pressure drop across the air and gas path.
      g. Check and, if required, correct air heater leakage and choking.
      h. Check for no air ingress through manholes, peepholes, and so on.
      i. Check and, if required, correct leakage in the gas path.
      j. Check availability of sampling points for fuel, bottom ash, fly ash, pulverizer rejects, steam, water, and so on.
      k. Operate steam generator at or near specified test conditions and note abnormalities, if any.
      l. Adjust airflow based on flue gas oxygen content.
      m. If necessary, shut down the steam generator for carrying out all corrective actions.
  II. Four hours prior to each test, all wall blowers and soot blowers will be operated to keep the heat surfaces clean.
 III. A performance test will consist of preferably two runs, each of about 2 h' operation.
 IV. A trial test of duration of about 2 h to check the working of all instruments and to train the test personnel for taking readings. Personnel, assigned for recording readings, should be available during the trial.
  V. All readings pertaining to the actual test shall be taken to make the test personnel conversant with the test.
 VI. The test will be preceded by at least 1 h of operation near the test load to establish steady-state conditions.
 VII. Problems encountered during recording readings and collecting samples shall be attended.
VIII. The frequency of recording readings generally conform to the following:
Flow measurements ≤ 5 min
Flue gas analysis ≤ 5 min
Pressure and temperature measurements ≤ 2 min
Sampling of water and steam ≤ 15 min

Note

Because of fuel variability, control system tuning, and other factors, variations in operational parameters are inevitable. To minimize the uncertainty, more measurements are taken during the test to reduce random errors in the data collected (Clause No. 4-3.3, P36) [1].

IX. When recording readings automatically through a data logger, readings will be more frequent than those listed in step VIII.

X. The average value of each measured parameter is calculated, which is used in the final computation.

XI. Fuel samples and ash samples are to be collected before, after, and at '1-h intervals in between tests' for analysis.

XII. Operating conditions at the time of the test may differ from the specified conditions that were used to establish design or guarantee performance levels. Correction factors (Section A.2.1) may be obtained from correction curves to take care of deviation in any operating condition. The correction curves must be supplied by the steam generator manufacturer prior to conducting the test.

XIII. Deviation in any operating condition during the test should comply with the requirement of Table A.1.

XIV. At the commencement of the test, the steam generator shall be operated under the following conditions:
   a. All blowdowns have to be kept closed.
   b. No oil support should be provided.
   c. Soot blowing shall be avoided.

XV. During the test, the required load on steam generator and operating conditions, as close to design as possible, are to be ensured by the customer.

XVI. The test report shall consist of the following:
   a. All recorded test data, duly countersigned by the customer, the engineer, the steam generator manufacturer, and the supplier (contractor)
   b. Fuel and ash sample analyses
   c. Calculation and analysis of test data
   d. Computation of performance test parameters
   e. Final findings, analysis of the findings, and recommendations based on final findings, if any

Table A.1 Permissible deviation in operating conditions

Sl. No.	Parameter	Instant Fluctuation	Average Fluctuation
1	Superheater outlet MS pressure	4% (Max. 175 kPa)	3% (Max. 140 kPa)
2	Superheater/reheater outlet MS temperature	11.0 K	5.5 K
3	Superheater/reheater attemperator spray water flow	40% spray flow or 2% MS flow	Not applicable
4	Feedwater flow (drum-type steam generator)	10%	3%

Continued

Table A.1 Permissible deviation in operating conditions—cont'd

Sl. No.	Parameter	Instant Fluctuation	Average Fluctuation
5	Steam flow (once-through steam generator)	4%	3%
6	Feedwater temperature	11.0 K	5.5 K
7	O_2 content in flue gas at APH inlet	1.0 (points of O_2)	0.5 (points of O_2)
8	CO	150 mL kL^{-1}	50 mL kL^{-1}

Source: Table 3-2-1, P 23 [1].

A.2.1 Correction Factors

Correction factors that are typically adopted in the industry due to deviation in various parameters, for making the measured efficiency of the steam generator equivalent to the guaranteed efficiency of the steam generator, are given below. Correction factors are shown in percentages. Hence, depending on whether the correction factor is positive or negative, it will be either added or subtracted from the measured efficiency of steam generator to arrive at the corrected efficiency of steam generator.

The following deviations are pertinent to tests carried out on an higher heating value (HHV) basis.

  i. Deviation in total moisture in fuel (Fig. A.1)
 ii. Deviation in ash content in as-received fuel (Fig. A.2)
iii. Deviation in H_2 content in fuel (Fig. A.3)
 iv. Deviation in HHV of fuel (Fig. A.4)
  v. Deviation in ambient air temperature at air preheater (APH) inlet (Fig. A.5)
 vi. Deviation in absolute humidity of ambient air (Fig. A.6)
vii. Deviation in feedwater temperature at economizer inlet (Fig. A.7)

A.2.2 Evaluation of Efficiency

From Chapter 17, it is known that the efficiency of a steam generator could be evaluated using either HHV of fuel (ASME PTC4 or BS EN 12952-15) or lower heating value (LHV) of fuel (DIN EN 12952-15). The following paragraphs discuss the detailed method of efficiency evaluation using both HHV and LHV of fuel. The starting point in both cases is the accurate analysis of coal.

A.3 ASME PTC 4

The procedure of the test is discussed in the following paragraphs.

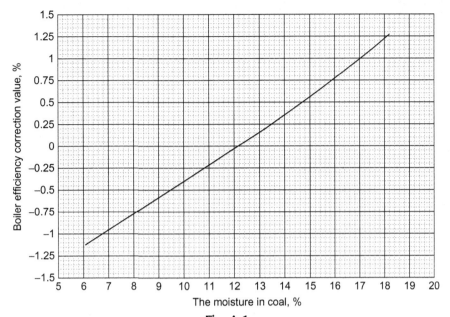

Fig. A.1
Efficiency of steam generator (SG) versus total moisture in fuel.

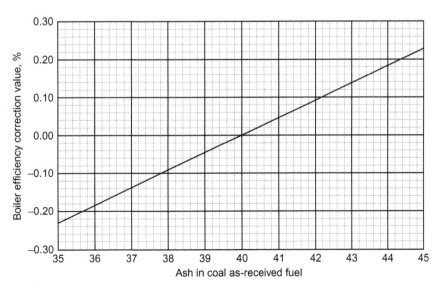

Fig. A.2
Efficiency of SG versus ash in as-received fuel.

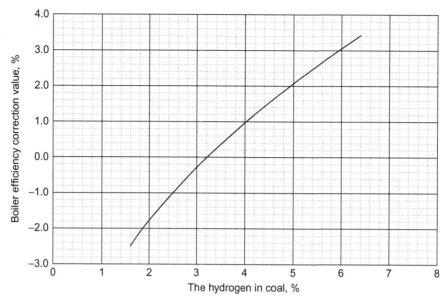

Fig. A.3

Efficiency of SG versus H_2 content in fuel.

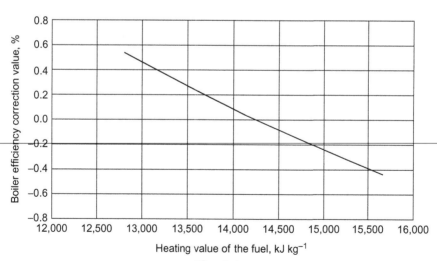

Fig. A.4

Efficiency of SG versus HHV of fuel.

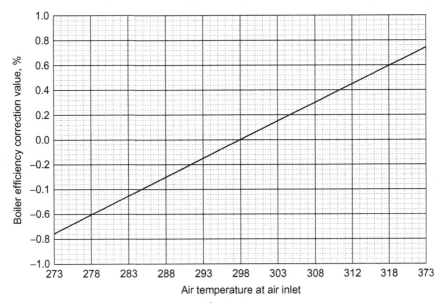

Fig. A.5
Efficiency of SG versus ambient air temperature at APH inlet.

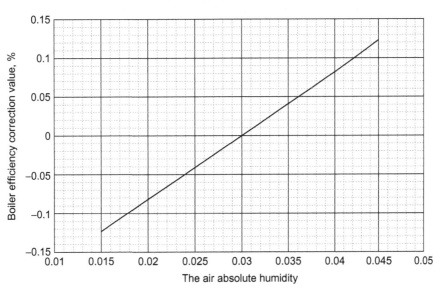

Fig. A.6
Efficiency of SG versus absolute humidity of ambient air.

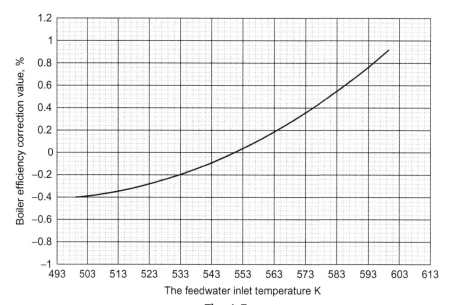

Fig. A.7

Efficiency of SG versus feedwater temperature at economizer inlet.

A.3.1 Measured Parameters

Parameters, which are measured while the test is in progress, are given in Table A.2.

Table A.2 Parameters to be measured

Sl. No.	Description	Unit	Symbol and Formula
1	APH outlet flue gas temperature	K	TFgLv
2	Air temperature (dry bulb) at the boundary of test envelope	K	Tdb
3	Wet bulb temperature at the boundary of test envelope	K	Twb
4	Reference temperature	K	TRe
5	APH outlet air temperature	K	TA
6	Ambient air/barometric pressure	Pa	Pa
7	APH outlet air pressure	Pa	PA
8	APH outlet flue gas pressure	Pa	PFg
9	Oxygen content of dry flue gas at APH inlet	%	$DVpO_2$
10	Carbon dioxide content in dry flue gas at APH outlet	%	$DVpCO_2$
11	Carbon monoxide content in dry flue gas	%	DVpCO
12	Bottom ash (slag) temperature	K	TBA
13	Fly ash (flue dust) temperature	K	TFA
14	MS temperature	K	TSt32
15	MS pressure	MPa	PSt32

Continued

Table A.2   Parameters to be measured—cont'd

Sl. No.	Description	Unit	Symbol and Formula
16	Hot reheat (reheater outlet) steam temperature	K	TSt34
17	Hot reheat steam pressure	MPa	PSt34
18	Cold reheat (reheater inlet) steam temperature	K	TSt33
19	Cold reheat steam pressure	MPa	PSt33
20	Feedwater temperature at economizer inlet	K	TW24
21	Feedwater pressure at economizer inlet	MPa	PW24
22	Spray water temperature (for MS)	K	TW25
23	Spray water pressure (for MS)	MPa	PW25
24	Spray water temperature (for reheat steam)	K	TW26
25	Spray water pressure (for reheat steam)	MPa	PW26
26	MS flow	$kg\ s^{-1}$	MrSt32
27	Reheat steam flow (evaluated from regenerative cycle)	$kg\ s^{-1}$	MrSt33
28	Superheat steam attemperator spray water flow	$kg\ s^{-1}$	MrW25
29	Reheat steam attemperator spray water flow	$kg\ s^{-1}$	MrW26
30	Mass flow rate of pulverizer rejects	$kg\ s^{-1}$	MrPr
31	HHV of pulverizer rejects	$kJ\ kg^{-1}$	HHVPr
32	Pulverizer outlet temperature	K	TPr
33	Fuel mass flow rate	$kg\ s^{-1}$	MrF
34	Total steam supplied to steam driven auxiliary equipment	$kg\ s^{-1}$	MrStX
35	Enthalpy of steam supplied to drive the auxiliaries	$kJ\ kg^{-1}$	HStEn
36	Enthalpy of steam exhaust from the auxiliaries	$kJ\ kg^{-1}$	HStLv
37	Energy input to electrically driven equipment (eg, pulverizer, gas recirculating fan, hot primary air fan, boiler circulating pump, etc.)	kWh	QX
38	Overall drive (steam/electric) efficiency	%	EX

A.3.2 Fuel and Ash Analysis

Samples of as-fired fuel and ash collected during the test are analyzed in the laboratory. Analyses of such samples are presented in Table A.3.

Table A.3 Laboratory analyzed (as-fired fuel) data

Sl. No.	Description	Unit	Symbol and Formula
1	Carbon content in fuel	%	MpCF
2	Hydrogen content in fuel	%	MpH_2F
3	Sulfur content in fuel	%	MpSF
4	Oxygen content in fuel	%	MpO_2F

Continued

Table A.3   Laboratory analyzed (as-fired fuel) data—cont'd

Sl. No.	Description	Unit	Symbol and Formula
5	Nitrogen content in fuel	%	MpN_2F
6	Moisture content in fuel	%	MpWF
7	Ash content in fuel	%	MpASF
8	HHV of fuel (bomb calorimeter)	$kJ\ kg^{-1}$	HHVFcv
9	Unburned combustible content of bottom ash (slag)	%	MpRsb
10	Unburned combustible content of fly ash (flue dust)	%	MpRsf

A.3.3 Standard Data

For the evaluation of steam generator efficiency it becomes essential to use certain standard data, per Table A.4.

Table A.4 Standard data

Sl. No.	Description	Unit	Symbol and Formula
1	Mean specific heat of flue gas	$kJ\ kg^{-1}\ K^{-1}$	Cpg (between TFgLv and 298 K)
2	Mean specific heat of bottom ash (slag)	$kJ\ kg^{-1}\ K^{-1}$	CpBA (between TBA and 298 K)
3	Mean specific heat of fly ash (flue dust)	$kJ\ kg^{-1}\ K^{-1}$	CpFA (between TFA and 298 K)
4	Mean specific heat of ash at ambient	$kJ\ kg^{-1}\ K^{-1}$	CpRa (between Tdb and 298 K)
5	Mean specific heat of pulverizer rejects	$kJ\ kg^{-1}\ K^{-1}$	CpPr (between TPr and 298 K)
6	Mean specific heat of dry air at ambient	$kJ\ kg^{-1}\ K^{-1}$	CpDA (between Tdb and 298 K)
7	Mean specific heat of water vapor at ambient	$kJ\ kg^{-1}\ K^{-1}$	CpWv (between Tdb and 298 K)
8	Mean specific heat of bituminous coal	$kJ\ kg^{-1}\ K^{-1}$	CpF (between TPr and 298 K)
9	Heating value of unburned combustibles	$kJ\ kg^{-1}$	HHVCRs
10	Heating value of carbon monoxide	$kJ\ kg^{-1}$	HHVCO
11	Molecular weight of CO	$kg\ mol^{-1}$	MwCO
12	Universal molar gas constant	$kJ\ mol^{-1}\ K^{-1}$	R (8.3145)

Note

Values of mean specific heat, as required under Sl. Nos. 1 through 8, may be gathered from Figs. 5-19-1, 5-19-2, 5-19-3, and 5-19-4 of *ASME* PTC 4-2008.

A.3.4 Inputs From Steam Table

In addition to the aforementioned there are certain conditions, the enthalpies of which are either obtained from the steam table or are calculated (Table A.5).

Table A.5 Parameters from steam table/calculated

Sl. No.	Description	Unit	Symbol and Formula
1	Enthalpy of steam at partial pressure of 6.895 kPa[a] and gas outlet temperature TFgLv	kJ kg^{-1}	HStLVcr
2	Enthalpy of water at partial pressure of 6.895 kPa and air inlet temperature Tdb	kJ kg^{-1}	HWRe
3	Enthalpy of steam at partial pressure of 6.895 kPa and gas outlet temperature TFgLv	kJ kg^{-1}	HStRe
4	Enthalpy of MS at TSt32 and PSt32	kJ kg^{-1}	HSt32
5	Enthalpy of feedwater at TW24 and PW24	kJ kg^{-1}	HW24
6	Enthalpy of MS spray water at TW25 and PW25	kJ kg^{-1}	HW25
7	Enthalpy of hot reheat steam at TSt34 and PSt34	kJ kg^{-1}	HSt34
8	Enthalpy of cold reheat steam at TSt33 and PSt33	kJ kg^{-1}	HSt33
9	Enthalpy of reheat steam spray water at TW26 and PW26	kJ kg^{-1}	HW26
10	Enthalpy of dry air entering steam generator	kJ kg^{-1}	HDAEn = CpDA × (Tdb − TRe)
11	Enthalpy of water vapor in dry air entering steam generator	kJ kg^{-1}	HWvEn = CpWv × (Tdb − TRe)
12	Enthalpy of the fuel at TPr	kJ kg^{-1}	HFEn = CpF × (TPr − TRe)

[a]6.895 kPa = 1 psi (5-19.5, P 111 [1]).

A.3.5 Miscellaneous Calculations

Tables A.2–A.5 provide enough information to calculate the efficiency of the steam generator. Yet one piece of information which is pertinent to proceeding further is the percentage distribution of ash content in fuel between bottom ash (Asb) and fly ash (Asf). Depending on the pulverizer grinding capability and the combustion efficiency of fuel in the furnace, the Asb:Asf ratio typically varies from 5:95 to 20:80. Detailed calculations of various properties, as furnished in Table A.6, are developed assuming Asb:Asf is 20:80.

Table A.6 Detailed calculations of various properties

Sl. No.	Reference [1]	Description	Unit	Symbol and Formula
1	Cl. No. 5-8.1 (eqn. 5-8-1)	Heating value of fuel	kJ kg^{-1}	$HHVF = HHVFcv + 2.644\ MpH2F$
2	Cl. No. 5-10.1 (eqn. 5-10-1)	Mass of residue	kg kg^{-1}	$MFrRs = MpAsF/(100 - MpCRs)$
3	Cl. No. 5-10.3 (eqn. 5-10-6)	Weighted average of unburned carbon in residue	%	$MpCRs = Asb \times MpRsb + Asf \times MpRsf$
4	Cl. No. 5-10.4 (eqn. 5-10-8)	Unburned carbon in fuel	%	$MpUbC = MpCRs \times MFrRs$
5	Cl. No. 5-10.5 (eqn. 5-10-9)	Actual percent of carbon in fuel that is burned	%	$MpCb = MpCF - MpUbC$
6	Cl. No. 5-11.2	Weight of water vapor per unit of dry air (from psychrometric chart)	kg kg^{-1}	$MFrWA$ (at Tdb and Twb)
7	Cl. No. 5-11.3 (eqn. 5-11-8)	Theoretical air (corrected) for combustion	kg kg^{-1}	$MFrThACr = 0.1151 \times MpCb + 0.3429 \times MpH2F + 0.0431 \times MpSF - 0.0432 \times MpO_2F$
8	Cl. No. 5-11.3 (eqn. 5-11-7)	Theoretical air (corrected) for combustion	kg kJ^{-1}	$MqThACr = MFrThACr/HHVF$
9	Cl. No. 5-11.3 (eqn. 5-11-9)	Moles of theoretical air	mol kg^{-1}	$MoThACr = MFrThACr/28.963$
10	Cl. No. 5-11.4.1 (eqn. 5-11-12)	Moles of dry products from the combustion of fuel	mol kg^{-1}	$MoDPc = MpCb/1201.1 + MpSF/3206.5 + MpN2F/2801.34$
11	Cl. No. 5-11.4.1 (eqn. 5-11-11)	Excess air	%	$XpA = 100 \times [\{DVpO_2 \times (MoDPc + 0.7905 \times MoThACr)\}/\{(20.95 - DVpO_2) \times MoThACr\}]$
12	Cl. No. 5-11.4.2 (eqn. 5-11-18)	Moles of dry gas	mol kg^{-1}	$MoDFg = MoDPc + MoThACr \times (0.7905 + XpA/100)$
13	$DVpO_2$, $DVpCO_2$, and $DVpCO$ are already measured (Tables A.2–A.11)			
14	Cl. No. 5-11.4.2 (eqn. 5-11-15)	Sulfur dioxide content in dry flue gas	%	$DVpSO_2 = MpSF/(32.065 \times MoDFg)$
15	Cl. No. 5-11.4.2 (eqn. 5-11-16)	Nitrogen content from fuel in dry flue gas	%	$DVpN_2F = MpN_2F/(28.0134 \times MoDFg)$
16	Cl. No. 5-11.4.2 (eqn. 5-11-17)	Atmospheric nitrogen content in dry flue gas	%	$DVpN_2a = 100 - DVpO_2 - DVpCO_2 - DVpSO_2 - DVpN_2F$
17	Cl. No. 5-11.4.3 (eqn. 5-11-21)	Moles of wet products from the combustion of fuel	mol kg^{-1}	$MoWPc = MoDPc + MpH_2F/201.59 + MpWF/1801.53$
18	Cl. No. 5-11.4.4 (eqn. 5-11-28)	Moles of wet gas	mol kg^{-1}	$MoFg = MoWPc + MoThACr \times \{0.7905 + XpA/100\}$
19	Cl. No. 5-11.5 (eqn. 5-11-29)	Dry air entering steam generator	kg kJ^{-1}	$MqDA = MqThACr \times (1 + XpA/100)$
20	Cl. No. 5-11.5 (eqn. 5-11-30)	Dry air entering steam generator	kg kg^{-1}	$MFrDA = MFrThACr \times (1 + XpA/100)$

Continued

Table A.6 Detailed calculations of various properties—cont'd

Sl. No.	Reference [1]	Description	Unit	Symbol and Formula
21	Cl. No. 5-11.6 (eqn. 5-11-31)	Total (wet) air for combustion	kg kJ^{-1}	MqA = (1 + MFrWA) MqDA
22	Cl. No. 5-11.6 (eqn. 5-11-32)	Total (wet) air for combustion	kg s^{-1}	MrA = MqA × MrF × HHVF
23	Cl. No. 5-11.7 (eqn. 5-11-35)	Molecular weight of wet air	kg mol^{-1}	MwA = (1 + MFrWA)/(1/28.9625 + MFrWA/18.0153)
24	Cl. No. 5-11.7 (eqn. 5-11-34)	Specific gas constant of wet air	kJ kg^{-1} K^{-1}	RkA = R/MwA
25	Cl. No. 5-11.7 (eqn. 5-11-33)	Density of air	kg m^{-3}	DnA = (Pa + PA)/(RkA × (273.2 + 297.1))
26	Cl. No. 5-12.1 (eqn. 5-12-1)	Wet gas from fuel	kg kJ^{-1}	MqFgF = (100 − MpAsF − MpUbC)/ 100 × HHVF
27	Cl. No. 5-12.2 (eqn. 5-12-2)	Moisture from H$_2$O in fuel	kg kJ^{-1}	MqWF = MpWF/100 × HHVF
28	Cl. No. 5-12.3 (eqn. 5-12-3)	Moisture from combustion of hydrogen in fuel	kg kJ^{-1}	MqWH2F = 8.937 × MpH2F/ 100 × HHVF
29	Cl. No. 5-12.6 (eqn. 5-12-6)	Moisture in air	kg kJ^{-1}	MqWA = MqDA × MFrWA
30	Cl. No. 5-12.8 (eqn. 5-12-9)	Total moisture in flue gas	kg kJ^{-1}	MqWFg = MqWF + MqWH2F + MqWA
31	Cl. No. 5-12.9 (eqn. 5-12-10)	Total wet flue gas flow weight	kg kJ^{-1}	MqFg = MqA + MqFgF
32	Cl. No. 5-12.10 (eqn. 5-12-12)	Dry flue gas weight	kg kJ^{-1}	MqDFg = MqFg − MqWF
33	Cl. No. 5-12.11 (eqn. 5-12-13)	Moisture in flue gas	%	MpWFg = 100 × MqWFg/MqFg
34	Cl. No. 5-12.13 (eqn. 5-12-19)	Molecular weight of dry flue gas	kg mol^{-1}	MwDFg = 0.31999 × DVpO$_2$ + 0.4401 × DVpCO$_2$ + 0.64063 × DVpSO$_2$ + 0.28013 × DVpN$_2$F + 0.28158 × DVpN$_2$a
35	Cl. No. 5-12.13 (eqn. 5-12-20)	Percent moisture in dry flue gas	%	DVpH$_2$O = 100 × (MoFg − MoDFg)/MoDFg
36	Cl. No. 5-12.13 (eqn. 5-12-18)	Molecular weight of wet flue gas	kg mol^{-1}	MwFg = (MwDFg + 0.18015 × DVpH$_2$O) × MoDFg/MoFg
37	Cl. No. 5-12.16	Specific gas constant of wet flue gas	kJ kg^{-1} K^{-1}	RkFg = R/MwFg
38	Cl. No. 5-12.13 (eqn. 5-12-15)	Density of wet flue gas	kg m^{-3}	DnFg = (Pa + PFg)/(RkFg × (273.2 + TFg))

Note

Higher heating value, HHVF, refers to the as-fired higher heating value on a constant pressure basis. For solid and liquid fuels, HHVFcv is determined in a bomb calorimeter, which is a constant volume device. Because fuel is burned in a steam generator under essentially constant pressure conditions, the bomb calorimeter values must be corrected to a constant pressure basis.

A.3.6 Calculation of Heat Losses

Based on the information presented in Tables A.2–A.6, various heat losses are given in Table A.7.

Table A.7 Heat losses

Sl. No.	Reference [1]	Description	Unit	Symbol and Formula
1	Cl. No. 5-14.1 (eqn. 5-14-3 & 5-19-29)	Dry gas heat loss	%	$QpLDFg = 100 \times MqDFg \times HDFgLvCr = 100 \times MqDFg \times Cpg \times (TFgLv - Tdb)$
2	Cl. No. 5-14.2.1 (eqn. 5-14-4)	Heat loss due to moisture formed from the combustion of hydrogen in the fuel	%	$QpLH2F = 100 \times MqWH_2F \times (HStLvCr - HWRe)$
3	Cl. No. 5-14.2.2 (eqn. 5-14-5)	Heat loss due to moisture content of the fuel	%	$QpLWF = 100 \times MqWF \times (HStLvCr - HWRe)$
4	Cl. No. 5-14.3 (eqn. 5-14-7)	Heat loss due to moisture in air	%	$QpLWA = 100 \times MqWA \times (HStLvCr - HstRe)$
5	Cl. No. 5-14.4.1 (eqn. 5-14-8)	Heat loss due to sensible heat of unburned carbon in residue	%	$QpLUb = MpUbC \times HHVCRs/HHVF$
6	Cl. No. 5-14.4.3 (eqn. 5-14-10)	Heat loss due to carbon monoxide in flue gas	%	$QpLCO = DvpCO \times MoDFg \times MwCO \times HHVCO/HHVF$
7	Cl. No. 5-14.4.4 (eqn. 5-14-12)	Mass flow rate of pulverizer rejects	kg kJ^{-1}	$MqPr = MrPr/(MrF \times HHVF)$
8	Cl. No. 5-14.4.4 (eqn. 5-19-29)	Enthalpy of pulverizer rejects	kJ kg^{-1}	$HPr = CpPr \times (TPr - 25)$
9	Cl. No. 5-14.4.4 (eqn. 5-14-11)	Heat loss due to pulverizer rejects	%	$QpLPr = 100 \times MqPr \times (HHVPr + HPr)$
10	Cl. No. 5-14.5 (eqn. 5-14-14)	Mass flow rate of bottom ash	kg kJ^{-1}	$MqRsba = \{(MpAsF/100) \times (Asb/100) \times (1 + MpRsb/100)\}/HHVF$
11		Mass flow rate of fly ash	kg kJ^{-1}	$MqRsfa = \{(MpAsF/100) \times (Asf/100) \times (1 + MpRsf/100)\}/HHVF$
12	Cl. No. 5-14.5 (eqn. 5-14-14)	Enthalpy of bottom ash	kJ kg^{-1}	$Hrba = CpBA \times (TBA - 25) - CpRa \times (Tdb - 25)$
13		Enthalpy of fly ash	kJ kg^{-1}	$Hrfa = CpFA \times (TFA - 25) - CpRa \times (Tdb - 25)$
14	Cl. No. 5-14.5 (eqn. 5-14-14)	Heat loss due to sensible heat of residue	%	$QpLRs = 100 \times (MqRsba \times Hrba + MqRsfa \times Hrfa)$
15	Cl. No. 5-14.9 (eqn. 5-14-19-too complicated)	Heat loss due to surface radiation and convection	%	$QrLSrc$ (from ABMA curve)
	Total heat losses		%	**QpL** $= QpLDFg + QpLH_2F + QpLWF + QpLWA + QpLUb + QpLCO + QpLPr + QpLRs + QrLSrc$

A.3.7 Calculation of Heat Credits

Heat credits are included in Table A.8.

Table A.8 Heat credits

Sl. No.	Reference [1]	Description	Unit	Symbol and Formula
1	Cl. No. 5-15.1 (eqn. 5-15-1)	Heat credit due to dry air entering steam generator	%	$QpBDA = 100 \times MqDA \times HDAEn$
2	Cl. No. 5-15.2 (eqn. 5-15-2)	Heat credit due to moisture in entering air	%	$QpBWA = 100 \times MqWA \times HWvEn$
3	Cl. No. 5-15.3 (eqn. 5-15-3)	Heat credit due to sensible heat in fuel	%	$QpBF = 100 \times HFEn/HHVF$
4	Cl. No. 5-15.5.1 (eqn. 5-15-5)	Heat credit due to steam-driven auxiliary equipment	kJ s^{-1} (kW)	$QrBX = MrStX \times (HStEn - HStLv) \times EX/100$
5	Cl. No. 5-15.5.2 (eqn. 5-15-6)	Heat credit due to electrically driven auxiliary equipment	kJ s^{-1} (kW)	$QrBX = QX \times EX/100$
	Total heat credits		%	**QpB** = QpBDA + QpBWA + QpBF + $100 \times QrBX \times MrF/HHVF$

Hence, the efficiency of the steam generator, $\textbf{Eff} = \textbf{100} - \textbf{QpL} + \textbf{QpB}$, %.

A.4 DIN 1942 (DIN EN 12952-15)

The efficiency test of the steam generator following the code DIN 1942 (DIN EN 12952-15) is conducted by the heat-loss method based on the LHV of fuel.

A.4.1 Measured Parameters

Parameters that are measured during the test are given in Table A.9.

Table A.9 Measured parameters

Sl. No.	Description	Unit	Symbol and Formula
1	Air-heater outlet flue gas temperature	K	t_G
2	Fuel temperature	K	t_B
3	Air temperature (dry bulb) at envelope boundary	K	t_L
4	Wet bulb temperature at envelope boundary	K	t_{WL}
5	Reference temperature	K	t_b
6	Ambient air/barometric pressure	Pa	p_L
7	Oxygen content of dry flue gas	m^3 m^{-3}	y_{O_2T}
8	Carbon monoxide content in dry flue gas	m^3 m^{-3}	y_{COT}

Continued

Table A.9 Measured parameters—cont'd

Sl. No.	Description	Unit	Symbol and Formula
9	Bottom ash (slag) temperature	K	t_S
10	Fly ash (flue dust) temperature	K	Same as t_G
11	MS temperature	K	t_D
12	MS pressure	MPa	p_D
13	Hot reheat (boiler outlet) steam temperature	K	t_{ZH}
14	Hot reheat steam pressure	MPa	p_{ZH}
15	Cold reheat (boiler inlet) steam temperature	K	t_{ZC}
16	Cold reheat steam pressure	MPa	p_{ZC}
17	Feedwater temperature at economizer inlet	K	t_{SP}
18	Feedwater pressure at economizer inlet	MPa	p_{SP}
19	Spray water temperature (for MS)	K	t_{ED}
20	Spray water pressure (for MS)	MPa	p_{ED}
21	Spray water temperature (for reheat steam)	K	t_{EZ}
22	Spray water pressure (for reheat steam)	MPa	p_{EZ}
23	MS flow	kg s^{-1}	$\dot{m}_D$
24	Reheat steam flow (evaluated from regenerative cycle)	kg s^{-1}	$\dot{m}_Z$
25	Superheat steam attemperator spray water flow	kg s^{-1}	$\dot{m}_{ED}$
26	Reheat steam attemperator spray water flow	kg s^{-1}	$\dot{m}_{EZ}$
27	Mass flow of fuel	kg s^{-1}	$\dot{m}_B$
28	Shaft power of feeders	kW	P_M

A.4.2 Fuel and Ash Analysis

Data related to the analysis of fuel and ash in the laboratory are presented in Table A.10.

Table A.10 Analysis of fuel and ash

Sl. No.	Description	Unit	Symbol and Formula
1	Carbon content in fuel	kg kg^{-1}	γ_C
2	Hydrogen content in fuel	kg kg^{-1}	γ_H
3	Sulfur content in fuel	kg kg^{-1}	γ_S
4	Oxygen content in fuel	kg kg^{-1}	γ_O
5	Nitrogen content in fuel	kg kg^{-1}	γ_N
6	Moisture content in fuel	kg kg^{-1}	γ_{H_2O}
7	Ash content in fuel	kg kg^{-1}	γ_A
8	Volatile matter content of fuel	kg kg^{-1}	$\gamma_{Fl.B}$
9	Gross calorific value (GCV) of fuel	kJ kg^{-1}	H_o
10	Unburned combustible content of bottom ash (slag)	kg kg^{-1}	u_S
11	Unburned combustible content of fly ash (flue dust)	kg kg^{-1}	u_F

A.4.3 Data From Steam Table

The required enthalpy of steam and water as obtained from the steam table is given in Table A.11.

Table A.11 Enthalpy from steam table

Sl. No.	Description	Unit	Symbol and Formula
1	Enthalpy of MS at p_D and t_D	kJ kg^{-1}	h_D
2	Enthalpy of feedwater at p_{SP} and t_{SP}	kJ kg^{-1}	h_{SP}
3	Enthalpy of MS spray water at p_{ED} and t_{ED}	kJ kg^{-1}	h_{ED}
4	Enthalpy of hot reheat steam at p_{ZH} and t_{ZH}	kJ kg^{-1}	h_{Z2}
5	Enthalpy of cold reheat steam at p_{ZC} and t_{ZC}	kJ kg^{-1}	h_{Z1}
6	Enthalpy of reheat steam spray water at p_{EZ} and t_{EZ}	kJ kg^{-1}	h_{EZ}

A.4.4 Standard Data

Table A.12 provides standard data, which are used in the calculations.

Table A.12 Standard data

Sl. No.	Description	Unit	Symbol and Formula
1	Standard density of dry air	kg m^{-3}	ρ_{nLT} (=1.2930)
2	Oxygen content of dry air	m^{-3}/m^{-3}	y_{O_2LT} (=0.20938)
3	Volatile matter content of ash	kg kg^{-1}	ν (=5) (pulverized coal-fired steam generator)
4	Specific heat of flue gas	kJ kg^{-1} K^{-1}	$\check{c}_{pG}$ (between t_G and t_b)
5	Specific heat of combustion air	kJ kg^{-1} K^{-1}	$\check{c}_{pL}$ (between t_L and t_b)
6	Mean specific heat of bottom ash (slag)	kJ kg^{-1} K^{-1}	$\hat{C}_S$ (=1.0) (dry-bottom furnace)
7	Mean specific heat of fly ash (flue dust)	kJ kg^{-1} K^{-1}	$\hat{C}_F$ (=0.84)
8	Mean specific heat of fuel	kJ kg^{-1} K^{-1}	$\hat{C}_B = (1 + 0.95 \times \gamma_{Fl.B}) \times 0.877$ (hard coal)
9	Heating value of unburned combustibles	MJ kg^{-1}	H_{uu} (=33.0) (hard coal)
10	Heating value of carbon monoxide (LHV)	MJ m^{-3}	$H_{uCOn} = 12.633$
11	Weight of water vapor per unit of dry air (from psychometric chart)	kg kg^{-1}	$\chi_{H_2 OLT}$ (at t_L and t_{WL})
12	Constant factor for radiation and convection		C (=0.0220) (hard coal)

For a pulverized coal-fired steam generator, the ratio of bottom ash (η_S) to fly ash (η_F) is 20:80, per this code.

A.4.5 Calculation of Efficiency

Based on the data provided in Tables A.9–A.12, calculations of the efficiency of the steam generator are furnished in Table A.13.

Table A.13 Calculations of efficiency

Description	Unit	Symbol and Formula
Theoretical air for combustion	kg kg^{-1}	$\mu_{LOT} = 11.5122 \times \gamma_C + 34.2974 \times \gamma_H + 4.3129 \times \gamma_S - 4.3212 \times \gamma_O$
Dry flue gas per unit of fuel at standard temperature and pressure (STP)	m^3 kg^{-1}	$V_{GOT} = 8.8930 \times \gamma_C + 20.9724 \times \gamma_H + 3.3190 \times \gamma_S - 2.6424 \times \gamma_O + 0.7997 \times \gamma_N$
Dry air entering steam generator	kg kg^{-1}	$\mu_{LT} = \mu_{LOT} + \rho_{nLT} \times V_{GOT} \times \left\{ y_{O_2T} / \left(y_{O_2LT} - y_{O_2T} \right) \right\}$
Total (wet) air for combustion	kg kg^{-1}	$\mu_L = \mu_{LT} \left(1 + \chi_{H_2\,OLT} \right)$
Wet flue gas flow	kg kg^{-1}	$\mu_G = \mu_L + 1 - \gamma_A \times (1 - \nu)$
Mass flow rate of bottom ash	kg s^{-1}	$\dot{m} = \eta_S \times \gamma_A \times \dot{m}_B$
Mass flow rate of fly ash	kg s^{-1}	$\dot{m}_F = \eta_F \times \gamma_A \times \dot{m}_B$
Ratio of unburned combustibles to supplied fuel	–	$l_u = \{ \gamma_A \times (1 - \nu)/(1 - \gamma_A - \gamma_{H2O}) \} \times [\{u_S \times \eta_S/(1 - u_S)\} + \{u_F \times \eta_F/(1 - u_F)\}]$
Enthalpy of bottom ash	kJ kg^{-1}	$J_S = \{ \gamma_A \times (1 - \nu)/(1 - l_u) \} \times \{ \eta_S \times h_S/(1 - u_S) \}$
Enthalpy of fly ash	kJ kg^{-1}	$J_F = \{ \gamma_A \times (1 - \nu)/(1 - l_u) \} \times \{ \eta_F \times h_F/(1 - u_F) \}$
Enthalpy of combustion air	kJ kg^{-1}	$J_L = \mu_L \times \check{c}_{pL} \times (t_L - t_b)$
Enthalpy of fuel	kJ kg^{-1}	$h_B = \check{c}_B \times (t_B - t_b)$
LHV of fuel at temperature t_B	kJ kg^{-1}	$H_u = H_o - 2442.5 \times (8.9370 \times \gamma_H + \gamma_{H2O})$
Overall LHV of fuel	kJ kg^{-1}	$H_{utot} = (H_u + h_B)/(1 - l_u) + J_L$
Heat input proportional to fuel burned	kW	$\dot{Q}_{zB} = \dot{m}_B \times H_{utot}$
Useful heat output	kW	$\dot{Q}_N = \dot{m}_D \times (h_D - h_{SP}) + \dot{m}_{ED} \times (h_D - h_{ED}) + \dot{m}_Z \times (h_{Z2} - h_{Z1}) + \dot{m}_{EZ} \times (h_{Z2} - h_{EZ})$
Dry gas heat loss	–	$l_{GB} = \mu_G \times \check{c}_{pG} \times (t_G - t_b)/H_{utot}$ (related to $\dot{Q}_{zB}$)
Heat loss due to carbon monoxide	–	$l_{COB} = V_{GT} \times y_{COT} \times H_{uCOn}/H_{utot}$ (related to $\dot{Q}_{zB}$)
Heat loss due to sensible heat of residue	–	$l_{SFB} = (J_S + J_F)/H_{utot}$ (related to $\dot{Q}_{zB}$)
Losses due to enthalpy and unburned combustibles in slag and flue dust	kW	$\dot{Q}_v = \dot{Q}_{SF} = \dot{m}_B \times (J_S + J_F)$
Heat loss due to radiation and convection	kW	$\dot{Q}_{St} = C \times \dot{Q}_N^{0.7}$
Heat credit due to auxiliaries	kW	$\dot{Q}_z = P_M$
Total loss resulting from burning fuel	–	$\Sigma l_B = l_{GB} + l_{COB} + l_{SFB}$
Total loss related to useful heat output	–	$\dot{Q}_{Vtot} = \left\{ \dot{Q}_{St} + \dot{Q}_v - \left(\dot{Q}_z \times \Sigma l_B \right) \right\}/\dot{Q}_N$
Boiler efficiency	%	$\eta_k = \left\{ (1 - \Sigma l_B)/\left(1 + \dot{Q}_{Vtot} \right) \right\} \times 100$

A.5 PG/PA Test of Steam Turbine Heat Rate

The PG/PA test of steam turbines is carried out in accordance with the latest version of ASME PTC 6/BS 752/DIN 1943/IEC 953-1 [IS 14198 (Part 1)]. The following paragraphs are developed based on the requirement of ASME PTC 6:2004.

This code recognizes that "the accurate determination of primary flow to the turbine is necessary to compute turbine heat rate or steam rate if the results are to be considered as a basis for turbine acceptance......and recommends measurement of water flow in the feedwater cycle" (Clause No. 4-8.1, P 29). Hence, in order to maximize accuracy of flow measurement, it is normal practice in the industry to measure "condensate flow" to the deaerator, from which feedwater flow and main steam (MS) flow are evaluated.

Against the aforementioned backdrop, sequences of activities that are usually adopted in the industry during the PG/PA test of steam turbines are:

I. Inspection and tuning of the steam turbine generator as follows:
    a. Verify availability of minimum 20D upstream and 10D downstream straight length of piping from the flow element used to measuring condensate flow (D corresponds to internal diameter of piping).
    b. Check availability of all instruments, including the DCS, required for the evaluation of the test.
    c. Check the calibration of all primary and secondary instruments.
    d. Check and correct steam leakage. As per international practice, steam leakage should be within 0.1%.
    e. Check for no or least air ingress into the turbine through turbine gland seals.
    f. Check for no or least air ingress into the condenser through the glands of valves located on interconnecting piping with the condenser, vacuum breaking valve, all atmospheric vent valves on the shell side (steam side) of the condenser, feedwater heater vents and drain connections, steam drain connections, makeup water connections, condensate spray connections, LP bypass dump line connections, air evacuation connections, and so on.
    g. Check the availability of sampling points for steam, water, and so on.
    h. Operate steam turbine generator at or near specified test conditions and note abnormalities, if any.
    i. Undertake any adjustment required to make operating parameters consistent.
    j. If necessary, shut down the steam turbine to carry out all corrective actions.

II. In order to eliminate probable errors in test results, the following extraneous flows need to be isolated/closed (Clause No. 3-5.5, P 10, [2]):
    1. HP/LP bypass systems and auxiliary steam lines
    2. bypass lines for primary flow measuring devices
    3. turbine sprays

4. drain lines on emergency stop valve (ESV), interceptor valve (IV), and contorl valves (CVs)

5. drain lines on MS, cold reheat, hot reheat, and extraction steam piping

6. interconnecting lines to other units, if any

7. chemical feed equipment using condensate

8. steam generator fill lines

9. steam generator vents

10. steam-operated soot blowers

11. steam for fuel oil atomization and heating

12. condensate and feedwater flow bypassing heaters

13. heater drain bypasses

14. heater shell drains

15. heater waterbox vents

16. hogging ejector

17. condenser waterbox priming vents

18. steam or water lines for station heating

19. steam or water lines installed for water washing the turbine

III. In addition to the previous items, it is preferable to measure the following extraneous flows to eliminate errors from test results (Clause No. 3-5.6, P 10 & 11, [2]):

1. Sealing and gland cooling flow on the following (both supply and return):

a. condensate extraction pumps

b. boiler feed pumps

c. boiler circulating pumps, if provided

d. heater drain pumps, if provided

e. turbines for turbine-driven pumps, if provided

2. superheater/reheater attemperator spray water flow

3. minimum flow recirculation line and balance flow line of boiler feed pumps

4. steam generator blowdowns

5. turbine gland sealing steam supply and leakage

6. makeup water, if necessary

7. pegging or sparging steam to deaerator

8. heater shell vents, are to be throttled to a minimum

9. deaerator overflow

10. deaerator vents, throttled to a minimum

11. water leakage into water-sealed vacuum breakers

12. water and steam sampling equipment

13. steam supply to APHs.

IV. The PG/PA test will consist of at least two runs, each of about 2 h' operation.

V. Conduct a trial test of a duration of about 2 h to check the working of all instruments and train the test personnel for taking readings. Personnel assigned for recording readings should be available during the trial.

VI. All readings pertaining to the actual test shall be taken to make the test personnel conversant with the test.

VII. Consistency of operating parameters must be ensured during the test. Parameters which appear to be nonconsistent are to be rejected.

VIII. The test will be preceded by operation near the test load for a duration until steady-state conditions are established.

IX. Problems encountered during recording readings and collecting samples shall be attended.

X. The frequency of recording readings generally conform to the following:
Flow measurements ≤ 1 min
Power output measurements ≤ 1 min
Pressure and temperature measurements ≤ 5 min
Water level measurements ≤ 10 min

XI. When recording readings automatically through a data logger, recording readings will be more frequent than those listed in number X.

XII. The average value of each measured parameter is calculated, and will be used in the final computation.

XIII. Operating conditions at the time of the test may differ from the specified conditions that were used to establish design or guarantee performance levels. Correction factors may be obtained from correction curves (Section A.5.1) to take care of deviation in any operating condition. The correction curves must be supplied by the steam turbine manufacturer prior to conducting the test.

XIV. Deviation in any operating condition during the test should comply with the requirements of Table A.14.

XV. At the commencement of the test, steam turbines shall be operated under the following conditions:
a. Alternate drain lines to and from HP and LP feedwater heaters closed
b. Steam to and from common auxiliary steam header, if applicable, closed
c. Makeup water supply to deaerator feedwater storage tank and condenser hotwell closed

XVI. During the test, required load on steam turbine generator and operating conditions, as close to design as possible, are to be ensured by the customer

XVII. The test report shall consist of the following:
a. All recorded test data, duly countersigned by the customer, the engineer, the steam turbine manufacturer, and the supplier (contractor)
b. Calculation and analysis of test data
c. Computation of performance test parameters
d. Final findings, analysis of the findings, and recommendations based on final findings, if any

Table A.14 Permissible deviation in operating parameters

Sl. No.	Parameter	Permissible Deviations for the Average of the Test Conditions from Design-Rated Conditions	Permissible Fluctuations During Any Test Run
1	MS pressure	±3.0% of the absolute pressure	±0.25% of the absolute pressure or 34.5 kPa, whichever is larger
2	Main and reheat steam temperature	±8 K when superheat is 15–30 K; ±16 K when superheat is in excess of 30 K.	±2 K when superheat is 15–30 K; ±4 K when superheat is in excess of 30 K.
3	Pressure drop through reheater	±50.0%	Not specified
4	Extraction pressure	±5.0%	Not specified
5	Extraction flow	±5.0%	Not specified
6	Temp. of FW leaving final heater	±6 K	Not specified
7	Exhaust pressure	±0.34 kPa or ±2.5% of the absolute pressure, whichever is larger	±0.14 kPa or ±1.0% of the absolute pressure, whichever is larger
8	Load	The corrected load will be within ±5.0% of the load specified for the test	±0.25%
9	Voltage	±5.0%	Not specified
10	Power factor	Not specified	±1.0%
11	Speed	±5.0%	±0.25%
12	Aggregate isentropic enthalpy drop of any one of the sections of an automatic extraction turbine	±10.0%	Not specified

A.5.1 Correction Factors

In line with ASME PTC 6, correction factors that are typically applied, due to deviation in various parameters, to measured turbine heat rate to make it equivalent to the guaranteed turbine heat rate are given next. Corrections to be applied follow the relationship given here:

$$HR_{\text{CORRECTED}} = HR_{\text{MEASURED}} \left(1 + \frac{\%}{100} \right)$$

i. Deviation in MS pressure (Fig. A.8)
ii. Deviation in MS temperature (Fig. A.9)
iii. Deviation in hot reheat steam temperature (Fig. A.10)
iv. Deviation in reheater pressure drop (Fig. A.11)
v. Deviation in condenser vacuum (Fig. A.12)
vi. Deviation in feedwater temperature at economizer inlet (Fig. A.13)

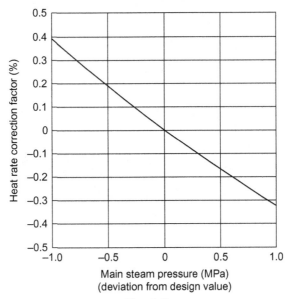

Fig. A.8
Turbine HR versus MS pressure.

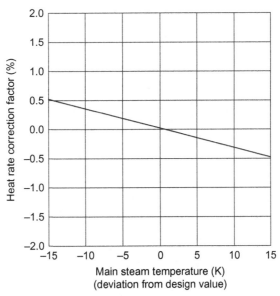

Fig. A.9
Turbine HR versus MS temperature.

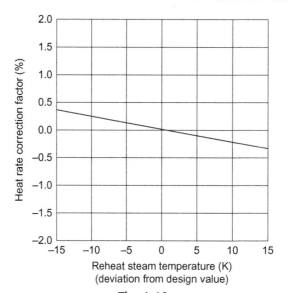

Fig. A.10

Turbine HR versus hot reheat steam temperature.

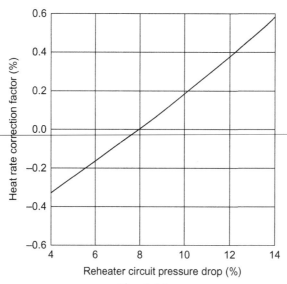

Fig. A.11

Turbine HR versus reheater pressure drop.

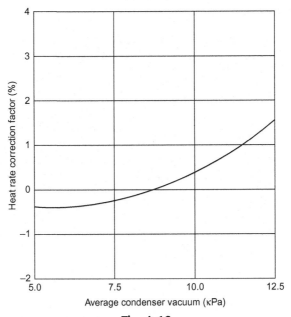

Fig. A.12
Turbine HR versus condenser vacuum.

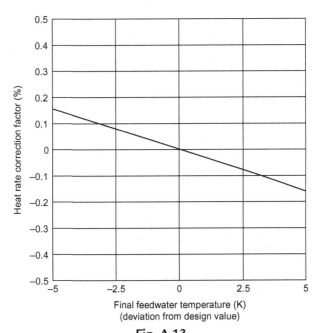

Fig. A.13
Turbine HR versus feedwater temperature at economizer inlet.

The following paragraphs discuss in detail the evaluation of the turbine heat rate.

A.5.2 Measured Parameters

For the computation of the turbine heat rate, it is essential to measure certain parameters, per Table A.15.

Table A.15 Measured parameters

Sl. No.	Description	Unit	Symbol and Formula
1	Superheater outlet MS pressure	MPa	P_{SHO}
2	Superheater outlet MS temperature	K	T_{SHO}
3	HP turbine inlet MS pressure	MPa	P_{HPI}
4	HP turbine inlet MS temperature	K	T_{HPI}
5	HP turbine exhaust steam pressure	MPa	P_{HPE}
6	HP turbine exhaust steam temperature	K	T_{HPE}
7	HP casing differential temperature	K	ΔT_{HPC}
8	IP turbine inlet steam pressure	MPa	P_{IPI}
9	IP turbine inlet steam temperature	K	T_{IPI}
10	IP turbine exhaust steam pressure	MPa	P_{IPE}
11	IP turbine exhaust steam temperature	K	T_{IPE}
12	IP casing differential temperature	K	ΔT_{IPC}
13	Control valve 1 position	%	Pos_{CV1}
14	Control valve 2 position	%	Pos_{CV2}
15	Control valve 3 position	%	Pos_{CV3}
16	Control valve 4 position	%	Pos_{CV4}
17	Reheater pressure drop (Sl. No. 3–Sl. No. 6)	MPa	ΔP_{RH}
18	Extraction 1 steam pressure	MPa	P_{EX1}
19	Extraction 1 steam temperature	K	T_{EX1}
20	Extraction 2 steam pressure	MPa	P_{EX2}
21	Extraction 2 steam temperature	K	T_{EX2}
22	Extraction 3 steam pressure	MPa	P_{EX3}
23	Extraction 3 steam temperature	K	T_{EX3}
24	Extraction 4 steam pressure	MPa	P_{EX4}
25	Extraction 4 steam temperature	K	T_{EX4}
26	Extraction 5 steam pressure	MPa	P_{EX5}
27	Extraction 5 steam temperature	K	T_{EX5}
28	Extraction 6 steam pressure	MPa	P_{EX6}
29	Extraction 6 steam temperature	K	T_{EX6}
30	Extraction 7 steam pressure	Pa	P_{EX7}
31	Extraction 7 steam temperature	K	T_{EX7}
32	Extraction 8 steam pressure	Pa	P_{EX8}
33	Extraction 8 steam temperature	K	T_{EX8}
34	LP turbine exhaust steam pressure	Pa	P_{LPE}

Continued

Table A.15   Measured parameters—cont'd

Sl. No.	Description	Unit	Symbol and Formula
35	LP turbine exhaust steam temperature	K	T_{LPE}
36	Condenser vacuum	Pa	P_{COND}
37	HPH 1 outlet feedwater pressure	MPa	P_{HPH1O}
38	HPH 1 outlet feedwater temperature	K	T_{HPH1O}
39	HPH 1 inlet feedwater pressure	MPa	P_{HPH1I}
40	HPH 1 inlet feedwater temperature	K	T_{HPH1I}
41	HPH 1 drain temperature	K	T_{HPH1DR}
42	HPH 2 outlet feedwater pressure	MPa	P_{HPH2O}
43	HPH 2 outlet feedwater temperature	K	T_{HPH2O}
44	HPH 2 inlet feedwater pressure	MPa	P_{HPH2I}
45	HPH 2 inlet feedwater temperature	K	T_{HPH2I}
46	HPH 2 drain temperature	K	T_{HPH2DR}
47	HPH 3 outlet feedwater pressure	MPa	P_{HPH3O}
48	HPH 3 outlet feedwater temperature	K	T_{HPH3O}
49	HPH 3 inlet feedwater pressure	MPa	P_{HPH3I}
50	HPH 3 inlet feedwater temperature	K	T_{HPH3I}
51	HPH 3 drain temperature	K	T_{HPH3DR}
52	Economizer inlet feedwater pressure	MPa	P_{ECOI}
53	Economizer inlet feedwater temperature	K	T_{ECOI}
54	Superheater attemperator spray water inlet pressure	MPa	P_{SSP}
55	Superheater attemperator spray water inlet temperature	K	T_{SSP}
56	Reheater attemperator spray water inlet pressure	MPa	P_{RSP}
57	Reheater attemperator spray water inlet temperature	K	T_{RSP}
58	Superheater attemperator spray water flow	kg s^{-1}	M_{SSP}
59	Reheater attemperator spray water flow	kg s^{-1}	M_{RSP}
60	Makeup flow to hotwell	kg s^{-1}	M_{MU}
61	Hotwell temperature	K	T_{HOT}
62	Hotwell level at the beginning of PG/PA test	mm	L_{HOTB}
63	Hotwell level at the end of PG/PA test	mm	L_{HOTE}
64	Condensate inlet flow to deaerator	kg s^{-1}	M_{COND}
65	Condensate pressure at deaerator inlet	MPa	P_{COND}
66	Condensate temperature at deaerator inlet	K	T_{COND}
67	Deaerator pressure	MPa	P_{DEA}
68	Deaerator feedwater temperature	K	T_{DEA}
69	Dearator feedwater storage tank level at the beginning of PG/PA test	mm	L_{DEAB}
70	Deaerator feedwater storage tank level at the end of PG/PA test	mm	L_{DEAE}

Continued

Table A.15   Measured parameters—cont'd

Sl. No.	Description	Unit	Symbol and Formula
71	Boiler drum level at the beginning of PG/PA test (applicable drum type steam generator)	mm	L_{BDB}
72	Boiler drum level at the end of PG/PA test (applicable drum type steam generator)	mm	L_{BDE}
73	Extraction flow to BFP turbine (if provided)	kg s^{-1}	M_{BFPT}
74	Turbine shaft vibration 2X/2Y	μ	VIB_{SH1}
75	Turbine shaft vibration 2X/2Y	μ	VIB_{SH2}
76	Turbine shaft vibration 3X/3Y	μ	VIB_{SH3}
77	Turbine shaft vibration 4X/4Y	μ	VIB_{SH4}
78	Turbine shaft vibration 5X/5Y	μ	VIB_{SH5}
79	Turbine shaft vibration 6X/6Y	μ	VIB_{SH6}
80	Turbine bearing vibration 1X/1Y	μ	VIB_{BRG1}
81	Turbine bearing vibration 2X/2Y	μ	VIB_{BRG2}
82	Turbine bearing vibration 3X/3Y	μ	VIB_{BRG3}
83	Turbine bearing vibration 4X/4Y	μ	VIB_{BRG4}
84	Turbine bearing vibration 5X/5Y	μ	VIB_{BRG5}
85	Turbine bearing vibration 6X/6Y	μ	VIB_{BRG6}
86	Differential expansion HP-IP	mm	ΔEXP_{HP-IP}
87	Differential expansion LP	mm	ΔEXP_{LP}
88	Axial shift	mm	AXSH
89	Turbine rotor speed	rpm	N
90	Turbine bearing drain oil temperature 1	K	T_{BDO1}
91	Turbine bearing drain oil temperature 2	K	T_{BDO2}
92	Turbine bearing drain oil temperature 3	K	T_{BDO3}
93	Turbine bearing drain oil temperature 4	K	T_{BDO4}
94	Turbine bearing drain oil temperature 5	K	T_{BDO5}
95	Turbine bearing drain oil temperature 6	K	T_{BDO6}
96	Thrust bearing metal temperature	K	T_{THB}
97	Lube oil pressure	MPa	P_{LO}
98	Oil cooler inlet oil temperature	K	T_{OCI}
99	Oil cooler outlet oil temperature	K	T_{OCO}
100	Gross generator output	kW	P_G
101	Auxiliary power	kW	P_{AUX}
102	Power factor	—	PF
103	Frequency	Hz	F
104	Bus voltage	kV	V_{BUS}
105	Generator winding temperature	K	T_{GWDG}
106	Main exciter air temperature	K	T_{EXA}

Continued

Table A.15   Measured parameters—cont'd

Sl. No.	Description	Unit	Symbol and Formula
107	Hydrogen pressure in generator	MPa	P_{HYD}
108	Cold hydrogen gas temperature	K	T_{CHYD}
109	Seal oil pressure	MPa	P_{SO}
110	Cold seal oil temperature	K	T_{CSO}
111	Stator cooling water pressure	MPa	P_{STCW}
112	Cold stator cooling water temperature	K	T_{CSTCW}
113	Stator winding cooling water flow	kg s^{-1}	M_{STCW}

Note

Strike out whichever parameter is not applicable.

A.5.3 Calculated Parameters

Over and above measurement, certain flows need to be calculated. Theses parameters are presented in Table A.16.

Table A.16   Calculated parameters

Sl. No.	Description	Unit	Symbol and Formula
1	MS Flow	kg s^{-1}	M_{MS}
2	Reheat steam flow	kg s^{-1}	M_{HR}
3	Feedwater flow	kg s^{-1}	M_{FW}
4	Unaccounted-for leakage flow	kg s^{-1}	M_{UNL}
5	Extraction 1 steam flow	kg s^{-1}	M_{EX1}
6	Extraction 2 steam flow	kg s^{-1}	M_{EX2}
7	Extraction 3 steam flow	kg s^{-1}	M_{EX3}
8	Extraction 4 steam flow	kg s^{-1}	M_{EX4}
9	Extraction 5 steam flow	kg s^{-1}	M_{EX5}
10	Extraction 6 steam flow	kg s^{-1}	M_{EX6}
11	Extraction 7 steam flow	kg s^{-1}	M_{EX7}
12	Extraction 8 steam flow	kg s^{-1}	M_{EX8}
13	HP turbine front and rear gland seal leakage steam flow	kg s^{-1}	M_{GL}

A.5.4 Inputs From Steam Table

In addition to the measured and calculated parameters, there are certain conditions, the enthalpies of which are obtained from the steam table or are calculated (Table A.17).

Table A.17 Parameters from steam table

Sl. No.	Description	Unit	Symbol and Formula
1	Enthalpy of MS at superheater outlet at P_{SHO} and T_{SHO}	$kJ\ kg^{-1}$	H_{SHO}
2	Enthalpy of MS at HP turbine (HPT) inlet at P_{HPI} and T_{HPI}	$kJ\ kg^{-1}$	H_{HPI}
3	Enthalpy of cold reheat steam at HPT exhaust at P_{HPE} and T_{HPE}	$kJ\ kg^{-1}$	H_{CR}
4	Enthalpy of hot reheat steam at IP turbine inlet at P_{IPI} and T_{IPI}	$kJ\ kg^{-1}$	H_{HR}
5	Enthalpy of feedwater at economizer inlet at P_{ECOI} and T_{ECOI}	$kJ\ kg^{-1}$	H_{FWE}
6	Enthalpy of feedwater at HPH 1 outlet at P_{HPH1O} and T_{HPH1O}	$kJ\ kg^{-1}$	H_{FWH1O}
7	Enthalpy of feedwater at HPH 1 inlet at P_{HPH1I} and T_{HPH1I}	$kJ\ kg^{-1}$	H_{FWH1I}
8	Enthalpy of extraction steam 1 at HPH 1 inlet at P_{EX1} and T_{EX1}	$kJ\ kg^{-1}$	H_{EX1}
9	Enthalpy of drain at HPH 1 outlet at P_{EX1} and $T_{HPH1DRI}$	$kJ\ kg^{-1}$	H_{DR1}
10	Enthalpy of feedwater at HPH 2 outlet at P_{HPH2O} and T_{HPH2O}	$kJ\ kg^{-1}$	H_{FWH2O}
11	Enthalpy of feedwater at HPH 2 inlet at P_{HPH2I} and T_{HPH2I}	$kJ\ kg^{-1}$	H_{FWH2I}
12	Enthalpy of extraction steam 2 at HPH 2 inlet at P_{EX2} and T_{EX2}	$kJ\ kg^{-1}$	H_{EX2}
13	Enthalpy of drain at HPH 2 outlet at P_{EX2} and T_{HPH2DR}	$kJ\ kg^{-1}$	H_{DR2}
14	Enthalpy of feedwater at HPH 3 outlet at P_{HPH3O} and T_{HPH3O}	$kJ\ kg^{-1}$	H_{FWH3O}
15	Enthalpy of feedwater at HPH 3 inlet at P_{HPH3I} and T_{HPH3I}	$kJ\ kg^{-1}$	H_{FWH3I}
16	Enthalpy of extraction steam 3 at HPH 3 inlet at P_{EX3} and T_{EX3}	$kJ\ kg^{-1}$	H_{EX3}
17	Enthalpy of drain at HPH 3 outlet at P_{EX3} and T_{HP3DR}	$kJ\ kg^{-1}$	H_{DR3}
18	Enthalpy of superheater attemperator spray water at P_{SSP} and T_{SSP}	$kJ\ kg^{-1}$	H_{SSP}
19	Enthalpy of reheater attemperator spray water at P_{RSP} and T_{RSP}	$kJ\ kg^{-1}$	H_{RSP}
20	Enthalpy of extraction steam 4 at deaerator inlet at P_{EX4} and T_{EX4}	$kJ\ kg^{-1}$	H_{EX4}
21	Enthalpy of feedwater at deaerator outlet at P_{DEA} and T_{DEA}	$kJ\ kg^{-1}$	H_{DO}
22	Enthalpy of extraction steam 5 at P_{EX5} and T_{EX5}	$kJ\ kg^{-1}$	H_{EX5}
23	Enthalpy of extraction steam 6 at P_{EX6} and T_{EX6}	$kJ\ kg^{-1}$	H_{EX6}
24	Enthalpy of extraction steam 7 at P_{EX7} and T_{EX7}	$kJ\ kg^{-1}$	H_{EX7}
25	Enthalpy of extraction steam 8 at P_{EX8} and T_{EX8}	$kJ\ kg^{-1}$	H_{EX8}

A.5.5 Evaluation of Heat Rate

From Chapter 17, combining Eqs. (17.7), (17.8) we get

$$HR = \frac{(M_{MS} \times H_{MS} - M_{FW} \times H_{FW}) + (M_{HR} \times H_{HR} - M_{CR} \times H_{CR}) - (M_{SSP} \times H_{SSP} + M_{RSP} \times H_{RSP})}{P_G}, kJkWh^{-1}$$

(A.1)

We also get from Eq. (17.9) that

$$M_{MS} = M_{FW} + M_{SSP} - M_{UNL}, \ kgs^{-1}$$

(A.2)

It is evident from the previous two equations that while the values of enthalpy, H_{MS}, H_{FW}, H_{HR}, H_{CR}, H_{SSP}, and H_{RSP}, could be gathered from Table A.17, values of flows, M_{MS}, M_{FW}, M_{HR}, M_{CR}, M_{SSP}, M_{RSP}, and M_{UNL}, essentially have to be calculated. The following paragraphs provide details of various flow calculations.

A.5.6 Computation of Feedwater Flow (M_{FW})

In order to compute the value of feedwater flow it becomes apparent to carry out heat balance across HP heater (HPH) 1, HPH 2, HPH 3, and the deaerator (Fig. A.14; Table A.18).

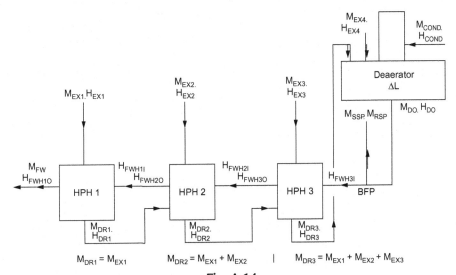

Fig. A.14

Schematic diagram of extraction steam flow to HPHs and deaerator, HPH drain flows, and feedwater flow.

For steady-state condition from Fig. A.14, we get

$$M_{DR1} = M_{EX1} \quad M_{DR2} = M_{EX1} + M_{EX2} \quad M_{DR3} = M_{EX1} + M_{EX2} + M_{EX3}$$

Table A.18 Heat balance across HPHs and deaerator

Item	Calculation	Factors
HPH 1	$$M_{EX1}(H_{EX1} - H_{DR1}) = M_{FW}(H_{FWH1O} - H_{FWH1I})$$ Or, $$M_{EX1} = M_{FW}\frac{(H_{FWH1O} - H_{FWH1I})}{(H_{EX1} - H_{DR1})} = K_1 \cdot M_{FW} \quad (A.3)$$	$$K_1 = \frac{(H_{FWH1O} - H_{FWH1I})}{(H_{EX1} - H_{DR1})},$$
HPH 2	$$M_{EX2}(H_{EX2} - H_{DR2}) + M_{EX1}(H_{DR1} - H_{DR2}) = M_{FW}(H_{FWH2O} - H_{FWH2I})$$ Or, $$M_{EX2}(H_{EX2} - H_{DR2}) + K_1 \cdot M_{FW} \cdot (H_{DR1} - H_{DR2}) = M_{FW}(H_{FWH2O} - H_{FWH2I})$$ Or, $$M_{EX2} = \frac{(H_{FWH2O} - H_{FWH2I}) - K_1(H_{DR1} - H_{DR2})}{(H_{EX2} - H_{DR2})} M_{FW}$$ $$M_{EX2} = (K_2 - K_1 K_3) \cdot M_{FW} \quad (A.4)$$	$$K_2 = \frac{(H_{FWH2O} - H_{FWH2I})}{(H_{EX2} - H_{DR2})}$$ $$K_3 = \frac{(H_{DR1} - H_{DR2})}{(H_{EX2} - H_{DR2})},$$
HPH 3	$$M_{EX3}(H_{EX3} - H_{DR3}) + (M_{EX1} + M_{EX2})(H_{DR2} - H_{DR3}) = M_{FW}(H_{FWH3O} - H_{FWH3I})$$ Or, $$M_{EX3}(H_{EX3} - H_{DR3}) + \{K_1 + (K_2 - K_1 \cdot K_3)\} M_{FW}(H_{DR2} - H_{DR3}) = M_{FW}(H_{FWH3O} - H_{FWH3I})$$ Or, $$M_{EX3} = \frac{(H_{FWH3O} - H_{FWH3I}) - \{K_1 + (K_2 - K_1 \cdot K_3)\}(H_{DR2} - H_{DR3})}{(H_{EX3} - H_{DR3})} M_{FW}$$ $$M_{EX3} = (K_4 - K_1 K_5 - K_2 K_5 + K_1 K_3 K_5) M_{FW} \quad (A.5)$$	$$K_4 = \frac{(H_{FWH3O} - H_{FWH3I})}{(H_{EX3} - H_{DR3})},$$ $$K_5 = \frac{(H_{DR2} - H_{DR3})}{(H_{EX3} - H_{DR3})},$$
Deaerator	$$M_{EX4}(H_{HEX4} - H_{DO}) + M_{DR3}(H_{DR3} - H_{DO}) + M_{COND}(H_{COND} - H_{DO}) = \pm \Delta L \cdot H_{DO}$$ Or, $$M_{EX4}(H_{HEX4} - H_{DO}) + (M_{EX1} + M_{EX2} + M_{EX3})(H_{DR3} - H_{DO}) + M_{COND}(H_{COND} - H_{DO}) = \pm \Delta L \cdot H_{DO}$$ Or, $$M_{EX4}(H_{HEX4} - H_{DO}) + \{K_1 + (K_2 - K_1 K_3) + (K_4 - K_1 K_5 - K_2 K_5 + K_1 K_3 K_5)\} \times M_{FW}(H_{DR3} - H_{DO}) + M_{COND}(H_{COND} - H_{DO}) = \pm \Delta L \cdot H_{DO}$$ Or, $$M_{EX4}(H_{HEX4} - H_{DO}) = \pm \Delta L \cdot H_{DO} - \{K_1 + (K_2 - K_1 K_3) + (K_4 - K_1 K_5 - K_2 K_5 + K_1 K_3 K_5)\} \times M_{FW}(H_{DR3} - H_{DO}) - M_{COND}(H_{COND} - H_{DO})$$ Or, $$M_{EX4} = \frac{\pm \Delta L \cdot H_{DO}}{(H_{HEX4} - H_{DO})} - \{K_1 + (K_2 - K_1 K_3) + (K_4 - K_1 K_5 - K_2 K_5 + K_1 K_3 K_5)\} \times$$ $$M_{FW} \cdot \frac{(H_{DR3} - H_{DO})}{(H_{HEX4} - H_{DO})} - M_{COND}\frac{(H_{COND} - H_{DO})}{(H_{HEX4} - H_{DO})}$$ $$M_{EX4} = \pm \Delta L \cdot K_6 - \{K_1(1 - K_3 - K_5 + K_3 K_5) + K_2(1 - K_5) + K_4\} K_7 \cdot M_{FW} - K_8 \cdot M_{COND} \quad (A.6)$$	$$K_6 = \frac{H_{DO}}{(H_{HEX4} - H_{DO})},$$ $$K_7 = \frac{(H_{DR3} - H_{DO})}{(H_{HEX4} - H_{DO})}$$ $$K_8 = \frac{(H_{COND} - H_{DO})}{(H_{HEX4} - H_{DO})}$$

Feed water at deaerator outlet, $M_{DO} = M_{COND} + M_{EX4} + M_{DR3} \pm \Delta L$
$$= M_{COND} + M_{EX1} + M_{EX2} + M_{EX3} + M_{EX4} \pm \Delta L$$

Feed water at HPH 1 outlet, $M_{FW} = M_{DO} - M_{SSP} - M_{RSP}$

Hence,

$$M_{FW} = M_{COND} + M_{EX1} + M_{EX2} + M_{EX3} + M_{EX4} - M_{SSP} - M_{RSP} \pm \Delta L \qquad (A.7)$$

A.5.6.1 Calculation of feedwater flow

Putting the values from Eqs. (A.3)–(A.6) in Eq. (A.7), we get

$$
\begin{aligned}
M_{FW} = \ & K_1 \cdot M_{FW} + (K_2 - K_1 K_3) \cdot M_{FW} + (K_4 - K_1 K_5 - K_2 K_5 + K_1 K_3 K_5) M_{FW} + \pm \Delta L \cdot \\
& K_6 - (K_1 K_7 + K_2 K_7 - K_1 K_3 K_7 + K_4 K_7 - K_1 K_5 K_7 - K_2 K_5 K_7 + K_1 K_3 K_5 K_7) \cdot \\
& M_{FW} - K_8 \cdot M_{COND} + M_{COND} - M_{SSP} - M_{RSP} \pm \Delta L
\end{aligned}
$$

Or,

$$
\begin{aligned}
& M_{FW}(1 - K_1 - K_2 + K_1 K_3 - K_4 + K_1 K_5 + K_2 K_5 - K_1 K_3 K_5 \\
& + K_1 K_7 + K_2 K_7 - K_1 K_3 K_7 + K_4 K_7 - K_1 K_5 K_7 - K_2 K_5 K_7 \\
& + K_1 K_3 K_5 K_7) = \pm \Delta L \cdot K_6 - K_8 \cdot M_{COND} + M_{COND} - M_{SSP} \\
& - M_{RSP} \pm \Delta L
\end{aligned}
$$

Solving above,

$$M_{FW} = \frac{\{\pm \Delta L \, (K_6 + 1) + M_{COND}(1 - K_8) - M_{SSP} - M_{RSP}\}}{K} \qquad (A.8)$$

where $K = 1 - (K_1 + K_2 + K_4 - K_1 K_3 - K_1 K_5 - K_2 K_5 + K_1 K_3 K_5)\,(1 - K_7)$.

Example A.1

Table A.19 furnishes values of enthalpy across HPH 1, HPH 2, HPH 3, and the deaerator. Condensate flow to the deaerator is 436.63 kg s^{-1}. If the level in the deaerator between the start and end of the test remains unchanged and there was no spray water requirement in the superheater and reheater attemperator during the test, determine the feedwater flow at the economizer inlet and extraction steam flows to HPHs and the deaerator.

To determine desired parameters let us proceed as under.

From Tables A.18 and A.19, Table A.20 is prepared.

Table A.19 Enthalpy across HPH 1, HPH 2, HPH 3, and deaerator

Item	Parameter	Unit	Value
HPH 1	H_{EX1}	kJ kg^{-1}	3159.21
	H_{DR1}	kJ kg^{-1}	1226.93
	H_{FWH1O}	kJ kg^{-1}	1361.30
	H_{FWH1I}	kJ kg^{-1}	1193.44

Continued

Continued

Example A.1—cont'd

Table A.19　Enthalpy across HPH 1, HPH 2, HPH 3, and deaerator—cont'd

Item	Parameter	Unit	Value
HPH 2	H_{EX2}	kJ kg^{-1}	3045.35
	H_{DR2}	kJ kg^{-1}	1031.44
	H_{FWH2O}	kJ kg^{-1}	1193.44
	H_{FWH2I}	kJ kg^{-1}	1011.77
HPH 3	H_{EX3}	kJ kg^{-1}	3448.47
	H_{DR3}	kJ kg^{-1}	836.79
	H_{FWH3O}	kJ kg^{-1}	1011.7
	H_{FWH3I}	kJ kg^{-1}	826.33
Deaerator	H_{EX4}	kJ kg^{-1}	3172.60
	H_{DO}	kJ kg^{-1}	785.30
	H_{DR3}	kJ kg^{-1}	836.79
	H_{COND}	kJ kg^{-1}	637.12

Table A.20　Values of factors

Factor	Equation	Value
K_1	$\dfrac{(H_{FWH1O} - H_{FWH1I})}{(H_{EX1} - H_{DR1})} = \dfrac{167.86}{1932.28}$	0.0868715
K_2	$\dfrac{(H_{FWH2O} - H_{FWH2I})}{(H_{EX2} - H_{DR2})} = \dfrac{181.67}{2013.91}$	0.0902076
K_3	$\dfrac{(H_{DR1} - H_{DR2})}{(H_{EX2} - H_{DR2})} = \dfrac{195.49}{2013.91}$	0.0970699
K_4	$\dfrac{(H_{FWH3O} - H_{FWH3I})}{(H_{EX3} - H_{DR3})} = \dfrac{185.37}{2611.68}$	0.0709773
K_5	$\dfrac{(H_{DR2} - H_{DR3})}{(H_{EX3} - H_{DR3})} = \dfrac{194.65}{2611.68}$	0.0745306
K_6	$\dfrac{H_{DO}}{(H_{HEX4} - H_{DO})} = \dfrac{785.30}{2387.30}$	0.3289490
K_7	$\dfrac{(H_{DR3} - H_{DO})}{(H_{HEX4} - H_{DO})} = \dfrac{51.49}{2387.30}$	0.0215683
K_8	$\dfrac{(H_{COND} - H_{DO})}{(H_{HEX4} - H_{DO})} = \dfrac{-148.18}{2387.30}$	-0.0620701
$K_2 - K_1 K_3$	$0.0902076 - 0.0868715 \times 0.0970699$	0.0817750
$K_4 - K_1 K_5 - K_2 K_5$ $+ K_1 K_3 K_5$	$0.0709773 - 0.0868715 \times 0.0745306 -$ $0.0902076 \times 0.0745306 +$ $0.0868715 \times 0.0970699 \times 0.0745306$	0.0584080

Continued

Example A.1—cont'd

Table A.20   Values of factors—cont'd

Factor	Equation	Value
$K_1(1 - K_3 - K_5 + K_3 K_5)$ $+ K_2(1 - K_5) + K_4$	$0.0868715\,(1 - 0.0970699 - 0.0745306 +$ $0.0970699 \times 0.0745306) +$ $0.0902076\,(1 - 0.0745306) + 0.0709773$	0.2270545
K	$1 - 0.2270545(1 - 0.0215683)$	0.7778427

From the given conditions $\Delta L = 0$, $M_{SSP} = 0$, and $M_{RSP} = 0$.
Hence, from Eq. (A.8),

$$\text{Feed water flow, } M_{FW} = M_{COND}(1 - K_8)/K = 436.63\{1 - (-0.0620701)\}/$$
$$0.7778427 = 596.18\,\text{kgs}^{-1}$$

From Eqs. (A.3)–(A.6), extraction steam flows are

$$\text{To HPH 1, } M_{EX1} = K_1 \cdot M_{FW} = 0.0868715 \times 596.18 = 51.79\,\text{kgs}^{-1}$$

$$\text{To HPH 2, } M_{EX2} = (K_2 - K_1 K_3) M_{FW} = 0.0817750 \times 596.18 = 48.75\,\text{kgs}^{-1}$$

$$\text{To HPH 3, } M_{EX3} = (K_4 - K_1 K_5 - K_2 K_5 + K_1 K_3 K_5) M_{FW} = 0.0584080 \times$$
$$596.18 = 34.82\,\text{kgs}^{-1}$$

$$\text{To Deaerator, } M_{EX4} = \pm \Delta L \cdot K_6 - \{K_1(1 - K_3 - K_5 + K_3 K_5) + K_2(1 - K_5) + K_4\}$$
$$K_7 \cdot M_{FW} - K_8 \cdot M_{COND} = 0 - 0.2270545 \times 0.0215683 \times 596.18 - (-0.0620701) \times 436.63$$
$$= 24.18\,\text{kgs}^{-1}$$

A.5.6.2 Unaccounted-for leakage (M_{UNL})

From Eq. (A.2), it is evident that while M_{SSP} is measurable and M_{FW} can be calculated from Eq. (A.8), the term M_{UNL} also needs to be calculated, as given here:

$$M_{UNL} = \pm M_{\Delta LHOT} \pm M_{\Delta LDEA} \pm M_{\Delta LBD} - M_{ML} - M_{SAMP}, \text{kg s}^{-1} \qquad (A.9)$$

where $M_{\Delta LHOT}$: equivalent mass flow rate corresponding to change in hotwell level during the test;

$M_{\Delta LDEA}$: equivalent mass flow rate corresponding to change in deaerator feedwater storage tank level during the test;

$M_{\Delta LBD}$: equivalent mass flow rate corresponding to change in boiler drum level during the test (applicable to drum-type steam generator);

Description	Increase	Decrease
Boiler drum level	$(-)$ ve	$(+)$ ve
Deaerator feedwater storage tank level	$(-)$ ve	$(+)$ ve
Hotwell level	$(-)$ ve	$(+)$ ve
Net storage water level	$(-)$ ve	$(+)$ ve

M_{ML}: measured leakage flow, if any;

M_{SAMP}: cumulative flow of samples from the cycle.

In Eq. (A.9), change in level could be either positive or negative per the following conditions:

Having measured the change in level, equivalent change in mass flow rate is calculated as

$$M_{\Delta L} = \frac{l \cdot \left[\left(R^2 \cdot \sin^{-1}\frac{L_E}{R} + L_E \cdot \sqrt{R^2 - L_E^2} \right) - \left(R^2 \cdot \sin^{-1}\frac{L_B}{R} + L_B \cdot \sqrt{R^2 - L_B^2} \right) \right]}{t \cdot v}, \text{kg s}^{-1}$$
(A.10)

where $M_{\Delta L}$: equivalent mass flow rate corresponding to change in level;

l: length of the straight portion of the vessel, m;

R: inside radius of the vessel, m;

L_E: height of water level from the center line of the vessel at the end of the test, m;

L_B: height of water level from the center line of the vessel at the beginning of the test, m;

t: duration of the test, s;

v: specific volume of water in the vessel, $\text{m}^3 \text{ kg}^{-1}$.

A.5.7 Computation of Flow of Cold Reheat Steam to Reheater

One more parameter of Eq. (A.1) that needs to be calculated is cold reheat steam flow to the reheater, which is given by

$$M_{CR} = M_{MS} - M_{GL} - M_{CV} - M_{EX1} - M_{AUX}, \text{kg s}^{-1}$$
(A.11)

where M_{GL}: cumulative leakage steam flow from HP turbine gland seals;

M_{CV}: cumulative leakage steam flow from HP control valves;

M_{EX1}: extraction steam flow from cold reheat line to HPH 1;

M_{AUX}: auxiliary steam flow from cold reheat line, kg s^{-1}.

Note

In the event that it is difficult to measure the values of M_{GL} and M_{CV}, values supplied by the steam turbine manufacturer may be used in the calculation.

A.5.8 Computation of Flow of Hot Reheat Steam to IP Turbine Inlet

Another parameter is hot reheat steam flow, calculated as under

$$M_{HR} = M_{CR} + M_{RSP} \tag{A.12}$$

All parameters on the right-hand side of Eq. (**A.1**) are now available through either calculation or measurement, by which turbine heat rate can be determined.

A.5.9 Correction Toward Degradation in Turbine Performance Due to Aging

ASME PTC 6 recommends "the acceptance test should be scheduled as soon as practicable, preferably within eight weeks, after the turbine is first loaded.....Adjusting of heat rate test results for the effects of aging is not permitted" (Clause No. 3-3.1. P8).

Notwithstanding the previous stipulation, in the event that the PG/PA test gets delayed beyond eight weeks for reasons attributable to the customer, then either of the following two methods, as agreed between the customer and the turbine manufacturer, may be adopted to prove the guaranteed heat rate after applying the correction factor for performance degradation due to aging.

a. One of the methods is [3]:

$$\%\text{Deterioration} = \frac{BF}{\log MW} f\left(\frac{P}{16.55}\right)^{0.5} \tag{A.13}$$

where *BF*: the base factor, as under

N, months of normal operation	0	12	24	36	48
BF, base factor (%)	0	1.0	1.5	1.9	2.2

MW: rated power output of the turbine;
f: 1.0 for fossil units;
P: MS pressure at the inlet to HP turbine, MPa.

b. Another method recommends following deterioration in heat rate [4]:

Turbine Rating (MW)	Months of Normal Operation After Synchronization		
	2–12 Months	**12–24 Months**	
$\leq$150	0.10	0.06	
>150	$0.10\left(\dfrac{150}{MW}\right)^{0.5}$	$0.06\left(\dfrac{150}{MW}\right)^{0.5}$	% Per month

A.6 PG/PA Test of Electrostatic Precipitator (ESP)

Principal elements of an ESP are collecting electrodes and discharge electrodes. Collecting electrodes comprise rows of electrically grounded vertical parallel plates between which flue gas flows. Discharge electrodes, located between each pair of parallel plates, typically are wires through which high-voltage direct current is supplied from an external source. Thus a unidirectional, nonuniform strong electric field is generated between collecting electrodes and discharge electrodes. While the intensity of this field near the wires (discharge electrodes) is considerably high, it is very feeble on the plates (collecting electrodes). When the voltage supplied to wires is high enough, a corona is produced around them, which indicates generation of negatively charged flue gas ions that travel from discharge electrodes to grounded collecting electrodes. When flue gas dust or fly ash builds up on plates, build-up ash is removed periodically by hitting/rapping collecting electrodes with hammers. The remaining quantity of fly ash particles (suspended particulate matter) that escape to the atmosphere along with the flue gas must be restricted in accordance with the local environmental pollution control norm.

The World Bank stipulates that the guaranteed suspended particulate matter downstream of ESP (dust concentration at ESP outlet—*DCO*) for solid fuels must not exceed 50 mg Nm^{-3} at 100% TMCR unit load. Failure to meet this requirement would attract liquidated damage (Sl. No. 5, Table 17.4: Parameters included under category A).

The collection efficiency of an ESP is given by

$$Eff = \frac{DCI - \dfrac{DCO}{1000}}{DCI} \times 100 \tag{A.14}$$

where *Eff*: collection efficiency of ESP, %;

DCI: dust concentration at ESP inlet, g Nm^{-3}.

Having found out the efficiency of ESP at 100% TMCR unit load per Eq. (A.14), it must be verified that the efficiency is not less than the efficiency specified by the customer, lest this will also attract liquidated damage (Sl. No. 6, Table 17.4: Parameters included under category A).

The test code that is generally followed in the industry for evaluation of the aforementioned ESP parameters is either EPA 17 (method-17 of EPA) or ASME PTC 38 (Sl. No. 11, Table 17.7: Codes followed for PG/PA tests of some major equipment).

Another parameter of ESP, which is not guaranteed but is demonstrated, is "maximum pressure drop across ESP at the guarantee point flow condition." This parameter must not exceed 200 Pa (Sl. No. 19, Table 17.3: Parameters to be demonstrated during reliability run test).

A.6.1 Collection of Samples

The most important part of ESP performance evaluation is sample collection from flue gas passing through the inlet/outlet duct of the ESP. In order to ascertain a representative collection of fly ash sample from flue gas, the method used is termed isokinetic sampling, which means the velocity of flue gas entering the sampling nozzle is equal to the velocity of flue gas at the sampling point across the cross section of the flue gas duct at each inlet and outlet of ESP. Another factor that influences a representative collection of the sample is the number of points for collecting samples across the cross section of ducts. This number of points for the sampling collection depends on the internal equivalent diameter (De) of ducts and the extent of straight length available from a flow disturbance (eg, bends, location of temperature-sensing elements, expansion, contraction, divergence duct at inlet to ESP/convergence duct at outlet from ESP, etc.).

For a circular duct the equivalent diameter is equal to the internal diameter of the duct, while for a rectangular duct the equivalent diameter is given by

$$De = \frac{2 \times L \times B}{L + B} \qquad (A.15)$$

where L: length of the cross section of the duct;

B: breadth of the cross section of the duct.

The straight length in the duct upstream of the measurement site of the sampling probe (that is downstream from flow disturbance) must be greater than 8 De, and the straight length in the duct downstream of the measurement site of the sampling probe (that is upstream from flow disturbance) must be greater than 2 De. Typical sampling point traverse locations are shown in Figs. A.15 and A.16.

In the event that adequate straight length in duct upstream/downstream of the measurement site of the sampling probe is not available, the number of sampling points must be increased in accordance with the requirement shown in Fig. A.17. The distance of each traverse point from the inner wall of the circular duct to ensure equal area has to conform to the requirement shown in Table A.21 multiplied by the corresponding equivalent diameter. For a rectangular duct, the location of each traverse point is determined by dividing length L by row number N and breadth B by column number n (Fig. A.18) to arrive at the number of traverse points, per Fig. A.17.

As explained previously, samples are collected from flue gas flowing through the inlet/outlet duct of an ESP with the help of a sampling probe. The nozzle of this probe must be of unique diameter, which depends on the capacity of equipment collecting samples (gas meter) and the maximum average velocity encountered during the traverse of the sampling probe.

Fig. A.19 depicts the detailed arrangement of a typical sampling probe.

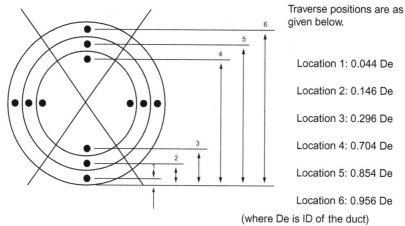

Fig. A.15

Location of traverse points in a circular duct divided into 12 equal areas. *Source: Figure 1-3, P 205 [5]*

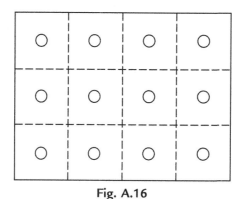

Fig. A.16

Location of traverse points in a rectangular duct divided into 12 equal areas. *Source: Figure 1-4, P 205 [5]*

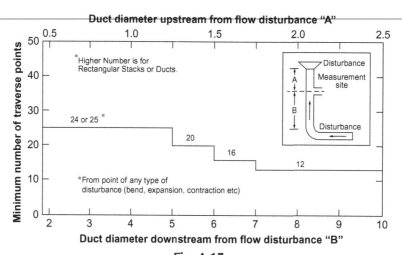

Fig. A.17

Minimum number of traverse points versus duct diameter. *Source: Figure 1-1, P 201 [5]*

Table A.21 Distance of the sampling point from the inner wall of circular flue gas duct

Serial No. of Sampling Point	Number of Concentric Circles				
	1	2	3	4	5
1	0.146	0.067	0.044	0.033	0.022
2	0.854	0.250	0.146	0.105	0.082
3		0.750	0.294	0.195	0.145
4		0.933	0.706	0.321	0.227
5			0.854	0.679	0.344
6			0.956	0.805	0.656
7				0.895	0.773
8				0.967	0.855
9					0.918
10					0.978

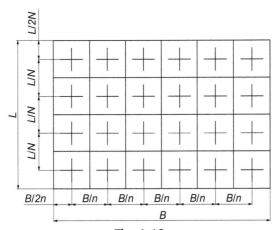

Fig. A.18
Location of traverse points in rectangular duct.

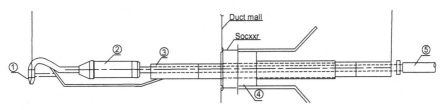

1: Nozzle tip   2: Filter holder   3: Sampling tubes   4: Probe holder   5: Rubber tube

Fig. A.19
Sampling probe.

A.6.2 Selection of Sampling Probe Nozzle

The maximum average velocity of flue gas through the duct (v_{max}), as encountered by the sampling probe, is measured by a manometer along with a pitot tube (Fig. A.20) and is calculated as

$$v_{max} = \sqrt{\frac{2 \times g \times H \times \rho_l \times K_m}{\rho_g}} \tag{A.16}$$

where g: acceleration due to gravity (9.81 m s^{-2});

H: dynamic head of liquid (eg, mercury, water, or ethyl alcohol) column in manometer, mm;

K_m: manometer constant;

ρ_l: density of manometer liquid, kg m^{-3};

ρ_g: density of flue gas at operating pressure and temperature, kg m^{-3}.

To find out the capacity of the gas meter, it is essential to calculate the largest attainable sampling constant (K_n) per the equation given here:

$$K_n = \frac{Q}{v_{max} \times \dfrac{\rho_g}{\rho_n} \times \sqrt{\dfrac{101.33}{BP}}} \tag{A.17}$$

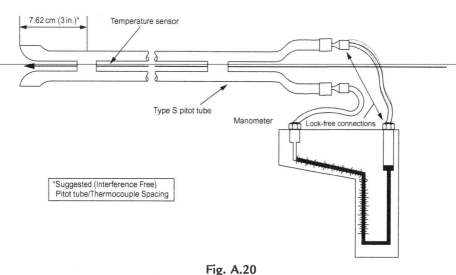

Fig. A.20

Type S pitot tube manometer assembly. *Source: Figure 2-1, P 244 [6]*

where Q: assumed flue gas flow rate through duct based on analysis of as-fired coal; v_{max} and ρ_g: as explained previously;

ρ_n: density of flue gas at 101.3 kPa and 273 K;

BP: barometric pressure, kPa.

Based on the values of v_{max} and K_n, the nozzle diameter of the sampling probe is selected. This nozzle diameter varies from project to project, usually ranges from 5 to 12 mm, and is recommended by the ESP manufacturer.

A schematic diagram of a typical sample collection system is depicted in Fig. A.21.

A.6.3 Measured Parameters

Salient parameters, which are to be recorded during the test, are presented in Tables A.22 and A.23.

A.6.4 Laboratory Analyzed Data

Analyses of as-fired coal samples are presented in Table A.24.

A.6.5 Sequence of Activities

The sequence of activities to be followed during the test is discussed here:

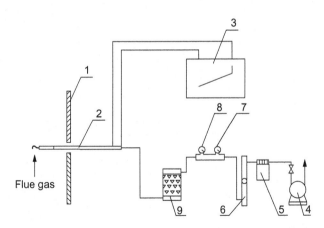

1: Flue gas duct 2: Sampling probe 3: Manometer 4: Vacuum pump

5: Integrating flowmeter   6: Rotameter 7: Vacuum gauge 8: Temperature gauge 9: Dryer

Fig. A.21
Scheme of sample collection system.

Table A.22 Measured parameters

Sl. No.	Parameters	Unit	Time		
1	Unit load	MW			
2	MS flow at superheater outlet (SHO)	$kg\,s^{-1}$			
3	MS pressure at SHO	MPa			
4	MS temperature at SHO	K			
5	Economizer inlet feedwater pressure	MPa			
6	Economizer inlet feedwater temperature	K			
7	Number of pulverizers in service	—			
8	Coal flow (W_C)	$kg\,s^{-1}$			
9	Total airflow (TA)	$kg\,s^{-1}$			
10	Oxygen content of dry flue gas at AH inlet	%			
11	Carbon dioxide content in dry flue gas at AH outlet	%			
12	Carbon monoxide content in dry flue gas	%			
13	AH outlet secondary air temperature	K			
14	AH outlet primary air temperature	K			
15	ID fan motor current	A			
16	FD fan motor current	A			

Table A.23 Parameters to be measured at inlet and outlet ducts at each test point

Sl. No.	Parameters	Unit	Time		
1	Barometric pressure	Pa			
2	Dry bulb temperature	K			
3	Wet bulb temperature	K			
4	Temperature at gas meter	K			
5	Flue gas pressure at inlet/outlet duct to ESP	Pa			
6	Flue gas temperature at inlet/outlet duct to ESP	K			
7	Manometer reading	mm			

I. Inspection and tuning of the ESP as narrated here:
   a. Check availability of all instruments, including the DCS, required for evaluation of the test.
   b. Check the calibration of all primary and secondary instruments.
   c. Check for no air ingress through manholes, peepholes, and so on.

Table A.24 Laboratory analyzed (as-fired coal) data

Sl. No.	Description	Unit
1	Carbon content in coal, C	%
2	Hydrogen content in coal, H	%
3	Sulfur content in coal, S	%
4	Oxygen content in coal, O	%
5	Nitrogen content in coal, N	%
6	Moisture content in coal, M	%
7	Ash content in coal, A	%
8	HHV of coal (bomb calorimeter)	$kJ\ kg^{-1}$

    d. Check and, if required, correct leakage in gas path.

    e. Verify that the locations of sampling points conform to code requirements as discussed.

    f. Operate steam generator at or near specified test conditions and note abnormalities, if any.

  II. Four hours prior to each test, all wall blowers and soot blowers will be operated to keep the heat surfaces clean.

  III. The performance test will commence for 2 h of operation.

  IV. A trial test must be conducted to check the working of all instruments and to train the test personnel for taking readings.

  V. All readings pertaining to the actual test shall be taken to make the test personnel conversant with the test.

  VI. Problems encountered during recording readings and collecting samples shall be attended.

  VII. The frequency of recording readings should be more frequent than at 5-min intervals.

  VIII. When recording readings automatically through a data logger, the readings will be more frequent than those in step VII.

  IX. The average value of each measured parameter is calculated, and is used in the final computation.

  X. The commencement of the test, the steam generator shall be operated under the following conditions:

    a. All blow downs have to be kept closed.

    b. No oil support should be provided.

    c. Soot blowing shall be avoided.

  XI. The test report shall consist of the following:

    a. All recorded test data, duly countersigned by the customer, the engineer, the ESP manufacturer, and the supplier (contractor)

    b. Calculation and analysis of test data

    c. Computation of performance test parameters
    d. Final findings, analysis of the findings, and recommendations based on final findings, if any

A.6.6 Evaluation of ESP Performance

In order to find out the collection efficiency of an ESP, the steps described next are generally followed.

i. *Calculation of density of dry flue gas*
From average measured values of $\%O_2$ and $\%CO_2$ content in dry flue gas by volume, density of dry flue gas at 101.33 kPa and 273 K is calculated as

$$\rho_{DFG} = \frac{1}{22.4}\left[\frac{32 \times O_2 + 44 \times CO_2 + \{100 - (O_2 + CO_2)\}28}{100}\right], kg\,Nm^{-3} \qquad (A.18)$$

ii. *Calculation of density of wet flue gas*
Knowing the measured values of dry bulb temperature and wet bulb temperature (Table A.22) of flue gas, moisture content in total air supplied (M_{DA}, kg kg^{-1}) can be found out from the psychometric chart.
From as-fired coal analysis (Table A.23) and moisture content in total air, we can find out the total moisture content in flue gas (M_{FG}):

$$M_{FG} = \frac{8.937 \times H + M}{100} + \frac{TA}{Wc} \times M_{DA}, kg\,kg^{-1} \qquad (A.19)$$

where H: hydrogen content in coal, %;
M: moisture content in coal, %;
TA: total airflow, kg s^{-1} (Sl. No. 9, Table A.22);
Wc: coal flow, kg s^{-1} (Sl. No. 8, Table A.22).
Hence, in line with the guideline of equation 5-12-18, Cl. No. 5-12.13, [1], the density of wet flue gas at 101.33 kPa and 273 K is

$$\rho_{WFG} = \rho_{DFG} + \frac{M_{FG}}{22.4}, kg\,Nm^{-3} \qquad (A.20)$$

iii. *Calculation of sampling flue gas volume*
The average velocity of sampling (v_{avg}) is

$$v_{avg} = K_p \times K_m \sqrt{2g \times \frac{\rho_1}{\rho_{WFG}} \times H \times \frac{101.3}{BP - P_S} \times \frac{T}{273}} \qquad (A.21)$$

where K_p: pitot constant;

K_m: manometer constant;

g: gravitational constant (9.81 m s^{-2});

ρ_l: density of manometer fluid;

ρ_{WFG}: density of flue gas from Eq. (A.20);

H: dynamic head in manometer, mm;

BP: barometric pressure, kPa;

P_S: static pressure of flue gas, kPa;

T: temperature of flue gas, K.

So the volume of flue gas sample is calculated as

$$Q_{FG} = v_{avg} \times A, \text{m}^3\,\text{s}^{-1} \tag{A.22}$$

where A: cross-sectional area of sampling nozzle, m^2.

iv. *Collected sample of fly ash*

For collection of fly ash samples, a filtering cartridge termed "thimble" is used inside the sampling probe. The number of thimbles required for conducting a test varies from project to project and shall conform to the recommendation of the equipment manufacturer. Prior to inserting a thimble it must be made free of any moisture. Hence, all the thimbles are put together in an oven and heated to a temperature of 378–473 K, as recommended. When all the moisture is driven out, take out the thimbles and put them in a desiccator for cooling.

Measure the weight of each dry thimble prior to putting it inside the probe in a precision balance. Insert the probe in sampling locations and cover all traverse points. Once the traverse is complete in all the points across the duct cross section, take out the thimbles and measure their weight again in the precision balance.

Dust concentration in flue gas is calculated as

Weight of all dry thimbles before traverse: W_1

Weight of all thimbles after traverse: W_2

The difference between W_2 and W_1 represents the quantity of fly ash content in flue gas flowing through the duct. This quantity of fly ash divided by flue gas volume yields the dust concentration in flue gas. The process of finding out the dust concentration in flue gas is repeated in both inlet and outlet ducts; the collection efficiency of ESP is then calculated following Eq. (A.14).

A.6.7 Correction Factors

Factors that influence the efficiency of ESPs are given next. During normal running of ESP, if any of these factors increases considerably from its design values, the efficiency of ESP will deteriorate. Hence, the measured value of efficiency must be divided by the corrective values obtained from the following figures to arrive at the corrected value of efficiency. This corrected value must not be less than the guaranteed efficiency; otherwise, this will lead to probable violation of local SPM emission regulation.

i) Volume of gas flow (Fig. A.22)
ii) Dust resistivity (Fig. A.23)
iii) Inlet dust concentration (Fig. A.24)

Notwithstanding the aforementioned factors, which directly influence the performance of an ESP, there are other factors that are indirectly responsible for degrading the performance of an ESP as discussed in following paragraphs. If ESP manufacturers intend to apply any correction factor for deviation in any or all of these conditions, then applicable correction curves must be supplied by the concerned manufacturer prior to conducting the PG/PA test of ESP.

i) Velocity of gas flow: If the actual velocity exceeds the design velocity, there would be apprehension of reentrainment of suspended particles, resulting in a rise in ESP outlet dust concentration.
ii) Temperature of flue gas: When this temperature reaches the dew-point temperature, moisture will form and cause dust particles to solidify into cakes, with consequent impairment of dust removal. If the flue-gas temperature exceeds a high limit, that will result in reduction in collection efficiency.

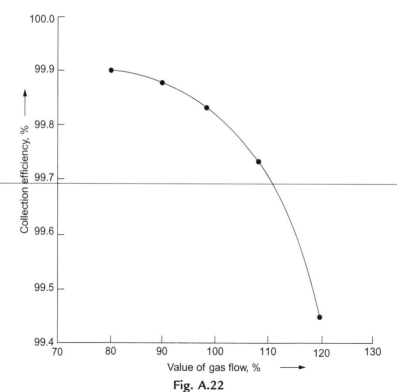

Fig. A.22

Collection efficiency versus volume of gas flow. *Source: Figure, 14.9, P 499 [7]*

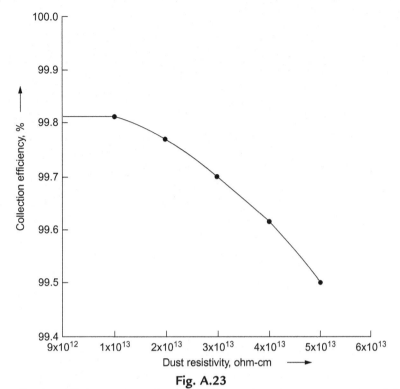

Fig. A.23

Collection efficiency versus dust resistivity. *Source: Figure, 14.10, P 501 [7]*

iii) Gas distribution: Flue gas distribution must be uniform from the inlet up to its outlet. In sections where velocity is high, ash may show up in the stack as a visible plume. At low velocity sections, overall collection efficiency may reduce.

iv) Weight of dust per unit volume of gas: In the event that this value exceeds design value, ESP inlet dust concentration would increase, causing a fall in collection efficiency.

v) Disposal of collected ash: If ash is not removed regularly from hoppers, it can build up to a point where it will bridge high-voltage frame, causing a short circuit and loss of an electrical section, with consequent failure of discharge electrodes (wires).

vi) Sulfur content in as-fired coal: If the sulfur content in coal is low, that will lead to high resistivity of dust (fly ash).

A.7 PG/PA Test of Cooling Tower

Looking at Section 14.1, it is gathered that in a recirculating type condenser cooling water system, circulating water (CW) from a CW sump is supplied to the condenser with the help of CW pumps. Hot water coming through condenser outlets is carried to the hot water section of the cooling tower, where the downward flow of water is cooled by the upward flow of air,

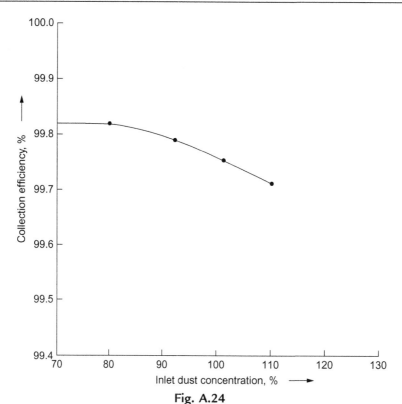

Fig. A.24

Collection efficiency versus inlet dust concentration. *Source: Figure, 14.12, P 503 [7]*

thereby rejecting heat to the atmosphere. Cooled water collected at the cold-water basin of the cooling tower then travels to the CW sump. Because the cooling water travels through a closed cycle, only the makeup water is replenished from the clarified water reservoir to the CW sump.

There are two different types of cooling towers—an induced draft cooling tower (IDCT) (Fig. A.25) and a natural draft cooling tower (NDCT) (Fig. A.26)—that are generally used in large thermal power plants.

Test codes which are used for PG/PA tests of cooling towers are CTI Code ATC 105/ASME PTC 23/BS 4485-PART 2 (Sl. No. 15, Table 17.7). It is recommended that performance tests of cooling towers are conducted within 12 months of structural completion of towers. The following parameters related to the water cooling capacity of cooling towers are usually guaranteed for evaluating tower performance:

i.   Range of cooling (cooling tower inlet hot water temperature—cold water temperature at cooling tower basin), K
ii. Power consumption by fan motors, kW (applicable to IDCT)

The aforementioned parameters are guaranteed at the following operating conditions:

  i.   specified maximum CW flow rate
 ii.   specified maximum ambient air temperature
iii.   specified inlet wet bulb temperature

A.7.1 Sequence of Activities

Prior to conducting PG/PA tests, it is recommended to adopt the following guidelines:

  I.   Verify that the protocol for the Erection Completion Checklist of mechanical and electrical equipment is available.
 II.   Verify also that calibration certificates of all test instruments are available.
III.   Inspect the cooling tower as follows:
    1.   Ensure that fills are free of foreign materials (eg, oil, tar, scale, algae, etc.) that may impede normal air and water flow, or alter heat transfer characteristics (Cl. No. 3-4 (d)) [8].

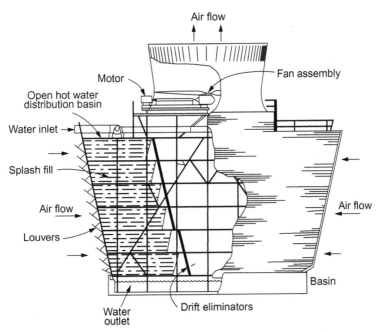

Fig. A.25
An induced draft cooling tower.

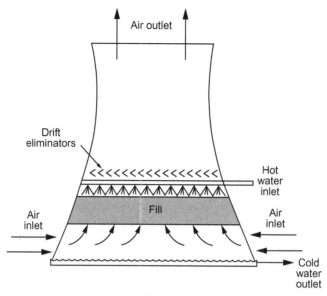

Fig. A.26
A natural draft cooling tower.

2. Ensure that the water distribution system is essentially clean and free of foreign material, without indication of any obstruction that may impair normal water flow (Cl. No. 3-4 (a)) [8].

3. Ensure that the water is distributed to all operating cells and/or parts of the tower as recommended by the manufacturer (Cl. No. 3-4 (a)) [8].

4. Ensure that drift eliminators are essentially free of foreign material such as oil, tar, scale, algae, and other deposits, the presence of which may impair normal airflow (Cl. No. 3-4 (c)) [8].

5. Ensure that normal water level in cold water basin is maintained with the least variation in level.

6. Ensure that fans are adjusted for proper rotation in the correct direction, at the correct speed, with uniform angle such that motor loading is within $\pm 10\%$ of the design motor power at design thermal conditions (Cl. No. 3-4 (b)) [8].

IV. Close blowdown from and makeup water supply to cooling tower basin.

V. Conduct the test at 90%, 100%, and 110% of the design CW flow, each of about 1-h operation for IDCT (Cl. No. 3-5.1.1 [8]) and about 6-h operation for NDCT (Cl. No. 3-5.1.2 [8]).

VI. A trial test may be conducted to check the working of all instruments and to train the test personnel for taking readings, making necessary adjustments, and establishing steady-state conditions. All readings pertaining to the actual test shall be taken to make the test personnel conversant with the test.

VII. Calculate the average value of each measured parameter, which will be used in the final computation.

VIII. During the test, the customer must ensure the required load on steam turbine and operating conditions, as close to design as possible.

IX. The test report shall consist of the following:

    a. All recorded test data, duly countersigned by the customer, the engineer, and the cooling tower manufacturer

    b. Calculation and analysis of test data

    c. Computation of performance test parameters

A.7.2 Operating Conditions

The test may be conducted with the following variations in operating parameters (Cl. No. 3-8 [8]):

1. Barometric pressure: ± 3.5 kPa
2. Fan power: $\pm 10\%$
3. Average wind velocity: within 4.5 m s^{-1}, with transient rise in velocity up to about 7.0 m s^{-1} for about a minute
4. Variation in wet bulb temperature: ± 8 K of design value
5. Variation in dry bulb temperature: ± 14 K of design value
6. Heat load: $\pm 20\%$
7. Cooling range: $\pm 20\%$ of design value
8. CW flow: $\pm 10\%$ of design value, with reasonably uniform water distribution in all operating cells
9. Quality of cooling water conforms to the following:

    i. Total dissolved solids are within 1.1 times the design concentration or 5000 mg kg^{-1}, whichever is higher

    ii. Concentration of oil, tar, or fatty material is within 10 mg kg^{-1}

A.7.3 Deviation in Operating Conditions

In order to achieve accurate PG/PA test results, it is necessary that all parameters remain reasonably steady. Permissible deviation in any operating condition during the test complies with the following (Cl. No. 3-6 [8]):

    i. $\pm 2\%$ variation in CW flow per hour

    ii. $\pm 5\%$ variation in heat load per hour

    iii. $\pm 5\%$ variation in range per hour

    iv. ± 1 K variation in average wet bulb temperature per hour

    v. ± 3 K variation in average dry bulb temperature per hour

Table A.25 Measured parameters

Sl. No.	Parameter	Unit	Type of Instrument	Time
1	Barometric pressure	kPa	Aneroid barometer	
2	Wind velocity at an open and unobstructed location within 30 m of and to windward of the tower; avoid direct sunlight	$m\ s^{-1}$	Rotating cup anemometer	
3	CW flow	$m^3\ s^{-1}$	Pitot tube/ ultrasonic flow meter	
4	Hot CW temperature at the inlet to cooling tower	K	Thermometer	
5	Cold CW temperature at cooling tower basin	K	Thermometer	
6	Inlet air dry bulb temperature at a location the same as below	K	Psychrometer	
7	Inlet air wet bulb temperature at a location 1500 mm above basin curb elevation, within 1200 mm of air intakes and at three equally spaced stations along each air intake side of the tower	K	Psychrometer	
8	Power consumed by fan motors (applicable to IDCT)	kW	Voltmeter, Ammeter, Power factor meter	
9	Total dissolved solids in cooling water	$mg\ kg^{-1}$		
10	Concentration of oil, tar, or fatty material in cooling water	$mg\ kg^{-1}$		

A.7.4 Frequency of Taking Readings

The frequency of taking readings generally conforms to the following (Cl. No. 3-7 [8]):

Wet bulb temperature ≤ 5 min
Cold water temperature ≤ 5 min
Hot water temperature ≤ 5 min
Wind velocity continuous
CW flow ≤ 15 min

Power consumption of the fan motor may be taken once during the test.

A.7.5 Measured Parameters

Parameters to be measured during the test are narrated in Table A.25.

A.7.6 Evaluation of Test Results

Test results are calculated as

i. Adjusted circulating water flow $=$ Measured circulating waterflow

$$\times \left(\frac{\text{Design fan power}}{\text{Measured fan power}} \right)^{\frac{1}{3}}$$

(Cl. No. 5-10.2.2 [8])

ii. Tower capability $= \dfrac{\text{Adjusted circulating water flow}}{\text{Predicted circulating water flow}} \times 100, \%$

(Cl. No. 5-10.3 [8])

References

[1] ASME PTC 4-2008: Fired Steam Generators—Performance Test Codes. American Society of Mechanical Engineers. An American National Standard. Date of Issuance: January 9, 2009.
[2] ASME PTC 6-2004: Steam Turbines—Performance Test Codes. American Society of Mechanical Engineers. An American National Standard. February 2006.
[3] ANSI/ASME PTC 6 Report 1985 (R1997)—Guidance for Evaluation of Measurement Uncertainty in Performance Tests of Steam Turbines. American Society of Mechanical Engineers. An American National Standard. 1997.
[4] IEC 953-2{IS 14198 (Part 2)}. Rules for Steam Turbine Thermal Acceptance Tests: Part 2 Method B—Wide Range of Accuracy for Various Types and Sizes of Turbines. International Electrotechnical Commission. Publication Date Dec. 1, 1990 IS 14198 (Part 2). 1994
[5] m-01. Method 1—Sample and Velocity Traverses for Stationary Sources. Environmental Protection Agency.

[6] m-02. Method 2—Determination of Stack Gas Velocity and Volumetric Flow Rate (Type S Pitot Tube). Environmental Protection Agency.

[7] D.K. Sarkar, Thermal Power Plant—Design and Operation, Elsevier, Amsterdam, Netherlands, 2015.

[8] ASME PTC 23-2003: Atmospheric Water Cooling Equipment—Performance Test Codes. American Society of Mechanical Engineers. An American National Standard. Date of Issuance: November 10, 2003.

General Safety Guidelines

B.1 Introduction

Safety is a fundamental necessity for operating any plant. It is ensured by resorting to safety education, adopting safe working conditions, utilizing safety equipment, abating environmental pollution, and most important, preventing an accident. Safety measures should comply with statutory obligations and also fulfill management and personnel aspects. Safety must be achieved from intrinsically safe design and from the trained conduct of an individual. Awareness of safety must come out of a sense of priority for safety from within oneself. One should not accept statements that the equipment one would work on or around is safe. One must assure oneself that it is safe and consider all apparatus to be in service or under pressure at all times unless properly blocked with DANGER tags placed on it. One with a correct attitude will always return home without suffering any bodily injury or causing any damage to any equipment. Plant authorities must enforce that all plant personnel cooperate in observing and exercising safety measures.

The objective of any thermal power plant is to develop and maintain an efficient, coordinated, and economical system of power supply. To meet this objective, operational procedures must ensure the following:

1. Personnel and plant are safeguarded against danger.
2. Plant is available for operation when required.
3. Plant is operated as efficiently as possible.

To ensure safe, reliable, and efficient operation of the plant, an adequate staff of properly qualified and trained personnel is required on duty at all times. These personnel should be fully knowledgeable of their responsibilities as well as of the effects of their actions upon other personnel with whom they relate in the operation/maintenance of the plant. They should also be aware of approved operating and administrative procedures. These procedures control all personnel, plant operation, and maintenance for ensuring coordinated and integrated operation of the plant. Adequate precautions must be taken before, during, and after any action is taken, to protect personnel who are not directly involved in the action, or who would not be directly informed, and thereby ensure their safety and the safety of other equipment and systems.

In order to ensure adequate communications between the plant and its operators, generally all equipment and systems are equipped with local and remote instrumentation, alarms, automatic controls, and a central data acquisition system in the central control room (CCR). With the help of local and remote instrumentation, the plant and equipment performance trends are carefully monitored. These operating devices and aids facilitate the anticipation of operational changes due to equipment deterioration. These devices, together with the plant communication system, inform and alert both CCR and roving operators regarding the healthiness of the plant condition. Continuous surveillance by both CCR operators and roving operators provides early warnings of problems, and will allow timely corrective actions to prevent unsafe conditions for personnel and equipment.

Notwithstanding the aforementioned benefits of instrumentation and control, they are unable to identify certain abnormal or unusual conditions, as delineated next, for which it is imperative to resort to visual inspection of the plant by trained roving operators. This will complement the control and instrumentation systems, as well as verify the information available in CCR.

A visual inspection is made for any intentional damage or any damage caused by plant operations during transient or emergencies, indicated by such things as:

a. Cracks in concrete foundations, deformed steel structures, and supports
b. Displaced equipment
c. Loose anchor fixings, fasteners, cabling, and tubing
d. Broken valve, gauge glass, and wires
e. Gnawing marks on insulation of cables and piping
f. Nests of animals, birds, and insects

Prior to taking any plant in service, it is essential that the plant is commissioned properly to achieve safe, trouble-free, efficient operation. Commissioning is a process that includes the activities necessary for the transition from constructional phase to operational phase of a plant. Hence it is necessary to ensure that all equipment is completely erected, and that the sequence of commissioning equipment follows the sequence of construction activities. During commissioning each equipment/system is kept under surveillance, both locally as well as from CCR, to establish that none of them goes beyond safe operating conditions.

While a plant is in operation or under maintenance it is necessary to observe effective safety measures. Maintenance of equipment, during plant operation or when the plant is under shutdown, requires careful planning and prepared procedures in achieving safe execution of the work with minimum interruption. Such practices will also provide efficient service to achieve high plant availability at the cheapest cost, keeping plant downtime at a minimum. Before undertaking maintenance, it is to be ensured that all administrative controls are properly authorized, issued, and understood by all concerned, so that operation of the plant can continue efficiently and all maintenance activities can be completed with assured safety to personnel and plant. During maintenance of any plant or equipment, rigid, foolproof safety measures are essential to prevent wrong operation of the concerned electrical switchgear, steam valve, air valve, and so on, in order to safeguard the people working on the equipment as well as the equipment itself.

Prior to taking any equipment or system under maintenance the operation department will shut down the system or the particular piece of equipment at a scheduled time, as required and agreed upon. They will isolate the equipment concerned by shutting down the valves, opening and drawing out the electrical switchgears, releasing the pressure trapped anywhere, and so on to make it safe for working. They are to take all possible precautions in order to safeguard the crew working on the equipment under shutdown to prevent any operation by mistake. Caution tags bearing clear instruction shall be put on each and every relevant valve, switch, switchgear, and so on.

The personnel of the operation department entrusted for shutdown of the equipment will hand over the plant for maintenance by issuing a "permit to work". On completion of the work, the person holding the permit shall surrender the same to the operation department, who will remove the tags and restore power supply, wherever applicable.

Prior to starting any rotating equipment, following typical prestart checks by the operation department would help in maintaining the safety of the equipment:

1. Make a general inspection of the equipment and ensure that there is nothing around it which may restrict free movement or effective operation.
2. Get clearance from the maintenance or erection group as to the proper alignment of the equipment and soundness of the foundation bolts.
3. Remove all tools, tackles, consumables, and other miscellaneous material lying near the equipment.
4. Check oil level or grease of each bearing.
5. Check normal cooling water flow.
6. Make sure that all valves are in correct starting position, and vents and drains are lined up.
7. In the case of a pump, check that it is primed and vented.
8. Be sure to know the design limits of pressure, temperature, flow, level of vibration, current, and voltage of the driving motor to be able to note any abnormal deviation from operating norms.
9. Check availability of power supply to the driving motor.
10. After completing all checks, the local operator should inform the CCR operator of the readiness of equipment to start.

B.2 General Safety Requirements

For effective implementation of safety rules it would be prudent to form a safety committee, whose primary function will be to enforce among all concerned that "All safety rules must be obeyed. Failure to do so will result in strict disciplinary action." In this respect the committee may take cognizance of Occupational Safety & Health Administration (OSHA) Safety Standards to formulate the specific plant safety guidelines so as to ensure that:

   i.   Safety procedures and rules are developed and implemented.

   ii.   Routine surveillance in workplace is in force.

  iii.   Safety trainings are conducted at regular intervals.

  iv.   The work environment is safe.

   v.   All concerned are provided with personal protective equipment (PPE).

  vi.   Proper PPE necessary for executing a specific job are in use.

 vii.   Basic training in first aid skills is provided.

viii.   Fire extinguishers are provided in fire-prone areas.

  ix.   Fire drills are conducted at regular intervals.

   x.   Fire lanes, escape lanes, escape stairways, and ladders are in place.

In continuation of the aforementioned guidelines, some general safety guidelines are narrated here:

1.   Before operating any equipment or system, all concerned personnel are to be informed of the actions being undertaken and the procedures being followed, so that no harm is caused to personnel or equipment. A check is made that all administrative controls are satisfactory for the operation and no out-of-service maintenance or testing work is in progress on the required equipment or any associated services or systems.

2.   Safety tags (Appendix C) are to be used to maintain the system integrity of a piping system, an electrical bus, or a piece of equipment in order to establish physical isolation of the system to prevent personnel from injury and/or equipment from damage.

3.   Proper ventilation, with no indication of fumes, gases, or high humidity, and temperature shall be ensured at the workplace to provide reasonable conditions for comfort to workers. Any sign of fumes, smoke, or fire from unusual places (eg, equipment, pipelines, cables, etc.) shall be reported immediately.

4.   Sufficient and suitable lighting shall be ensured at work areas and passages.

5.   Every rotating part of machinery shall be fenced/guarded securely.

6.   Ensure proper coupling guards for the rotating equipment. They should be strong enough to provide complete protection yet allow easy access to the coupling for inspection.

7.   Avoid standing on moving or operating equipment.

8.   Use proper valve wrenches when opening/closing valves.

9.   All platforms, staircases, and ladders shall be properly maintained without obstruction.

10.   All trenches/drains shall be properly covered.

11.   The welding machine, its cables (without joints), and power supply points shall be in safe condition and shall be properly earthed.

12.   Lighting cables or any other electrical temporary cable shall be in healthy condition without joints and shall be properly safeguarded against damage.

13.   Electrical junction boxes/panels and power sockets shall be covered properly.

14.   Use proper gas masks when entering a gas-laden area, chlorine cylinder locations, and so on.

15.   Report any leakage of water, steam, oil, gas, and so on immediately.

16.   Do not smoke in the vicinity of an acid tank or use equipment that can produce a spark.

17.   Use hand gloves, face mask, and protective clothing when working with acids or caustics.

18. If contact is made with acids or caustics, immediately flush with fresh water.
19. When free-blowing/flushing steam, air, or water lines, be sure that all personnel are quite away and no one else may be affected by the jet/blow-out material.
20. Keep all apparatus thoroughly cleaned. Do not allow accumulation of coal dust/dirt/combustible matter on or in equipment, controls, switchgears, pipelines, and so on, especially in a high-temperature region.
21. Where combustible dust accumulations are encountered, care must be taken to provide adequate ventilation, and fire or open flame shall be avoided.
22. While using compressed air for cleaning, observe that the air is clean and dry. Carefully regulate airflow. Carefully clean around bearings. Do not direct air blast toward automatic equipment or controls.
23. Do not undertake welding works by standing in water/damp condition.
24. Safety in usage of lifting tools, tackles, and wire ropes and their healthiness shall be ensured during material handling.
25. A proper working platform/floor shall be ensured at the basement. A dewatering pump shall also be available for safe working condition.
26. While carrying out gas/arc cutting work, suitable protection shall be ensured such that molten hot metal does not fall on any person or on flammable material.
27. Gas/arc cutting work shall not be carried out in the vicinity of hydrogen/any explosive gas/material storage area. Acetylene gas cylinders shall always be kept in the upright position and they shall be strapped or chained.
28. While the cutting is in progress, use shields when in the vicinity of an electric arc.
29. Keep the floor clean and clear of electrode pieces, scraps of metals, and tools.
30. The welding of pipes shall be carried out following welding codes by qualified/certified welders.
31. A quick communication system shall be established between fire-prone areas to the firefighting station.
32. Protective coatings at chemical hazard areas shall be applied regularly.
33. Avoid breathing ammonia vapors, which may cause damage to lung tissue.
34. Discharge from fuel gas atmospheric vents shall be located in such a way that there would not be any possibility of the discharge gas presenting a fire hazard.
35. A statutory warning with a danger sign has to be displayed around the fuel oil storage, unloading, and handling area.
36. Smoking and carrying naked light around the fuel oil storage, unloading, and handling area is strictly forbidden.
37. Wear a hard hat while working within the plant area.
38. Entry of unauthorized persons shall be restricted.
39. Day-to-day plant and equipment cleaning is another important aspect. A clean and orderly plant is an inducement to operators for disciplined and orderly performance and will help to achieve a smooth and efficient running station.
40. Proper lubrication of running equipment needs special attention.

B.3 A Few Operational Norms

A few norms for good operating practice are discussed next to ensure a safe work environment.

B.3.1 Plant Leakages

Any kind of leak, whether in pipe line, equipment, or valve, is a costly waste. Constant vigilance should be kept to detect a leak as soon as it develops and repair the same at the first opportunity. This will not only save money, but will also increase availability.

B.3.2 Abnormal Noises

Any change in the prevailing noise pattern around any piece of equipment is the first warning of a fault developing in a machine. Operators should be trained to detect such changes. Timely detection and prompt investigation may save a costly breakdown.

B.3.3 Water Conditioning and Quality Control

Constant vigilance on the water quality and maintenance of chemical conditioning within the prescribed limits is a vital necessity to maintain that steam, condensate, and feed systems are free from corrosion.

B.3.4 Load Changes

Load changes, as far as practicable, should be gradual and according to the procedure laid down. Nameplate ratings should not be exceeded even if there is apparent margin as indicated by temperature rise figures, especially for the generator.

B.3.5 Vibration Limits

The limiting condition of vibration of any rotating machine is classified according to its operating speed. For machines operating around 3000 rpm, a 12 μm peak-to-peak vibration would indicate a very satisfactory performance. A safe operating limit would, however, be 12–50 μm. But this limit can be exceeded to some extent for transient vibration. If the steady-state vibration exceeds 50 μm, it would call for inspection and rectification at the first opportunity.

For equipment operating around 1400 rpm or below, the permissible vibration limit may be as high as 120 μm.

For reciprocating and low-speed machines, which are subjected to vibration due to the nature of their operation, no limiting value can be prescribed. These machines should be checked for abnormal noise and rubbing while in operation.

B.4 Use of Fire Extinguisher

Outbreak of fire in an area, where it is not required, will lead to destructive burning of building, equipment, and even personnel trapped in that area. Even if a plant is adequately provided with automatic fire-extinguishing equipment and systems, immediate arrest of the fire from further spreading out could become effective with timely use of portable fire extinguishers. All plant personnel must be trained to operate fire extinguishers. Besides, there must be routine fire drills to keep the trained personnel conversant with the operation of fire extinguishers as well as to train them in the manner of exit to be followed in case of fire outbreak. In the event that fire becomes uncontrollable even after using fire extinguishers, professional firefighters must be informed to fight fires with the help of motor-operated pumps to shoot water or a chemical solution at high pressure.

Table B.1 shows some of the typical fire extinguishers used in the industry.

Table B.1 Types of fire extinguishers

Type of Material	Material Used in Extinguisher	Extinguishing Effect
Flammable liquids	Chemical foam Dry chemical powder Carbon dioxide	Blanketing of fire
Flammable gases	Dry chemical powder Carbon dioxide INERGEN FM 200	Blanketing of fire
Combustible metals Organic materials Electrical installation	Dry chemical powder Soda acid (water) INERGEN FM 200	Blanketing of fire Striking and cooling of fire Blanketing of fire

Note

In accordance with NFPA 2001: Clean Agent Fire Extinguishing Systems, the fire hazard of electrical installations (motors, switchgears, control circuits, control panels, computers, electronic equipment, cable vault, etc.) are to be protected with INERGEN or FM 200 instead of carbon dioxide or HALON 1301/1201.

1. INERGEN is a mixture of 52% N_2, 40% A, and 8% CO_2.
2. FM 200 is Heptafluoropropane (C_3HF_7: CF_3—CHF—CF_3).
3. Both INERGEN and FM 200 are fast-acting, clean, harmless, nonconductive, and ozone friendly.
4. Montreal Protocol 1987 prohibits the use of HALON as a fire extinguisher due to its effect on the depletion of the ozone layer.
5. Chemical foam and soda acid (water) are strictly prohibited from being applied on electrical installations as a fire extinguisher.
6. In order to douse fire in pulverizers, never use a CO_2 fire extinguisher. Instead, use steam or nitrogen. CO_2 may lead to the formation of dry ice that eventually would crack grinding elements.

B.5 First Aid and Emergency Treatment

In the event that any person gets injured or falls sick, before professional medical treatment can be arranged, temporary relief is provided to the person in the form of first aid that comprises simple medical techniques using minimal equipment. This timely support is most critical to the victim and may at times result in life saving. The order of treatment varies greatly and is determined by physical condition and nature of the injury. With proper training any layperson can dispense first aid.

When a need arises for first aid, one should proceed as such:

  i. Keep calm, never panic, and avoid excitement.
 ii. If necessary, provide artificial respiration, control bleeding, apply tapes as temporary plastering for a broken limb, and so on without delay.
iii. A badly injured person must never be moved unless it is essential to provide fresh air to the person or to protect him or her from further injury.
 iv. Carefully examine the victim.
  v. Seek medical attention immediately.
 vi. Send for an ambulance if the victim is unable to move on his or her own.
vii. Notify the police, fire department, and concerned authorities.

B.6 Some DOs and DON'Ts

Next, we describe some of the DOs and DON'Ts applicable to steam power plants, which if followed could ensure a safe working environment for the plant, machinery, and personnel. Many of the following conditions are applicable to gas turbine power plants and diesel generator plants as well, and hence are not discussed separately.

B.6.1 DOs

B.6.1.1 General

1. Always employ trained personnel who are familiar with the equipment/system for inspection, cleaning, and repair of any equipment.
2. Maintain adequate communication with other responsible personnel to allow coordinated and integrated actions to be taken.
3. Be clear with the system and equipment.
4. Always bear in mind safety and availability.
5. At the beginning of every shift, go round the equipment; keenly watch for malfunctioning, defects, and failures.
6. Set right all defects immediately.
7. Tighten loose bolts, nuts, and cap screws. Replace missing components.

8. All steam leaks should be attended immediately.
9. Disconnect electric supply, remove spring or mechanical loading, release the pressure, and isolate from other system equipment prior to attempting service or dismantling any equipment.
10. Always gag a safety valve before making ring adjustments.
11. Ensure that electric/pneumatic supplies are available for all controlling devices, instruments, and alarms.
12. Check that all alarms and annunciations are in working condition.
13. Check the operation of open-close limit switches.
14. Check for abnormal increase of motor currents.
15. Check all safety interlock systems prior to start-up. A functional test of the interlocks should be carried out at regular intervals.
16. Be alert at all times for possible malfunction of automatic control equipment.
17. Have constant watch over the proper operating parameters, namely pressure, differential pressure, temperature, level, flow, and so on, of various flowing mediums and set right deviations in a timely manner.
18. Carefully monitor plant and equipment performance trends.
19. Check that positions of all valves are as per the valve operating schedule recommended by equipment manufacturers.
20. Check for vibration, bearing temperature, and abnormal noise of rotating equipment.
21. Check for proper lubrication of bearings. Lubricate with proper grade of lubricants.
22. Check the temperature rise of lubricants in the bearings.
23. Check flow of cooling mediums.

B.6.1.2 Steam generator

1. Move dampers from their set position at least once a week so as to ensure that damper spindles do not become seized in their bearings.
2. The correct working of drum level gauges should be tested once every 8 h or more often if considered necessary.
3. The drum level gauges should be well illuminated under all conditions.
4. Drum level high and low alarms should be tested at least once a week.
5. Should the water level in the drum drop out of sight in the gauge glass, all firings must be stopped immediately.
6. Be alert to detect the symptoms of water carryover from the steam generator.
7. Operate all igniters once every day to keep them in good working condition.
8. The flue gas O_2 analyzer, which is tied up with total airflow control, should be checked periodically to assure continuous and dependable operation.
9. Maintain airflow through pulverizers whenever the unit is being fired.
10. Check for mill rejects periodically.

11. If fire in the coal bunker or in the pulverizer is noticed, immediately take action for putting out the fire.
12. Operate soot blowers once every shift during starting.
13. Blow compressed air once every day to clear off coal dust over piping, cable trays, equipment, and so on.
14. Keep environments clean. Fuel leaks and accumulation can lead to fire hazard.

Note

Items 2 through 5 are applicable to drum-type steam generators.

B.6.1.3 Steam turbine

1. Check the level of oil in the main oil tank.
2. Check bearing and pedestals for cleanliness.
3. Check the differential pressure at the lube oil and control fluid filters.
4. Check the oil pressure in the bearings.
5. Check the vibration of the bearings. It should be within the prescribed limits. In case of any rising tendency, watch and trip the set if the readings exceed the permissible values as recommended by the turbine manufacturer. When raising the speed of the turbine from 0 to 3000 rpm, hold the turbine at the same speed where there is a change in the level of vibration. If necessary, reduce the speed and restart only after the vibration is stabilized.
6. Raising speed should be carried out at a uniform rate to minimize shaft vibration. However, the critical speeds of the turbine should be passed rapidly at a steady rate.
7. Check the differential expansion and if the values exceed permissible limits as recommended by the turbine manufacturer, trip the set.
8. Check the control fluid pressure and the suction pressure of the main oil pump.
9. Check auto operation of auxiliary oil pumps (AOPs) and emergency oil pump (EOP).
10. In case of total power failure observe the following:
    i. The turbine has tripped.
    ii. Check the closing of ESVs, IVs, and HP and IP control valves.
    iii. Check that the DC emergency oil pump has cut in and oil to bearings is available.
    iv. Check that the DC seal oil pump has cut in.
11. In case of sudden heavy load throw-off due to grid disturbances, and if there is no possibility of immediate loading of the unit, carry out the following:
    i. Trip the turbine by hand.
    ii. Check that the AOP has cut in. If the AOP fails to start automatically, start it manually. If the AOP does not start, the emergency oil pump should start automatically. If that also fails to start automatically, start it manually.
12. The main oil tank filters should be cleaned, as required.

13. Turbine oil should be analyzed before filling and also at regular intervals. Take oil samples for analysis when the turbine is operating.
14. Before start-up and after shutdown of turbines, check the tightness of closing of ESVs, IVs, and HP and IP control valves.
15. Constantly check the operation of the gland sealing system.
16. Regularly note the readings of all the turbine supervisory (turbovisory) instruments, and take measures for abnormalities. At least once a year, check the correct function of all turbovisory instruments mounted on the turbo set.

B.6.2 DON'Ts

B.6.2.1 General

1. Do not start operation without valid schematic and layout diagrams.
2. Do not take any equipment for maintenance without service manuals and cross-sectional drawings.
3. Do not overload any equipment.
4. Do not mix oils or lubricants of different quality.
5. Do not run any rotating equipment when heavy vibrations are present.
6. Do not run any equipment without checking interlocks, if any.
7. Do not run any rotating equipment with high bearing temperature.
8. Do not run equipment when foreign matter is left inside.
9. Do not run any equipment without ensuring proper lubrication.
10. Do not attempt repair beyond your capabilities.
11. Do not accept statements that the equipment you would work on or around is safe.
12. Do not assume that the instruments are indicating wrong readings; investigate if any abnormal deviation is noticed.
13. Do not isolate the unit protection devices.
14. Do not rely solely on the lamp indications for the operation of pumps, valves, and fans. A physical check also has to be resorted to.
15. Do not overlook any abnormality in any of the parameters.

B.6.2.2 Steam generator

1. Do not run a fan if any defect is noticed during a trial run.
2. Do not increase flow through a fan unless the motor current becomes steady.
3. Do not run two fans at low loads.
4. Do not start an ID fan unless a clear flow path is established.
5. Do not rotate the regenerative air heater in the wrong direction.
6. Do not put in service an air heater unless adequate means of extinguishing air heater fires is available.

7. Do not operate the air heater below the cold end temperature as recommended by the steam generator manufacturer.
8. Do not let any combustibles get deposited in the air heater elements.
9. Do not take out air heaters and draft fans from service until the air heater gas inlet temperature drops below the recommended temperature limit, typically 478 K.
10. Do not start the pulverizer with the discharge valve closed.
11. Do not start the pulverizer without seal air supply.
12. Do not continue operation of pulverizer with high pulverizer outlet temperatures.
13. Do not use steel hammers directly on grinding rolls.
14. Do not start the feeder without the pulverizer running.
15. Do not start the feeder without ensuring proper ignition energy in the furnace.
16. Do not vary the speed of the coal feeder when the feeder is idle.
17. Do not keep open the hot air valve while the feeder is in idle condition.
18. Do not leave the igniter spark rod in advance position.
19. Do not assemble or dissemble spark rod forcibly.
20. Do not clean the contacts with a dirty cloth or oily tools.
21. Do not open the igniter exciter without the facilities to repair it.
22. Do not allow oil soaking or coal dust accumulation on components.
23. Do not operate soot blowers with wet steam.
24. Do not extend the soot blower lance without a blowing medium when the steam generator is in operation.
25. Do not attend any maintenance work when the power and/or steam supply is ON.
26. Do not subject the safety relief valve to any sharp impact while handling.
27. Do not reduce the airflow below 30% MCR airflow until all fires are out and the unit is off the line.
28. Do not increase the firing rate such that the furnace exit gas temperature exceeds the recommended limit, typically 813 K, until the steam flow is established through the reheater.
29. Do not remove warm-up oil guns and igniters from service until two adjacent coal elevations are in service with feeder ratings on each associated pulverizer greater than 50%.
30. Do not fire coal at separated coal nozzle elevations without the support of warm-up oil.

B.6.2.3 Steam turbine

1. Do not start the turbine under any condition, cold, warm, and hot starts, if the inlet steam temperature before ESVs is not in line with the recommendations given in the start-up procedures supplied by the turbine manufacturer.
2. Do not start the turbine if the differential expansion is not within the recommended limits.
3. Do not allow the inlet oil temperature of the bearings to fall below the recommended limit, typically 308 K.

4. Do not increase the speed of the turbine if the vibrations are increasing.
5. While raising speed, do not hold the turbine at or in the range of critical speeds. These speeds have to be passed rapidly.
6. Do not allow the rate of increase of metal temperature to exceed the prescribed limits during loading for various start-ups.
7. Do not carry out on-load test of both HP and IP control valves simultaneously.
8. Do not operate the turbine if the condenser pressure is more than the pressure limit as recommended by the turbine manufacturer, typically 30 kPa.
9. Do not run the turbine if the bearing-drain oil temperature exceeds 345 K or the temperature limit recommended by the turbine manufacturer.
10. Do not continue to run the turbine under the following conditions:
    i. sudden appearance of excessive vibrations
    ii. sudden appearance of water hammer
    iii. oil ignition
    iv. drop-off oil level in the main oil tank below the lowest permissible value
    v. appearance of metallic noise in the turbine
    vi. excessive axial shift of rotor
    vii. abrupt rise in condenser pressure
11. Do not run the turbine while the generator is motoring.
12. Do not run the set continuously at low load or no load. Do not run the turbine for a long time at no load or low loads immediately after it has run for long duration at higher loads, to avoid quenching of the internal parts.
13. Do not allow the turbine to run if the LP exhaust temperature goes beyond 373 K or the temperature limit recommended by the turbine manufacturer.
14. Do not start the turbine with turbovisory instruments out of order.
15. Do not isolate the oil vapor extraction fans mounted on the oil tank.
16. Do not drain oil saturated with hydrogen vapors into the oil tank.
17. Do not synchronize the machine with the grid without checking the operation of load shedding relay.
18. Do not start the turbine without ensuring that standby oil pumps are healthy to operate.
19. Do not run turbine only on AOP, that is, without the main oil pump in operation.
20. Do not start the turbine without lagging or with wet lagging material on the cylinders to avoid a temperature difference between the inner and outer casing wall that may lead to thermal stresses.
21. Do not work on duplex oil filters for cleaning unless it is assured that the standby filter has been taken in operation.
22. Do not start the turbine with faulty instruments. Get them rectified before starting.

Tagging Procedure

C.1 Introduction

During the operation of a plant or when a plant is put under maintenance or kept under mothballed condition, certain valves/areas/systems purposefully need to be kept isolated. Any attempt to violate such isolation, even inadvertently, may lead to harm, injury, or major disaster of the plant and personnel. In order to obviate such an inadvertent attempt of violation, warning tags of various types are applied on isolated valves/areas/systems, as laid down in a tagging procedure.

A tagging procedure comprises a series of activities, which when followed assure that no personnel get injured and no equipment gets damaged due to inadvertent energizing of a system or equipment. As construction and erection of equipment progresses, a stage is reached when a certain portion of a system or a certain piece of equipment is completed to the extent that it may be energized either electrically or mechanically, even if complete construction of the system or the equipment is awaited. The tagging procedures provide guidelines that prohibit the system or the equipment from operating before it is completely serviced and ready to be operated.

The prevention mechanisms are of various types of physical devices, such as blanking off pipe lines with spades or blind flanges, locking pins, gagging devices, lock-out switches, electrical isolators, locking of access hatchways, or a combination of two or more of these mechanisms. Equipment that may start automatically without prior intimation needs special attention for isolation by providing preventive devices overriding autoenergization. The tagging out of a piece of equipment may include opening/closing of valves or dampers, opening and racking out of circuit breakers, and so on, but is not necessarily restricted to these devices only. It should be understood that the tagging is dependent on the needs of the person requesting the tag.

Tagging procedures apply to all personnel of the customer, the engineer, equipment manufacturers, and suppliers (contractors and subcontractors) working in a plant. Every endeavor is to be made to ensure that good communications are established among all these personnel to see that accurate records of all mechanical and electrical tagging are followed. It is mandatory that all personnel working in the plant cooperate in observing and enforcing tagging procedures. Any violation of tagging procedures must be reported to the concerned plant authority (eg, plant superintendent/manager) entrusted with controlling these procedures.

C.2 Procedure

The tagging procedure described next applies to ensuring the safety of a plant and the equipment of a running plant, and is not applicable to other areas of the plant, such as store inventory control, and so on. The purpose of adopting a tagging procedure is to ensure that there are adequate controls to eliminate risks arising from inadvertent release of energy during cleaning, servicing, repairing, or alteration work of any equipment/system of a plant.

As a prelude to enforcing a tagging procedure, the plant superintendent/manager or his or her representative must identify and authorize a competent person who should be responsible for assessing the risk involved in a job and enforcing necessary steps to ensure safety. This person must possess the following qualities:

(a) He or she is dependable and capable of imparting responsibility through experience and/or proper training.
(b) He or she is familiar with the equipment/system and is aware of its relationship with plant operation.
(c) He or she is able to lock out, tag, and isolate the equipment/system for the purposes of cleaning, servicing, repairing, or alteration.
(d) He or she is capable of supervising the various aspects of safety.

Prior to working on running equipment or a energized system especially for testing and/or for undertaking maintenance, this authorized person must carry out the following steps:

   i. Identify the types of hazards and areas of locations, where a working person may get injured.
  ii. Assess the extent of risk involved in a particular job.
 iii. Shut down the concerned equipment/system.
  iv. Isolate the power supply (eg, electrical, mechanical, hydraulic, pneumatic, gaseous, chemical, thermal, etc.) to the concerned equipment/system by locking out the supply source.
   v. The lock-out switch shall be such that the key can be removed when the switch is in locked condition.
  vi. Release pressure, if any, within the equipment or the associated system.
 vii. Ensure preventing inadvertent switching of power supply to the equipment and/or building up of pressure in the associated system.
viii. Confirm that the isolation of power supply/pressure source is foolproof and that it is safe to work on the equipment/system. Attach applicable Danger Tags (Section C.3).
  ix. Ensure that personnel working on the equipment/system are competent, and that they have proper knowledge of all attributes of the activity.

x. Ensure also that these personnel are fully briefed on safe work procedures and are aware of the types of risks involved in the job.

xi. If more than one person/group is involved in completing the job, then assure that each person/group attaches their own personal Danger Tag while the work is in progress.

xii. Enforce that on completion of their portion of the job, they remove the Danger Tag installed by them.

xiii. Once all Danger Tags, installed by various persons/groups, are removed, the authorized person may unlock the power/pressure supply source and confirm that it is safe to put the equipment/system in service.

xiv. In the event that it is observed that the equipment/system fails to come into service, the authorized person must put an Out of Service tag on the equipment/system, narrating the proper reasons of such status.

xv. He or she must keep a record of the sequence of activities involved in an isolation process and the removal of the isolation.

xvi. Once satisfied, he or she must give clearance to place the equipment/system in service.

Note

An isolating device is one which physically prevents inadvertent transmission or release of energy (eg, valves, dampers, lock-out switches, circuit breakers, etc.).

C.3 Types of Tags

In a running plant, wherever no tag is used, normally it must be construed that those equipment/systems are either in service or are energized. The equipment/system that is idle or not in operation must be provided with the proper safety tags. For the convenience of readers, various types of safety tags generally used in the industry are shown next.

C.3.1 Red Tag

This tag is placed on any equipment which has been checked out and placed in automatic operation. These tags shall have red lettering reading:

WARNING: THIS EQUIPMENT IS AUTOMATICALLY CONTROLLED. IT MAY START AT ANY TIME WITHOUT PRIOR INTIMATION.

These tags serve as a warning to personnel working around equipment tagged as such.

C.3.2 White and Red Tag

This tag is provided on the equipment/system which is lock-out and must not be operated. This tag may be inscribed with:

DANGER. EQUIPMENT LOCK-OUT. DO NOT OPERATE.

This tag indicates that the push-button/control switch, contactor/circuit breaker, or mechanical equipment/system to which it is attached is under a clearance for the purpose of de-energizing the associated circuit or the equipment. Figs. C.1–C.3 represent different type of tags followed in the industry.

Fig. C.1
Danger. Equipment lock-out, do not operate.

Fig. C.2
Danger. Do not operate.

Fig. C.3
Danger. Use lock-out before working on equipment.

C.3.3 Yellow and Black Tag

A yellow tag with a black stripe and the lettering OUT OF SERVICE indicates that the equipment/system to which it is attached is mechanically or electrically defective or unsafe and shall not be operated until properly replaced or repaired.

This tag shall be used only by the personnel authorized by the plant superintendent/ manager or his or her representative. Two different types of such tags are shown in Figs. C.4 and C.5.

Fig. C.4
Caution. Out of service. Do not operate.

Fig. C.5
Out of service. This must not be used or operated.

C.3.4 Green Tag

A green tag with the red lettering ENERGIZED will be placed on electrical equipment (eg, switchgears, MCCs, power distribution panels, transformers, etc.) which is energized. The tag serves as a warning to personnel working on or around the equipment/system tagged as such.

C.3.5 Blue Tag

A blue tag with the white inscription UNDER TEST indicates that the push-button/control switch, contactor/circuit breaker, or mechanical equipment/system to which the tag is attached is under clearance for the purpose of a test or inspection. These items should be considered alive and must not be operated except on the order or by the person for whom the tag is attached.

Conversion Factors

Parameter	SI Units	Metric System of Units	Imperial and US System of Units
Length	1 m	100 cm	3.281 ft
			39.37 in.
		1000 mm	1.094 yd
	0.01 m	1 cm	0.3937 in.
	0.0254 m	2.54 cm	1 in.
	0.3048 m	30.48 cm	12 in.
			1 ft
	0.9144 m	91.44 cm	3 ft
			1 yd
	1.828 m	1.828 m	6 ft
			1 fathom
	1000 m	1 km	0.621 miles
			0.539 nautical miles
	1609 m	1.609 km	1760 yd
			1 mile
	1853 m	1.853 km	1 nautical mile
Area	$1\ m^2$	$10{,}000\ cm^2$	$1550\ in.^2$
			$10.764\ ft^2$
		$10^6\ mm^2$	$1.196\ yd^2$
	$0.0929\ m^2$	$929.03\ cm^2$	$144\ in.^2$
			$1\ ft^2$
	$0.836\ m^2$	$8361\ cm^2$	$9\ ft^2$
			$1\ yd^2$
	$10^{-4}\ m^2$	$1\ cm^2$	$0.155\ in.^2$
	$10^4\ m^2$	1 ha	2.47 acres
	$4049\ m^2$	0.405 ha	1 acre
			$4840\ yd^2$
	$1\ mm^2$	$10^{-2}\ cm^2$	$0.00155\ in^2$
	$645.2\ mm^2$	$6.45\ cm^2$	$1\ in^2$

Continued

Parameter	SI Units	Metric System of Units	Imperial and US System of Units
	10^6 m^2	1 km^2	1,196,836 yd^2
	2.59×10^6 m^2	259 ha	1 mile^2
			640 acres
Volume	1×10^{-3} m^3	1000 mL	0.2200 Imp. gallon
		1000 cc	0.2642 US gallon
	3.785×10^{-3} m^3	3.785 L	0.833 Imp. gallon
			1 US gallon
	4.55×10^{-3} m^3	4.55 L	1 Imp. gallon
			1.2 US gallon
	1 m^3	1000 L	35.31 ft^3
			1.308 yd^3
	10^{-6} m^3	1 cc	0.0610 in^3
	0.02832 m^3	28.32 L	1728 in^3
			1 ft^3
	0.7646 m^3	764.6 L	27 ft^3
			1 yd^3
	0.02685 N m^3	26,850 N cc	1 scf
	1 N m^3	10^6 N cc	37.244 scf
Mass	1 kg	1000 g	2.205 lb
	0.4536 kg	453.6 g	1 lb
			16 oz
			7000 gr
	64.8×10^{-6} kg	64.8 mg	1 gr
	6.350 kg	6.350 kg	14 lb
			1 stone
	1×10^{-3} kg	1 g	0.03528 oz
	0.0283 kg	28.34 g	1 oz
	100 kg	1 quintal	220.5 lb
	1000 kg	1 tonne	2205 lb
			0.984 ton
	907 kg	907 kg	2000 lb
			1 short ton
	1016 kg	1016 kg	2240 lb
			1 ton
			20 cwt (hundredweight)
Specific volume	1 m^3 kg^{-1}	1000 cc/g	16.02 ft^3/lb
	0.06243 m^3 kg^{-1}	62.43 cc/g	1 ft^3/lb

Continued

Parameter	SI Units	Metric System of Units	Imperial and US System of Units
Specific weight	1 kg m^{-3}	0.001 g/cc	0.0624 lb/ft^3
	1000 kg m^{-3}	1 g/cc	6.24 lb/ft^3
	160.26 kg m^{-3}	0.1602 g/cc	1 lb/ft^3
	0.022886 kg m^{-3}	2.2886×10^{-5} g/cc	1 gr/ft^3
Pressure and stress	1 Pa	0.102 mmwg	1.45×10^{-4} psi
		0.102 kg/m^2	
	1 N m^{-2}	7.5×10^{-3} mmHg	
	100 kPa	1 Bar (b)	14.5 psi
		1.02 kg/cm^2	
		750 mmHg	
	101.3 kPa	1013 mb	14.7 psi
		1.033 kg/cm^2	29.92 in.Hg
		760 mmHg	
	98.1 kPa	1 kg/cm^2	14.22 psi
		10 mwg	28.94 in.Hg
		736 mmHg	
	1 kPa	102 kg/m^2	0.145 psi
		102 mmwg	
		7.503 mmHg	0.295 in.Hg
	0.1333 kPa	1 mmHg	0.193 psi
	6.895 kPa	0.0703 kg/cm^2	1 psi
	9.81 Pa	1 mmwg	0.001422 psi
	249 Pa	0.00254 kg/cm^2	1 in.wg
		25.4 mmwg	
	1 MPa	10 b	145 psi
	1 N mm^{-2}	10.2 kg/cm^2	
	15.44 N mm^{-2}	1.575 kg/mm^2	1 Ton/in.^{2}
	9.80 N mm^{-2}	1 kg/mm^2	0.635 Ton/in.^{2}
Temperature	273 K	0°C	32°F/492 R
	0 K	−273°C	−460°F/0 R
	255.22 K	−17.78°C	0°F/460 R
Heat, power, work, and CV	1 W	0.2389 cal/s	9.478×10^{-4} Btu/s
	1 J s^{-1}		
	1 kW s	0.2389 kcal	0.948 Btu
	1 kJ	102 m kgf	738 ft lbf
	1 kW h	860 kcal	3413 Btu
	3600 kJ		
	4.186 kJ	1 kcal	3.9686 Btu
		427 m kgf	3088 ft lbf

Continued

Parameter	SI Units	Metric System of Units	Imperial and US System of Units
	1.055 kJ	0.252 kcal	1 Btu
		107.6 m kgf	778 ft lbf
	9.80 J	2.342×10^{-3} kcal	9.29×10^{-3} Btu
		1 m kgf	7.23 ft lbf
	1 kJ kg^{-1}	0.2389 kcal/kg	0.430 Btu/lb
	4.186 kJ kg^{-1}	1 kcal/kg	1.8 Btu/lb
	2.326 kJ kg^{-1}	0.556 kcal/kg	1 Btu/lb
	1 kJ m^{-3}	0.2389 cal/L	0.02684 Btu/ft^3
	4.186 kJ m^{-3}	1 cal/L	0.11235 Btu/ft^3
	37.26 kJ m^{-3}	8.90 cal/L	1 Btu/ft^3
	39.302 KJ Nm^{-3}	9.389 kcal/N m^3	1 Btu/scf
	1 kJ m^{-3} K^{-1}	0.2389 kcal/m^3°C	0.0149 Btu/ft^3°F
	4.186 kJ m^{-3} K^{-1}	1 kcal/m^3°C	0.06237 Btu/ft^3°F
	67.116 kJ m^{-3} K^{-1}	16.03 kcal/m^3°C	1 Btu/ft^3°F
	1 W m^{-2}	23.89×10^{-6} cal/cm^2 s	0.316 Btu/ft^2 h
	41.876 kW m^{-2}	1 cal/cm^2 s	13,233 Btu/ft^2 h
	3.165 W m^{-2}	75.57×10^{-6} cal/cm^2 s	1 Btu/ft^2 h
	1 kJ m^{-2}	0.2389 kcal/m^2	0.0881 Btu/ft^2
	4.186 kJ m^{-2}	1 kcal/m^2	0.3687 Btu/ft^2
	11.35 kJ m^{-2}	2.712 kcal/m^2	1 Btu/ft^2
	1 kJ m^{-2} K^{-1}	0.2389 kcal/m^2°C	0.0489 Btu/ft^2°F
	4.186 kJ m^{-2} K^{-1}	1 kcal/m^2°C	0.2048 Btu/ft^2°F
	20.44 kJ m^{-2} K^{-1}	4.883 kcal/m^2°C	1 Btu/ft^2°F
Specific heat	1 kJ kg^{-1} K^{-1}	0.2389 kcal/kg°C	0.2389 Btu/lb°F
	4.186 kJ kg^{-1} K^{-1}	1 kcal/kg°C	1 Btu/lb°F
Thermal conductivity	1 W m^{-1} K^{-1}	2.389×10^{-3} cal/cm s°C	0.578 Btu/ft h°F
	418.41 W m^{-1} K^{-1}	1 cal/cm s°C	241.84 Btu/ft h°F
	1.73 W m^{-1} K^{-1}	4.135×10^{-3} cal/cm s°C	1 Btu/ft h°F
Heat transfer coefficient	1 W m^{-2}K^{-1}	2.389×10^{-5} cal/cm^2 s°C	0.176 Btu/ft^2 h°F
	41858.5 W m^{-2} K^{-1}	1 cal/cm^2 s°C	7367 Btu/ft^2 h°F
	5.682 W m^{-2} K^{-1}	1.357×10^{-3} cal/cm^2 s°C	1 Btu/ft^2 h°F
Velocity	1 m s^{-1}	3.6 km/h	196.86 fpm
	0.278 m s^{-1}	1 km/h	54.68 fpm
	5.08×10^{-3} m s^{-1}	0.0183 km/h	1 fpm
	1 m s^{-1}	3.6 km/h	2.236 m/h
	0.278 m s^{-1}	1 km/h	0.621 m/h
	0.447 m s^{-1}	1.609 km/h	1 m/h

Continued

Parameter	SI Units	Metric System of Units	Imperial and US System of Units
Flow	$1\ m^3\ s^{-1}$	$3600\ m^3/h$	2118.6 cfm
	$2.78 \times 10^{-4}\ m^3\ s^{-1}$	$1\ m^3/h$	0.589 cfm
	$4.72 \times 10^{-4}\ m^3\ s^{-1}$	$1.698\ m^3/h$	1 cfm
	$1\ N\ m^3\ s^{-1}$	$3600\ N\ m^3/h$	2234.64 scfm
	$2.778 \times 10^{-4}\ N\ m^3\ s^{-1}$	$1\ N\ m^3/h$	0.6207 scfm
	$4.475 \times 10^{-4}\ N\ m^3\ s^{-1}$	$1.611\ N\ m^3/h$	1 scfm
	$6.31 \times 10^{-5}\ m^3\ s^{-1}$	$0.2271\ m^3/h$	1 US gpm
	$7.59 \times 10^{-5}\ m^3\ s^{-1}$	$0.273\ m^3/h$	1 Imperial gpm
	$1\ kg\ s^{-1}$	3.6 tph	7938 lb/h
	$0.278\ kg\ s^{-1}$	1 tph	2205 lb/h
	$1.26 \times 10^{-4}\ kg\ s^{-1}$	4.535×10^{-4} tph	1 lb/h
	$1\ m^3\ s^{-1}$	$3600\ m^3/h$	18.975 Imperial MGD
	$2.78 \times 10^{-4}\ m^3\ s^{-1}$	$1\ m^3/h$	5.275×10^{-3} Imperial MGD
	$0.0527\ m^3\ s^{-1}$	$189.58\ m^3/h$	1 Imperial MGD
	$1\ m^3\ s^{-1}$	$3600\ m^3/h$	22.827 US MGD
	$2.78 \times 10^{-4}\ m^3\ s^{-1}$	$1\ m^3/h$	6.341×10^{-3} US MGD
	$0.0438\ m^3\ s^{-1}$	$157.71\ m^3/h$	1 US MGD
Miscellaneous	$0.43\ kg\ GJ^{-1}$	1.8×10^{-3} g/kcal	1 lbm/MBtu

D.1 Basic SI Units (Systeme International D'Unites)

Sl. No.	Physical Quantity	Unit	Symbol	Definition of Unit
1	Length	Meter	m	—
2	Mass	Kilogram	kg	—
3	Time	Second	s	—
4	Temperature	Kelvin	K	—
5	Electric current	Ampere	A	—
6	Luminous intensity	Candela	cd	—
7	Amount of substance	Mole	mol	—
8	Plane angle	Radian	rad	—
9	Solid angle	Steradian	sr	—
10	Frequency	Hertz	Hz	s^{-1}
11	Energy	Joule	J	$kg\ m^2\ s^{-2}$
12	Force	Newton	N	$J\ m^{-1} = kg\ m\ s^{-2}$
13	Power	Watt	W	$J\ s^{-1} = kg\ m^2\ s^{-3}$
14	Pressure	Pascal	Pa	$N\ m^{-2} = J\ m^{-3} = kg\ m^{-1}\ s^{-2}$
15	Electric charge	Coulomb	C	$A\ s$

Continued

Sl. No.	Physical Quantity	Unit	Symbol	Definition of Unit
16	Electric potential difference	Volt	V	$W\,A^{-1}=J\,s^{-1}\,A^{-1}=kg\,m^2\,s^{-3}\,A^{-1}$
17	Electric resistance	Ohm	–	$V\,A^{-1}=W\,A^{-2}$ $J\,s^{-1}\,A^{-2}=kg\,m^2\,s^{-3}\,A^{-2}$
18	Electric capacitance	Farad	F	$C\,V^{-1}=kg^{-1}\,m^{-2}\,s^4\,A^2$
19	Magnetic flux	Weber	Wb	$V\,s=W\,s\,A^{-1}$ $J\,A^{-1}=kg\,m^2\,s^{-2}\,A^{-1}$
20	Magnetic flux density	Tesla	T	$Wb\,m^{-2}=V\,s\,m^{-2}=W\,s\,A^{-1}\,m^{-2}$ $J\,A^{-1}\,m^{-2}=kg\,s^{-2}\,A^{-1}$
21	Inductance	Henry	H	$V\,s\,A^{-1}=W\,s\,A^{-2}$ $J\,A^{-2}=kg\,m^2\,s^{-2}\,A^{-2}$
22	Conductance	Siemens	S	$V^{-1}\,A=W^{-1}\,A^2$ $J^{-1}\,s\,A^2=kg^{-1}\,m^{-2}\,s^3\,A^2$
23	Dynamic viscosity	Poiseuille	Pl	$N\,s\,m^{-2}=J\,s\,m^{-3}=kg\,m^{-1}\,s^{-1}$
24	Luminous flux	Lumen	lm	cd sr
25	Illumination	Lux	lx	$cd\,sr\,m^{-2}$

D.2 Prefixes Used in SI Units

Sl. No.	Prefix	Symbol	Factor
1	Tera	T	10^{12}
2	Giga	G	10^{9}
3	Mega	M	10^{6}
4	Kilo	k	10^{3}
5	Hecto*	h	10^{2}
6	Deca*	da	10^{1}
7	Deci*	d	10^{-1}
8	Centi*	c	10^{-2}
9	Milli	m	10^{-3}
10	Micro	μ	10^{-6}
11	Nano	n	10^{-9}
12	Pico	p	10^{-12}
13	Femto	f	10^{-15}
14	Atto	a	10^{-18}

Notes

1. Prefixes marked with * should normally be avoided.
2. It is preferable to write $J\,s^{-1}$, instead of J/s.

Index

Note: Page numbers followed by *f* indicate figures, *t* indicate tables, and *np* indicate footnotes.

Printed in the United States
By Bookmasters